AF577037

MATLAB® Numerical Methods with Chemical Engineering Applications

About the Author

Dr. Kamal I. M. Al-Malah, a portable, platform-independent professor of chemical engineering, is currently mounted at the University of Ha'il in Saudi Arabia. He holds B.S., M.S., and Ph.D. degrees in chemical/biochemical engineering. Dr. Al-Malah graduated from Oregon State University in 1993. His area of specialty during M.S. and Ph.D. programs there was protein interactions and behavior at interfaces in biological systems. He currently researches the modeling, simulation, and optimization aspects of physical/biophysical systems and the characterization of molecular properties within the dome of chemical, biochemical, pharmaceutical, and food engineering. In addition to his traditional and classical field of study, Dr. Al-Malah is a software developer, using Microsoft Visual Studio technology, and has created a bundle of Windows-based and MATLAB-based software for engineering applications (https://sites.google.com/site/almalahweb/software).

MATLAB® Numerical Methods with Chemical Engineering Applications

Kamal I. M. Al-Malah, Ph.D.

New York Chicago San Francisco
Athens London Madrid
Mexico City Milan New Delhi
Singapore Sydney Toronto

Library of Congress Cataloging-in-Publication Data

Al-Malah, Kamal I. M.
MATLAB numerical methods with chemical engineering applications / Kamal I. M. Al-Malah.
p. cm.
ISBN 978-0-07-183128-4 (hardback)
1. Chemical engineering—Data processing. 2. MATLAB. I. Title.
TP184.A4 2013
660.0285—dc23 2013018658

1 2 3 4 5 6 7 8 9 0 DOC/DOC 1 9 8 7 6 5 4 3

ISBN 978-0-07-183128-4
MHID 0-07-183128-2

Sponsoring Editor
Michael Penn

Editing Supervisor
Stephen M. Smith

Production Supervisor
Pamela A. Pelton

Acquisitions Coordinator
Amy Stonebraker

Project Manager
Vastavikta Sharma,
Cenveo® Publisher Services

Copy Editors
Mike and Lucia Read

Proofreader
Christine Andreasen

Indexer
Robert Swanson

Art Director, Cover
Jeff Weeks

Composition
Cenveo Publisher Services

Printed and bound by RR Donnelley.

This book is printed on acid-free paper.

Devotion

To my creator, who taught humans things they did not know before;

To the last prophet Mohammad and to the preceding prophets, peace be upon all of them;

To my first lovely teachers in this life, Mom and Dad;

To my wife Fadia, daughters Anwar and Lama, and sons Abdallah and Mohammad;

To my biological brothers and sisters;

To my first class teacher Raslan Al-Malah, mercy be upon him;

To my succeeding teachers and professors from my first class to my Ph.D.;

To my last professor (my Ph.D. supervisor, Dr. J. McGuire);

To the McGraw-Hill Education staff;

To Cenveo Publisher Services; and

Finally, to my brothers and sisters worldwide.

To the first, I say: I owe you everything and you are the first and last.

To the rest, I say: my sincere and warm greetings to all of you.

PRAISE IS TO ALLAH (GOD)

Kamal

Contents

Preface

MATLAB® is a comprehensive software system for mathematics and technical computing. It is an engineering tool and programming language that can be used to model and simulate any physical problem you may think of in the engineering realm. Thus, it contains concise explanations of essential MATLAB commands, as well as easily understood instructions for using MATLAB's programming features, graphical capabilities, and desktop interface. Written for MATLAB 7.11 (R2011), this book can also be used with earlier and later versions of MATLAB. This book contains successful examples of applications of MATLAB to interesting problems in mathematics and engineering.

Each chapter is a stand-alone entity and gives a separate dimension of a soft skill needed by engineers in the computational world. The last two chapters are written for specialized areas. Chapter 10 is exclusive for chemical engineering applications and Chap. 11 is exclusive for advanced users (i.e., developers), whether or not they are chemical engineers.

What Do You Get out of This Book?

The goal of writing this book is to get you started using MATLAB successfully and quickly. I point out the parts of MATLAB you need to know without overwhelming you with details. I do my best to avoid presenting cumbersome MATLAB features. I give you examples of real uses of MATLAB that you can refer to when performing your own exercises. When you are finished with this book you will be able to use MATLAB efficiently. You will also be ready to explore more MATLAB features on your own. You might not be a MATLAB expert when you finish this book, but you will be prepared to become one if that is your choice, although you are probably more interested in being an expert in your own specialty, whether science or engineering. Use MATLAB the way I do, as a tool. This book is designed to help you become a proficient MATLAB user as quickly as possible so you can get on with your own business.

Who Should Read This Book?

This book is useful to both beginning and experienced users. It is written for engineering students in general, who plan to take a course in numerical methods for engineers, and particularly for chemical engineers who would like to extend their knowledge of the modeling and programming aspects of chemical engineering applications in addition to numerical methods. Graduate engineers also can benefit from the book because it gives them an opportunity to replenish their skills in mastering this powerful tool. Once the engineer masters such a tool, he or she can use it in daily technical life. The book can be recommended to all engineering and chemical engineering unions, chapters, and organizations.

MATLAB is cumbersome for beginners. We need to facilitate the process of learning and later mastering such a powerful tool in a very short time without making their life miserable. This textbook is a recipe or cookbook type, and every step needed toward the final solution is algorithmically explained via snapshots of the MATLAB platform in parallel with the text.

Last Notes for Instructors

For a typical numerical methods course (junior-level) in engineering college, I recommend the sequence of the first eight chapters, except Chap. 4 (Image and Image Analysis). Chapter 9 (Statistics) will be a choice for the instructor whether to include in the course. Please keep in mind that the statistical background needed to describe the model goodness is covered in Sec. 5.4 (Statistical Definitions).

The need for image acquisition is on the rise and more applications require that users familiarize themselves with dealing with images as a source of input data in a typical experiment. Chapter 4 serves as an introduction to how MATLAB acquires different types of images. Moreover, Chap. 10 is truly chemical engineering territory, although parts of it may be applicable to mechanical engineering as well. Finally, Chap. 11 is written for advanced users and opens the door to those who wish to become applications developers or MATLAB-based Graphical User Interface (GUI) builders and dwellers.

The book presents problems at the end of each chapter, and there are additional problems for the instructor (available for download at www.mhprofessional.com/MATLABnumericalmethods, along with all M-files appearing in this book, all supplementary files, and software packages I have developed) in which he or she can adjust the numbers such that each student will have his or her own version on an exam or quiz. This will minimize the possibility of cheating among students if the instructor asks students to solve problems in a computer lab using MATLAB software. For example, the first or last five digits of a student's ID (alternatively, national ID or Social Security number) can be taken as an input, in the form of *abcde*, and plugged into the question statement such that each student ends up with his or her own version.

Conventions Used in This Book

Output from MATLAB is typeset in the Courier typewriter font. Commands that you type for interpretation by MATLAB are indicated by a boldface version of that font. These commands and responses are often displayed on separate lines, as they would be in a MATLAB session, as in the following example:

```
>> x = sqrt(2*pi + 2)
x =
2.8781
```

MATLAB®
Numerical Methods with Chemical Engineering Applications

1 MATLAB Basics

Chapter Outline

MATLAB® (MATrix LABoratory) is a high-performance interacting data-intensive software environment for high-efficiency engineering and scientific numerical calculations. Applications include heterogeneous simulations and data-intensive analysis of very complex systems and signals, comprehensive matrix and array manipulations in numerical analysis, finding roots of polynomials, 2D and 3D plotting and graphics for different coordinate systems, integration and differentiation, signal processing, control, identification, symbolic calculus, optimization, and so on. The goal of MATLAB is to enable users to solve a wide spectrum of analytical and numerical problems using matrix-based methods, attain excellent interfacing and interactive capabilities, compile with high-level programming languages, ensure robustness in data-intensive analysis and heterogeneous simulations, provide easy access to and straightforward implementation of state-of-the-art numerical algorithms, guarantee powerful graphical features, and so on. Because of high flexibility and versatility, the MATLAB environment has been significantly enhanced and developed during recent years. This provides users with advanced cutting-edge algorithms, enormous data-handling abilities, and powerful programming tools. MATLAB is based on a high-level matrix/array language with control flow statements, functions, data structures, input/output, and object-oriented programming features.

1.1 Starting MATLAB

You start MATLAB as you would start any other software application. On a Windows-based PC you access it via the Start menu, in Programs (All Programs in Vista, or Windows 7) under a folder such as MATLAB (followed sometimes by MATLAB R2011a subfolder), then click the icon, like MATLAB R2011a. Alternatively, you may have an icon (shortcut on the desktop) set up that enables you to start MATLAB with a simple double-click.

Once you start MATLAB, you will briefly see a window that displays the MATLAB logo as well as some MATLAB product information, and then a MATLAB Desktop window will launch. That window will contain a title bar, a menu bar, a tool bar, and four sub-windows. The largest and most important window is the *Command Window* in the middle. The other three windows are the *Workspace*, the *Command History*, and the *Current Folder* (*Directory*). For now we concentrate on the *Command Window* to get you started issuing MATLAB commands as quickly as possible. At the top of the *Command Window*, you may see some general information about MATLAB, perhaps some special instructions for getting started or accessing help, but most important of all, is a line that contains a prompt. The prompt will likely be a double caret (>>). If the *Command Window* is "active," its title bar will be blue colored, and the prompt will be followed by a cursor (a vertical line or box, usually blinking). That is the place where you will enter your MATLAB commands (see Chap. 2). If the Command Window is not active, then its title bar will become gray. Figure 1.1 contains an example of a newly launched MATLAB Desktop.

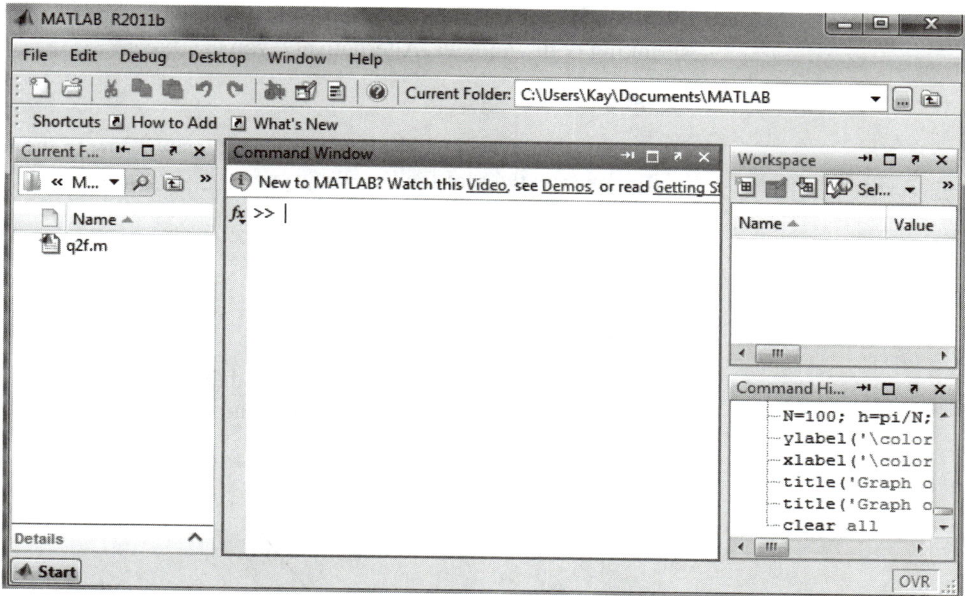

Figure 1.1 MATLAB Desktop. (The main window contains four sub-windows.)

Try typing **sqrt(9)**; then press ENTER or RETURN. Next try **factor (30)** and finally **sin(pi/2)**. Notice that **factor(N)** returns a vector containing the prime factors of N. Your MATLAB Desktop should look like Fig. 1.2.

Notice that many of the graphical examples in this book assume that the figure window is empty. To ensure an empty figure window, issue the command:

```
>> clf
```

clf stands for *clear figure*.

```
>> Sqrt (9)

ans =

     3

>> factor (30)

ans =

     2        3        5

>>  sin (pi/2)

ans =

    1
```

Figure 1.2 Some simple commands (i.e., MATLAB acts like a scientific calculator).

If you find that the figure window is obscured by your command window, try shrinking both windows. Alternatively, you can type:

```
>> shg
```

shg means (show graphic) to bring the graphics window to the front.

To generate the plot for $f(x) = x^2$, type:

```
>> ezplot('x^2');
```

On the other hand, the command:

```
>> ezplot('x^2',[-1 5]);
```

will plot the function within [–1 5] interval (Fig. 1.3).

More on plots and graphics can be found in Sec. 1.8.

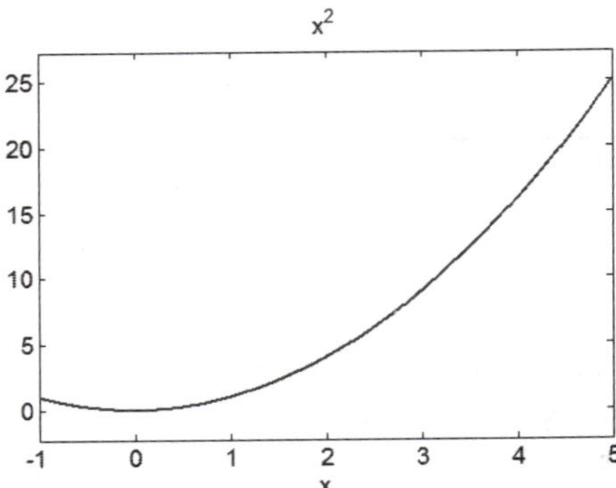

Figure 1.3 Plot of $f(x) = x^2$ using `ezplot('x^2')`.

NOTE An important note before you navigate deeper into the MATLAB ocean: As Fig. 1.1 shows, there is a Command History window shown at the right bottom corner. Unless you clear Command History, such a window keeps a record of all previous commands (that is why it is called a Command History) instructed to MATLAB by the user. There are points worth mentioning here. You may highlight (i.e., select any previous command) via left-mouse clicking and releasing the mouse. Once you select a given command from previous history, right-click your mouse and a shortcut pop-up menu appears, as shown in Fig. 1.4. With the highlighted command, you can copy the string (text) to the Windows clipboard where it can be pasted later in the Commands window; it can either be evaluated (immediately executed; no need for Supreme Court approval) or an M-file can be created from that command. An M-file is a scripting (programming) language, such as the C language, that can be understood by the MATLAB compiler. Regarding the last step, when creating an M-file, however, it is preferable that you highlight a list or set of commands rather than only one command, and then you can create an M-file out of the selected set. To highlight a list of consecutive commands (i.e., a block of commands) all that you need to do is highlight the starting command by clicking the left mouse and releasing it. Then press the Shift key and while in pressed mode, move your cursor downward to the last statement that you intend to execute or include in the M-file. Then click the left mouse, followed by the release

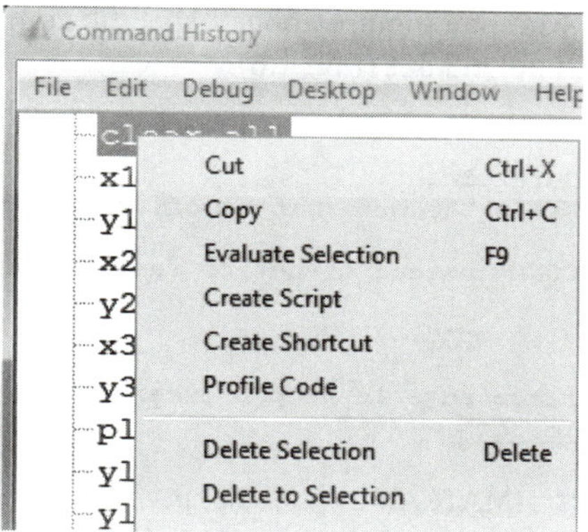

Figure 1.4 The shortcut pop-up menu that appears if you right-click a highlighted previous command found in Command History window. You can choose "Copy" and paste it later in the Commands window, immediately following "Evaluate Selection", or "Create M-File".

of the Shift key. On the other hand, using the ctrl key in pressed mode allows the user to selectively choose some specific commands and leave others.

1.2 Online Help

MATLAB has an extensive online help mechanism. In fact, using only this book and the online help, you should be able to become quite proficient with MATLAB.

You can access the online help in one of several ways. Typing help at the command prompt will reveal a long list of topics on which help is available. Just to illustrate, try typing **help general**. Now you see a long list of "general purpose" MATLAB commands (Fig. 1.5).

```
>> help general
  General purpose commands.
  MATLAB Version 7.13 (R2011b) 08-Jul-2011

  General information.
    syntax        - Help on MATLAB command syntax.
    demo          - Run demonstrations.
    ver           - MATLAB, Simulink and toolbox version information.
    version       - MATLAB version information.
    verLessThan   - Compare version of toolbox to specified version string.
    logo          - Plot the L-shaped membrane logo with MATLAB lighting.
    membrane      - Generates the MATLAB logo.
    bench         - MATLAB Benchmark.

  Managing the workspace.
    who           - List current variables.
    whos          - List current variables, long form.
    clear         - Clear variables and functions from memory.
    onCleanup     - Specify cleanup work to be done on function completion.
    pack          - Consolidate workspace memory.
    load          - Load workspace variables from disk.
    save          - Save workspace variables to disk.
    saveas        - Save Figure or model to desired output format.
    memory        - Help for memory limitations.
    recycle       - Set option to move deleted files to recycle folder.
    quit          - Quit MATLAB session.
    exit          - Exit from MATLAB.
```

Figure 1.5 A short list of "general purpose" MATLAB commands.

In particular, you may inquire about a specific command, such as **help det** to learn about the command **det** (Fig. 1.6).

```
>> help det
 det    Determinant.
    det(X) is the determinant of the square matrix X.

    Use COND instead of det to test for matrix singularity.

    See also cond.

    Reference page in Help browser
       doc det
```

Figure 1.6 MATLAB help for the specific command **det**, which evaluates the determinant of the square matrix **X**.

In every instance in the preceding, if more information than your screen can hold is supplied by MATLAB, your screen will then scroll by. There is a much more user-friendly way to access the online help, namely, via the MATLAB Help Browser. You can activate it in several ways; for example, typing either **helpwin** or **helpdesk** at the command prompt brings it up.

Alternatively, online help is available through the menu bar under Help. The question mark button on the tool bar will also invoke the Help Browser. Starting with 2008 releases, you can invoke a context-sensitive function browser, which lies on the left of the command prompt, and has the symbol f_x. You may type, for example, the command **plot**, and MATLAB will attempt to find all related MATLAB functions that include the word *plot* as part of their function names (Fig. 1.7).

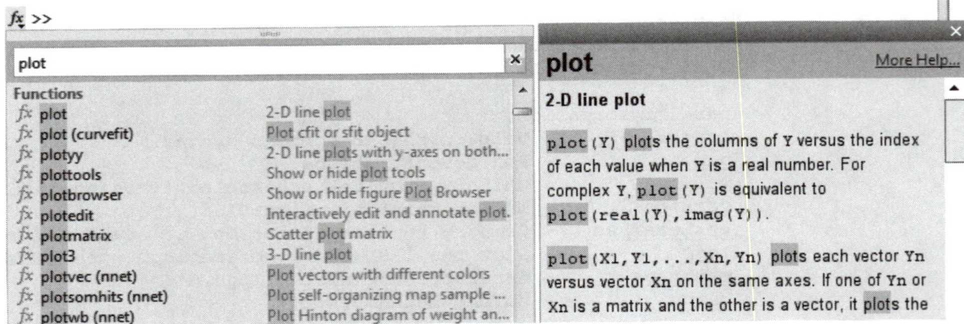

Figure 1.7 MATLAB context-sensitive functions browser, which lies on the left of the command prompt, and has the symbol f_x.

1.3 Simple Arithmetic Operations

In this section, you will start learning how to use MATLAB to do mathematics. You should read this chapter with your computer running MATLAB. Try the commands in the MATLAB Command Window as you go along. Feel free to experiment with the given examples; at the same time, you may also modify them to familiarize yourself with MATLAB. *The best way to find out how MATLAB responds to a command is to try it.*

You can use MATLAB to do arithmetic operations you would do with a calculator. You can use "+" for addition, "–" for subtraction, "*" for multiplication, "/" for division, and "^" for exponentiation. For example:

```
>> 4^2 + (5 + 4)/3 - 8*5
ans =
 -21
```

Notice that MATLAB prints the answer and stores the value to a *variable* called **ans**.

If you want to perform further calculations with the answer, you can recall the variable **ans** rather than retype the answer. Keep in mind that MATLAB retains the latest value in **ans**.

```
>> ans*(-1)
ans =
21
```

The new value of **ans** is no longer –21 but 21, as a result of the last carried operation. If the user desires to keep some intermediate values, then such values must be assigned or stored in separate variable "containers". For example:

```
>> my_ans=ans*2
my_ans =
  42
```

Now the variable **my_ans** will hold this value unchanged unless you carry out an operation that will assign a new value to it.

In the **Workspace** browser, the second type of MATLAB sub-window, you will notice that MATLAB shows a variable name called **my_ans** with a value of 42 and is separate from that of **ans,** which keeps changing as a result of successive mathematical operations.

NOTE MATLAB is case sensitive; i.e., **my_ans** ≠ **my_Ans** ≠ **My_ans** ≠ **My_Ans**.

If you carry out the following command:

```
>> My_Ans=ans*3
My_Ans =
  63
```

the Workspace browser will show the newly added variable.

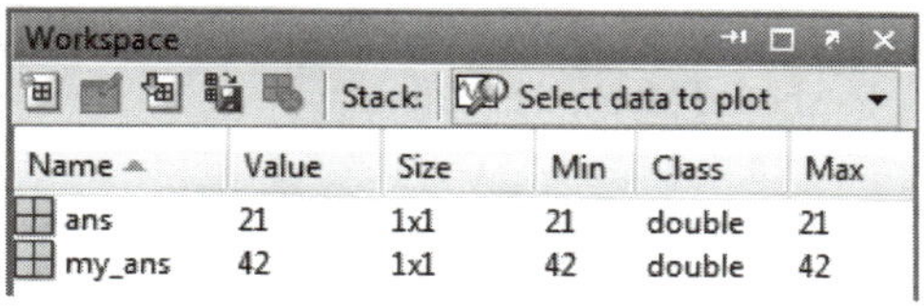

Workspace

Stack: Select data to plot

Name	Value	Size	Min	Class	Max
ans	21	1x1	21	double	21
my_ans	42	1x1	42	double	42

Figure 1.8 The Workspace browser shows the list of variables so far created by the user, each of which has an assigned or stored value as a result of mathematical operations instructed by the user.

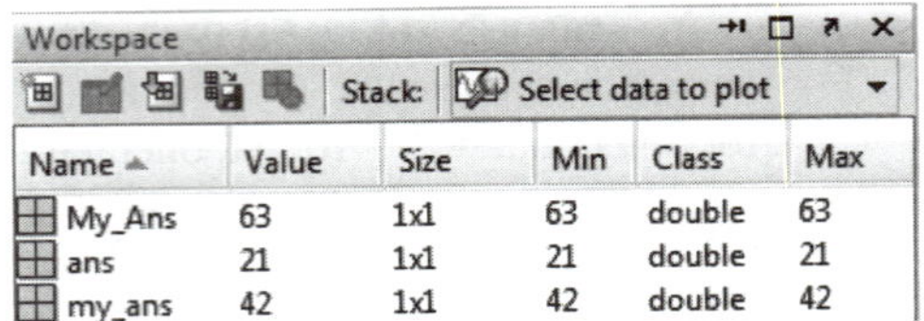

Workspace

Stack: Select data to plot

Name	Value	Size	Min	Class	Max
My_Ans	63	1x1	63	double	63
ans	21	1x1	21	double	21
my_ans	42	1x1	42	double	42

Figure 1.9 The updated list of variables as more operations are carried out. Notice that MATLAB is case sensitive in terms of naming a variable.

The **whos** command, shown in Fig. 1.10, also gives the user the same list that appears in Fig. 1.9.

```
>> whos
```

```
>> whos
  Name          Size            Bytes  Class

  My_Ans        1x1                 8  double
  ans           1x1                 8  double
  my_ans        1x1                 8  double
```

Figure 1.10 Results of applying the **whos** command that gives the same information about the list of variables appearing in the Workspace browser, shown in Fig. 1.9.

NOTE The **Size** and **Bytes** columns can also be shown in the Workspace browser simply by modifying what to show in the browser via right clicking on any column heading (name) where a shortcut menu pops up and the selection can be made of what entries to show in the Workspace browser.

The **clear** command rids MATLAB from variables. If you want to remove a specific variable, then invoke the clear command followed by the name of that variable. For example, if you want to get rid of **my_ans** variable then use

```
>> clear my_ans
```

NOTE Variable names in MATLAB must start with a letter and can be up to 31 characters long. The trailing characters can be numbers, letters, or underscores. (Some other characters are also available, but in this text we shall stick to these.) Hyphen (-), percent sign (%), and other such characters are not allowed in variable names. Also, reserved names should not be used as variable names. For example, **pi**, **i**, **j**, and **e** are reserved. Similarly, the names of functions and MATLAB commands should also be avoided.

1.4 MATLAB Built-in Library Functions

The user may use any of the MATLAB built-in library functions that are already defined and can be used in any command statement. MATLAB has a miscellany of functions. Some of these are standard functions, including trigonometric functions, and so on, and others are user-defined functions and third-party functions. All of these enable the user to easily carry out complex computational tasks. Examples are shown in Table 1.1.

Table 1.1 A list of common MATLAB built-in library functions.

	MATLAB Function	A Brief Explanation
		General Functions
1	**abs ( )**	Absolute value of the argument
2	**exp ( )**	Exponential of the argument
3	**log ( )**	Natural logarithm
4	**log10 ( )**	Common (base 10) logarithm
5	**sign ( )**	Sign ( ) returns 1 if the argument is greater than zero; 0 if it equals zero; and –1 if it is less than zero
6	**sqrt ( )**	Square root of the argument
		Complex Number Related Functions
7	**angle ( )**	Phase angle of a complex number; the answer (output) is in radians
8	**conj ( )**	Conjugate of a complex number
9	**imag ( )**	Imaginary part of a complex number
10	**real ( )**	Real part of a complex number
		Hyperbolic Trigonometric Functions
11	**acosh ( )**	Inverse hyperbolic cosine
12	**asinh ( )**	Inverse hyperbolic sine
13	**atanh ( )**	Inverse hyperbolic tan
14	**cosh ( )**	Hyperbolic cosine function
15	**sinh ( )**	Hyperbolic sine function
16	**tanh ( )**	Hyperbolic tan function
		Integer Related Functions
17	**ceil ( )**	Rounds toward plus infinity
18	**factor (N)**	Returns a vector containing the prime factors of N
19	**factorial (N)**	Factorial for a scalar argument, N, is the product of all the integers from 1 to N
20	**fix ( )**	Rounds toward zero
21	**floor ( )**	Rounds toward minus infinity
22	**round ( )**	Rounds to the nearest integer
		Trigonometric Functions
23	**acos ( )**	Arc-cosine (inverse cosine) of the arguement; the answer (output) is in radians
24	**acosd ( )**	Arc-cosine (inverse cosine) of the argument; the answer (output) is in degrees
25	**cos ( )**	Cosine of (argument in radians)
26	**cosd ( )**	Cosine of (argument in degrees)
27	**acot ( ), acotd ( ), cot ( ), and cotd ( )**	The same explanation as for cosine cases except you replace cosine by co-tangent
28	**asin ( ), asind ( ), sin ( ), and sind ( )**	The same explanation as for cosine cases except you replace cosine by sine
29	**atan ( ), atand ( ), tan ( ), and tand ( )**	The same explanation as for cosine cases except you replace cosine by tangent.

Here are some examples:

Example 1.1

```
>> var1=sin(pi/3)
var1 =
   0.8660
```

where **pi** is the ratio of the circumference to the diameter of a circle, $\pi = \frac{C}{D} =$ 3.142857.

```
>> var2=sind(60)
var2 =
   0.8660
```

Notice that **var1=var2** simply because $(\pi/3)$ *in radians* $= (180°/3) = 60°$ *in degrees.*

.

Example 1.2

```
>> var3=asin(1)
var3 =
   1.5708
>> var4=asind(1)
var4 =
   90
```

Notice that **var3=var4** simply because $90°$ *in degrees* $= (\pi/2) = (3.142857/2) =$ 1.5708 *in radians.*

Example 1.3

```
>> factorial (5)
ans =
     120.00
```

This means that $120 = 1 \times 2 \times 3 \times 4 \times 5$.

Example 1.4

```
>> factor (20)
ans =
      2.00      2.00      5.00
```

This means that $20 = 2 \times 2 \times 5$.

Example 1.5

```
>> fixed_value=fix(-6.5)
fixed_value =  -6
>> rounded_value=round(-6.5)
rounded_value =  -7
>> floor_value=floor(-6.5)
floor_value =  -7
>> ceiling_value=ceil(-6.5)
ceiling_value =  -6
```

1.5 Output Display in MATLAB

It should be kept in mind that the **format** command does not affect how MATLAB computations are done. Computations on float variables, namely, single or double, are done in appropriate floating point precision, no matter how those variables are displayed. Computations on integer variables are done natively in integer. MATLAB uses double-precision floating point arithmetic, which is accurate to approximately 15 digits and can work with the following data types (Table 1.2).

Table 1.2 Data types that can be dealt with by MATLAB.

Numbers	Details
Integer	Numbers without any fractional part and decimal point; e.g., 786
Real	Numbers with fractional part; e.g., 3.14159
Complex	Numbers having real and imaginary parts; e.g., 3 + 4i. MATLAB treats 'i' as well as 'j' to represent $\sqrt{-1}$
Inf	Infinity; e.g., 5/0 = Inf; log(0) = –Inf; Inf/0 = Inf; exp(1000) = Inf; and tand(90) = Inf. In general, any positive floating point number larger than **realmax** (equal to 1.7977e + 308) will overflow to single(Inf)
NaN	Not a number; e.g., 0/0 = NaN; inf-inf = NaN; inf/inf = NaN; and blank entries (i.e., cells in Excel sheet) imported to MATLAB are assigned NaN

MATLAB, however, displays only 5 digits by default. To display more digits, the user should type **format long**.

Example 1.6 >> **format long**

Now, all subsequent numerical output will have 15 digits displayed. If you try the previous command line, you will notice the difference.

```
>> var1=sin(pi/3)
var1 =
  0.866025403784439
```

There are other formats as well; let us see how **var1=sin(pi/3)** is displayed by MATLAB, depending on the active format.

Example 1.7 >> **format short e**

[Floating point format with 5 digits]

MATLAB displays **var1** = 8.6603e-001.

Example 1.8 >> **format long e**

[Floating point format with 15 digits for double and 7 digits for single]

MATLAB displays **var1** = 8.660254037844386e-001.

Example 1.9 >> **format short eng**

[Engineering format that has at least 5 digits and a power that is a multiple of three]

MATLAB displays **var1** = 866.0254e-003.

Example 1.10 >> **format long eng**

[Engineering format that has exactly 16 significant digits and a power that is a multiple of three]

MATLAB displays **var1** = 866.025403784439e-003.

Example 1.11 >> **format hex**

[Hexadecimal format]

MATLAB displays **var1** = 3febb67ae8584caa.

Example 1.12

```
>> format bank
```

[Fixed format for dollars and cents]

MATLAB displays **var1** = 0.87.

To return to the 5-digit display, type: ***format short.***

```
>> format short
```

1.6 Algebra

Using MATLAB's Symbolic Math Toolbox, you can carry out algebraic or symbolic calculations such as factoring polynomials or solving algebraic equations. Type **help symbolic** to make sure that the Symbolic Math Toolbox is installed on your system.

To perform symbolic computations, you must use **syms** or **sym** to declare the variables you plan to use to be symbolic variables. First, you have to introduce symbols to the MATLAB environment:

```
>> syms x y
```

Alternatively,

```
>> x=sym('x')
>> y=sym('y')
```

Now, let us consider some examples.

Example 1.13

```
>>term1= (x-3)*(x-5)
term1 =
(x - 3)*(x - 5)
```

If you expand **term1**, you will get:

```
>> myexpanded_term = expand (term1)
myexpanded_term =
x^2 - 8*x + 15
```

which is similar to:

$$x^2 - 8x + 15$$

You can revert to the compact term (i.e., bracketed terms) using the **factor** command:

```
>>mycompact_term = factor(myexpanded_term)
mycompact_term =
(x - 3)*(x - 5)
```

Example 1.14

```
>>term2= (x-3)*(y-5)
term2 =
(x - 3)*(y - 5)
>> expanded2=expand(term2)
expanded2 =
x*y - 3*y - 5*x + 15
```

which is similar to

$$xy - 3y - 5x + 15$$

Example 1.15

```
>> term3=(x-y)^3
term3 =
(x - y)^3
>> expanded3=expand(term3)
expanded3 =
x^3 - 3*x^2*y + 3*x*y^2 - y^3
```

which is similar to

$$x^3 - 3x^2y + 3xy^2 - y^3$$

Again, if you factor the previous answer, you will recover **term3,** as defined earlier:

```
>> compact3=factor(expanded3)
compact3 =
(x - y)^3
```

MATLAB has a command called **simplify**, which you can sometimes use to express a formula as simply as possible.

Example 1.16

```
>> simplified1=simplify((x^2-x-2)/(x+1))
simplified1 =
x - 2
>> simplified2=simplify((x^2-x+2)/(x+1))
simplified2 =
x + 4/(x + 1) - 2
```

Moreover, you can *symbolically* express some arguments rather than using their numeric values.

Example 1.17

```
>> sin(sym('pi/2'))
ans = 1
```

which is equivalent to:

```
>> sin(pi/2)
ans = 1
```

Notice that, in general,

```
>>S = sym(A)
```

constructs an object **S**, of class '**sym**', from **A**. If the input argument is a *string*, the result is a symbolic number or variable. If the input argument is a *numeric scalar or matrix*, the result is a symbolic representation of the given numeric values. If the input is a *function handle*, the result is the symbolic form of the body of the function handle. So, **x = sym('x')** creates the symbolic variable with name 'x' and stores the result in **x**.

Statements like

```
>>pi = sym('pi')
```

and

```
>>delta = sym('1/10')
```

create symbolic numbers that avoid the floating point approximations inherent in the values of **pi** and 1/10, respectively. The **pi** created in this way temporarily replaces the built-in numeric function with the same name. Thus:

```
>> delta+2/10
ans =
3/10
>> delta+delta
ans =
1/5
>>S = sym(A,flag)
```

converts a numeric scalar or matrix to symbolic form. The technique for converting floating point numbers is specified by the optional second argument, which may be 'f', 'r', 'e' or 'd'. The default is 'r'. First, let us introduce the **digits** command.

```
>> digits(D)
```

sets the digital accuracy, **digits,** to **D**, for subsequent calculations. **D** is an integer, or a string or sym representing an integer. Notice that the **digits** command is part of the Symbolic Math Toolbox.

```
>>digits(10)
>>sym(4/3,'d')
ans=
1.333333333
```

while with **digits(32)**,

```
>>sym(4/3,'d')
ans=
1.3333333333333332593184650249896
```

which does not end in a string of 3s but is an accurate decimal representation of the double-precision floating point number nearest to 4/3.

Similarly,

```
>>k = sym('k','positive')
```

makes **k** a positive (real) variable.

```
>>x = sym('x','real')
```

assumes that **x** is real, so that conj(x) is equal to **x**.

On the other hand,

```
>>x = sym('x','clear')
```

restores **x** to a formal variable with no additional properties (i.e., ensures that **x** is neither real nor positive).

1.7 Vectors

The basic object that MATLAB deals with is a matrix. A matrix is an array of numbers. On the other hand, vectors are reduced forms of matrices. In MATLAB, there are two types of vectors, the row vectors and the column vectors.

1.7.1 *The Row Vectors*

The row vectors are entities enclosed in pairs of square-brackets with numbers separated either by spaces or by commas. Let us define the two *row* vectors **X** and **Y** as:

```
>> X=[1 2 3]; Y=[4,5,6];
>> XYsum=X+Y
XYsum =
  5  7  9
```

The two row vectors are defined and then their sum **X+Y** is computed. The result is a row vector stored as **XYsum**. The usual operations with vectors can be easily carried out:

```
>> Result_4Xplus3Y=4*X+3*Y
Result_4Xplus3Y =
  16  23  30
```

The preceding case represents a linear combination of **X** and **Y**. One can also combine vectors to form another vector:

```
>> Z=[2*X,3*Y]
Z =
  2  4  6  12  15  18
```

The vectors **X** and **Y**, both of length 3, are combined to form a six-component vector **Z.** The results of row vector operations are shown in Fig. 1.11.

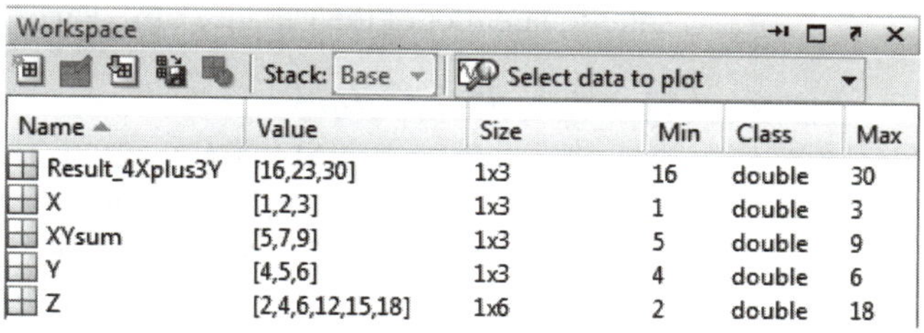

Workspace

Stack: Base | Select data to plot

Name ▲	Value	Size	Min	Class	Max
Result_4Xplus3Y	[16,23,30]	1x3	16	double	30
X	[1,2,3]	1x3	1	double	3
XYsum	[5,7,9]	1x3	5	double	9
Y	[4,5,6]	1x3	4	double	6
Z	[2,4,6,12,15,18]	1x6	2	double	18

Figure 1.11 Vectors formed as different combinations of **X** and **Y** vectors.

The following *statistical* operations can be carried out on row vectors as well as column vectors (see the following); where the row vector, **Z**, is treated as a distribution of discrete data:

```
>> min(Z)
ans =
  2
```

2 is the minimum value found in the row vector, **Z.**

```
>> max (Z)
ans =
  18
```

18 is the maximum value found in the row vector, **Z.**

```
>> mean(Z)
ans =
  9.5000
```

9.5 is the mean value of all entries in the row vector, **Z.**

```
>>median (Z)
ans =
  9
```

9 is the median value of the elements in **Z.** Notice that the value 9 lies in the middle between the third entry, **Z(3)**, and the fourth entry, **Z(4)**. Thus, the median represents the arithmetic mean of the two entries:

```
>> median=(Z(3)+Z(4))/2
median =
  9
>> std (Z)
ans =
  6.4420
```

6.4420 is the standard deviation of the Z-data distribution around their mean, 9.5.

```
>> var (Z)
ans =
 41.5000
```

41.5 is the variance of the Z-data distribution around their mean. Notice that **std (Z)** is the square root of the variance of Z, **var(Z)**.

1.7.2 The Column Vectors

The column vectors in MATLAB are formed by using a set of numbers in a pair of square brackets and separating them with a semicolon. Therefore, one can define two column vectors **X** and **Y** and add them to get **Z**, as follows:

```
>> X=[1;3;5]; Y=[2;4;6]; Z=X+Y
Z =
   3
   7
   11
```

Notice that all other standard operations with the row vectors can be carried out here, as well. For example, **min (Z), max (Z), mean (Z), median (Z), std (Z), and var (Z)** are applicable.

Exercise 1.1 Try to define a 2 × 2 matrix, say $A = \begin{bmatrix} 1 & 2 \\ 3 & 4 \end{bmatrix}$, in MATLAB using both the comma (,), or space ( ) and the semi-colon (;).

1.7.3 The Colon Notation

In order to form a vector as a sequence of numbers, one may use the colon notation, according to which **s:i:e** yields a sequence of numbers starting with **s**, and possibly ending with **e** in incremental steps of **i**. For example, **1:1:7** yields the following row vector:

```
>> 1:1:7
ans =
  1  2  3  4  5  6  7
```

Notice that in some cases, the upper limit may not be attainable. For example, in the case of **1:0.6:3**, the upper limit is not reached and the resulting vector in this case is:

```
>> 1:0.6:3
ans =
  1.0000  1.6000  2.2000  2.8000
```

On the other hand, if a decremental step is required, then the middle value should be negative. For example, the following statement assigns the integer values starting from 9 and down to 3 to the row vector, **A_Row_Vector**:

```
>> A_Row_Vector=9:-1:3
A_Row_Vector =
  9  8  7  6  5  4  3
```

1.7.4 Transpose

Of course, MATLAB does not entertain creating column vectors using the colon notation. But one can do it by first creating a row vector using the colon (range) notation and then transposing the resulting row vector into a column vector. The transpose is obtained with the apostrophe ' as shown in the following:

```
>> R=[1:5]; S=R'
S =
  1
  2
  3
  4
  5
```

Here, first a row vector [1 2 3 4 5] is formed which is called **R**. This vector is then transposed to form **S**, a column vector.

For complex vectors, the situation is a bit different. Let us take the following two examples.

```
>> E=[1+2i 1-2i]; F=E'
F =
 1.0000 - 2.0000i
 1.0000 + 2.0000i
```

Notice that **F** is not really a transported version of **E**; rather, it contains the conjugate values of the elements of **E**. To transpose without having the conjugate values, the following operation is carried out:

```
>> E=[1+2i 1-2i]; G=E.'
G =
 1.0000 + 2.0000i
 1.0000 - 2.0000i
```

Notice now that **G** is a transposed version of **E**.

1.8 Plots and Graphs

MATLAB offers powerful graphics and visualization tools. Let us start with some of the very basic graphics capabilities of MATLAB. The graph of the cosine function from $-\pi$ to π can be obtained in the following way:

```
>>N=100; h=pi/N; x= -pi:h:pi; y=cos(x); plot(x,y)
```

h represents the incremental step size which is equal to a hundredth of π (pi). The smaller the step size **h** is, the higher is the degree of resolution for showing the topology of $f(x)$, in general.

The command '**plot(x,y)**' generates the graph of this data and displays it in a separate window labeled Figure No. 1, which is shown in Fig. 1.12.

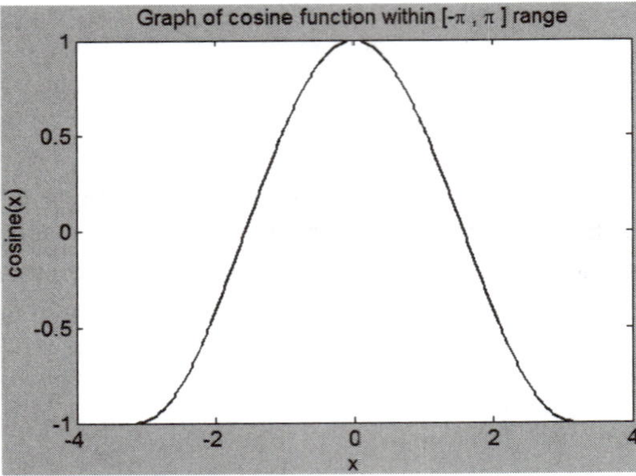

Figure 1.12 The plot of the cosine function over the interval [−π, π] using the plot command.

The graph title and x- and y-labels can be assigned using the following commands:

```
>> title('Graph of cosine function within [-\pi , \pi ] range')
>> xlabel('x')
>> ylabel('cosine(x)')
```

Notice that the backslash \ is used to tell MATLAB what comes next is to be displayed as a symbol. The word pi is spelled out after the backslash so that MATLAB displays π.

1.8.1 *Non-ASCII Character Display*

In general, you can define text that includes non-ASCII characters like symbols and Greek letters using the text function, assigning the character sequence to the String property of text objects. You can also include these character sequences in the string arguments of the title, xlabel, ylabel, and zlabel functions.

Other characters are shown in the following Tables 1.3a, 1.3b, and 1.3c.

1.8.2 *Font-Related Attributes*

You can also specify stream modifiers that control font type and color.

- \bf — **Bold** font
- \it — *Italic* font
- \sl — Oblique font (rarely available)
- \rm — Normal font
- \fontname{fontname}—Specify the name of the font family to use.
- \fontsize{fontsize}—Specify the font size in FontUnits.
- \color(colorSpec)—Specify color for succeeding characters.

Stream modifiers remain in effect until the end of the string or only within the context defined by braces { }.

For example, if the following function is to be plotted: $y = (\cos(x^2))^2 = \cos^2(x^2)$ over the interval [−π, π], then the following command will do the job:

```
>> N=100; h=pi/N; x= -pi:h:pi; y=(cos(x.^2)).^2; plot(x,y)
```

Table 1.3a The character sequence that should follow the backslash \ in order to display the corresponding mathematical symbol, Greek letter, or TeX character.

Character Sequence	Symbol	Character Sequence	Symbol	Character Sequence	Symbol
\alpha	α	\upsilon	υ	\sim	~
\angle	∠	\phi	Φ	\leq	≤
\ast	*	\chi	χ	\infty	∞
\beta	β	\psi	ψ	\clubsuit	♣
\gamma	γ	\omega	ω	\diamondsuit	♦
\delta	δ	\Gamma	Γ	\heartsuit	♥
\epsilon	ε	\Delta	Δ	\spadesuit	♠
\zeta	ζ	\Theta	Θ	\leftrightarrow	↔
\eta	η	\Lambda	Λ	\leftarrow	←
\theta	Θ	\Xi	Ξ	\Leftarrow	⇐
\vartheta	ϑ	\Pi	Π	\uparrow	↑
\iota	ι	\Sigma	Σ	\rightarrow	→

Table 1.3b The character sequence that should follow the backslash \ in order to display the corresponding mathematical symbol, Greek letter, or TeX character.

Character Sequence	Symbol	Character Sequence	Symbol	Character Sequence	Symbol
\kappa	κ	\Upsilon	ϒ	\Rightarrow	⇒
\lambda	λ	\Phi	Φ	\downarrow	↓
\mu	μ	\Psi	Ψ	\circ	°
\nu	ν	\Omega	Ω	\pm	±
\xi	ξ	\forall	∀	\geq	≥
\pi	π	\exists	∃	\propto	∝
\rho	ρ	\ni	∋	\partial	∂
\sigma	σ	\cong	≅	\bullet	•
\varsigma	ς	\approx	≈	\div	÷
\tau	τ	\Re	ℜ	\neq	≠
\equiv	≡	\oplus	⊕	\aleph	ℵ
\Im	ℑ	\cup	∪	\wp	℘

Table 1.3c The character sequence that should follow the backslash \ in order to display the corresponding mathematical symbol, Greek letter, or TeX character.

Character Sequence	Symbol	Character Sequence	Symbol	Character Sequence	Symbol
\otimes	⊗	\subseteq	⊆	\oslash	⊘
\cap	∩	\in	∈	\supseteq	⊇
\supset	⊃	\lceil	⌈	\subset	⊂
\int	∫	\cdot	·	\o	ο
\rfloor	⌋	\neg	¬	\nabla	∇
\lfloor	⌊	\times	x	\ldots	...
\perp	⊥	\surd	√	\prime	′
\wedge	∧	\varpi	ϖ	\0	∅
\rceil	⌉	\rangle	〉	\mid	|
\vee	∨			\copyright	©
\langle	〈				

The y-label should be written as follows:

```
>> ylabel('cos^2(x^2)')
```

Let us say that we would like to magnify the x-label to be a font size of 18, red color, and **boldface**. The following command should do the job:

```
>> xlabel('\color{red} \fontsize{18} \bf x')
```

In addition, if we are interested in making the title such that it says that x is bounded between the lower limit, $x_L = -\pi$ and the upper limit, $x_U = \pi$, then the following command should do the job:

```
>>title('Graph of cosine function within x_L= -\pi and x_U= \pi')
```

Notice that the subscript character _ in the title and the superscript character ^ in the y-label will modify the character or substring defined in braces immediately following.

Figure 1.13 shows new cosmetic changes, compared with Fig. 1.12, while changing y to be $y = (\cos(x^2))^2 = \cos^2(x^2)$.

1.8.3 *Plotted Curve (Line) Attributes*

MATLAB allows users to change the color as well as the line style of graphs using a third argument in the plot command. For example, using the same arguments used to generate the plot shown in Fig. 1.13 except the plot statement: **plot(x,y,'m-')** will generate the new plot shown in Fig. 1.14 (*top*) which plots x-y data using magenta (m) color and *solid line* style (-). On the other hand, using

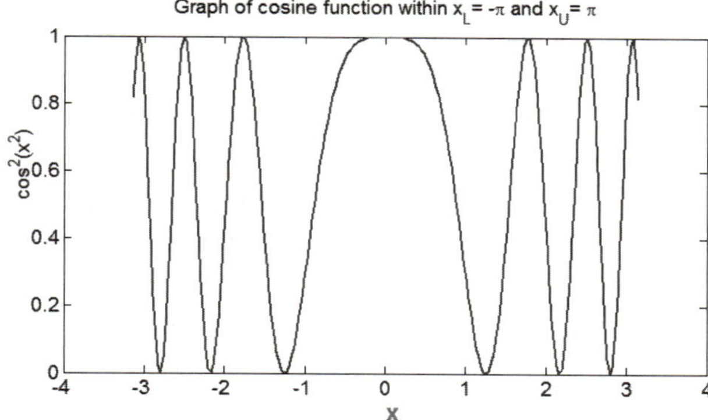

Figure 1.13 Compared with Fig. 1.12, the polished, cosmetic changes are shown as far as superscription, subscription, font size, and font color are concerned. $y = (\cos(x^2))^2 = \cos^2(x^2)$.

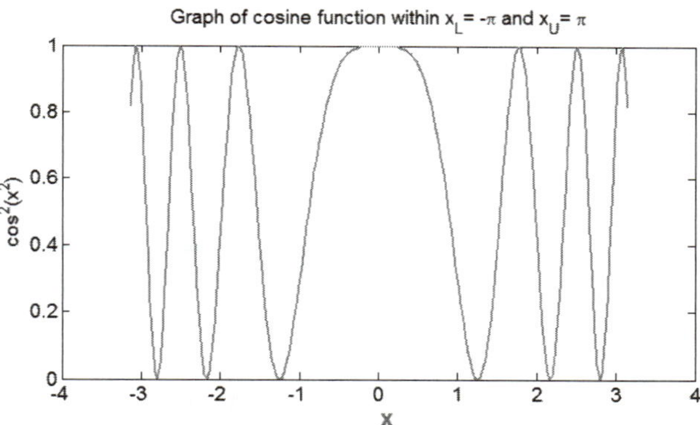

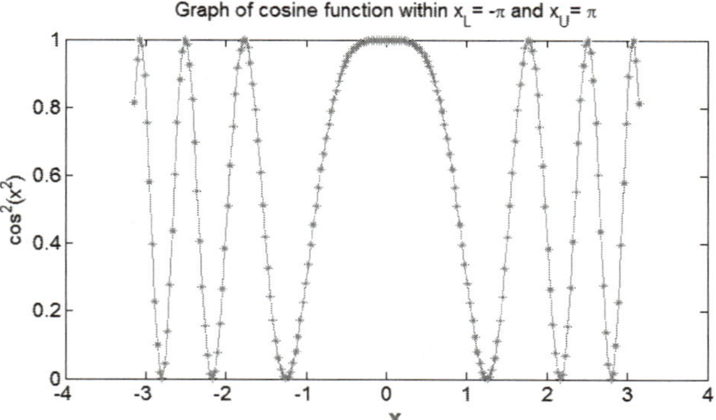

Figure 1.14 Re-plot of Fig. 1.13; *top*: Magenta-colored solid line using **plot(x,y,'m-')**. *Bottom*: Magenta-colored asterisks connected via magenta-colored solid line using **plot(x,y,'m*-')**. Further such options are given in Tables 1.4a, 1.4b, and 1.4c.

the same arguments used to generate the plot shown in Fig. 1.13, except the plot statement: **plot(x,y, 'm*-')** will now show asterisk points connected via a solid line with magenta color, as shown in Fig. 1.14 (*bottom*).

Table 1.4a Line style specifiers.

Specifier	Line Style
-	Solid line (default)
--	Dashed line
:	Dotted line
-.	Dash-dotted line

Table 1.4b Color specifiers.

Color Symbol	Color	Color Symbol	Color
k	Black	m	Magenta
b	Blue	r	Red
c	Cyan	y	Yellow
g	Green	w	White

Table 1.4c Marker specifiers.

Specifier	Marker Type
+	Plus sign
o	Circle
*	Asterisk
.	Point (see note below)
×	Cross
'square' or s	Square
'diamond' or d	Diamond
^	Upward-pointing triangle
v	Downward-pointing triangle
>	Right-pointing triangle
<	Left-pointing triangle
'pentagram' or p	Five-pointed star (pentagram)
'hexagram' or h	Six-pointed star (hexagram)

The grid lines on the graph can be switched on or off using the **grid** command. By issuing this command once, the grid will be turned on. Using it again, the grid will be turned off. Alternatively, it can be switched on and off using the following commands:

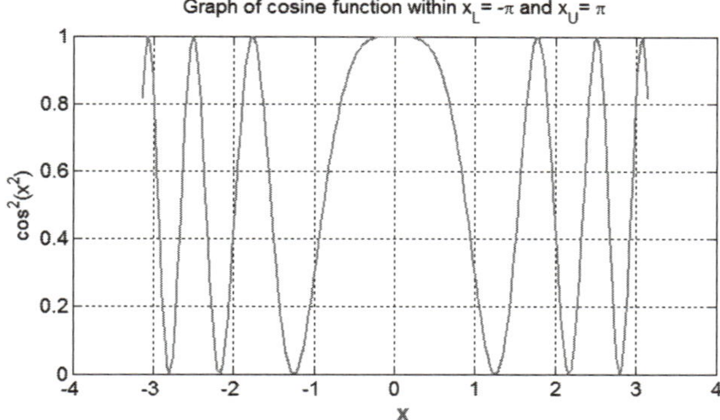

Figure 1.15 Re-plot of Fig. 1.14 (*top*) but with the "grid on" option enabled.

```
>> grid on
>>grid off
```

Figure 1.14 (*top*) is shown again in Fig. 1.15, but this time with the "grid on" feature enabled.

1.8.4 *Super-Positioning of Plots*

Let us say that we would like to have two functions, say, *sin(x)* and *cos(x)*, superposed one above the other. In this case, the plot statement will be:

```
>>N=100; h=pi/N; x= -pi:h:pi; y1=cos(x); y2=sin(x); plot(x,y1,
'm-',x,y2,'b--')
>>title('Graph of cosine & sine functions within x_L= -\pi
and x_U= \pi')
>>xlabel('\color{red}\fontsize{18} \bf x')
>>ylabel('cos(x) & sin(x)')
```

To resolve any ambiguity between the two functions, MATLAB provides what is called the legend, which tells what is what or maps the function with the symbols being used to display it.

```
>>legend('cosine','sine');
```

NOTE Entries of the legend statement are in harmony with the way the functions are introduced in the plot statement. Figure 1.16*a* shows the plot for two simultaneous functions; namely, cosine and sine. The legend box can be positioned anywhere within the plot region. Move the mouse over the legend box; click the left mouse button and while it is being pressed (i.e., the left mouse is down) you may drag the legend box anywhere you wish within the main frame dedicated for drawing; and after you decide on the proper location, just release the left mouse button.

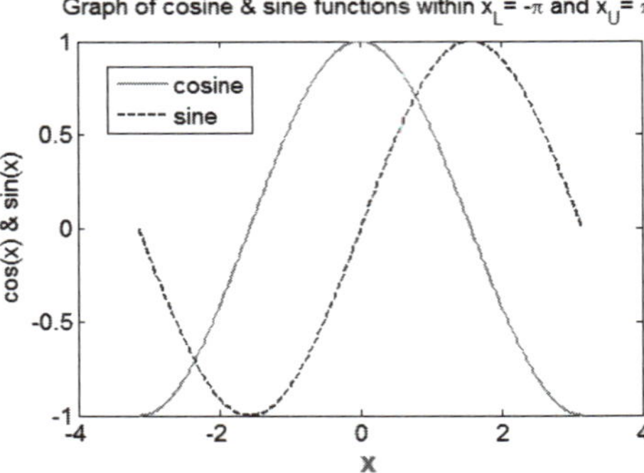

Figure 1.16(a) A plot of two simultaneous functions super-positioned on the same graph.

The number of inputs can also be extended. You can use the format **plot(x1,y1,x2,y2,...)** in which the rules mentioned in the preceding apply for each x and y pair. For example, if you want to plot three lines representing the data sets:

```
>>x1 = 0:.1:10;
y1 = cos(x1);
x2 = 1.5*x1;
y2 = 2*cos(x2);
x3 = 2*x1;
y3 = 2*sin(x3);
```

Then, you can use

```
>>plot(x1,y1,x2,y2,x3,y3)
>>ylabel('y1, y2, y3')
>>xlabel('x1, x2, x3')
>>legend('y1','y2','y3')
```

and Fig. 1.16*b* is generated.

NOTE If you do not define the color attributes, then by default MATLAB currently cycles through the colors blue, green, red, cyan, magenta, yellow, and black when creating multiple lines with a plot command, as shown in Fig. 1.16*b*.

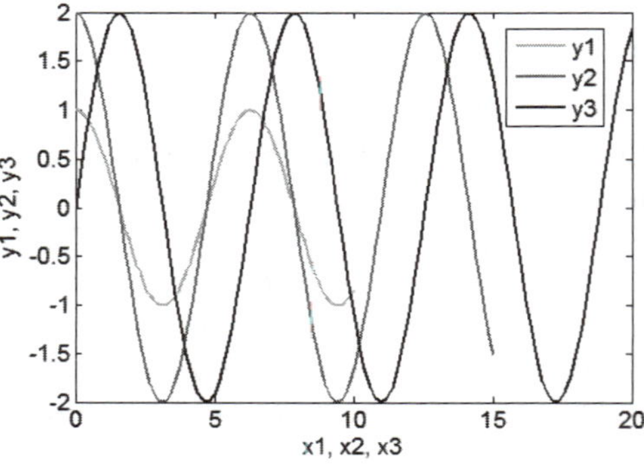

Figure 1.16(b) A plot of three simultaneous functions super-positioned on the same graph.

1.8.5 *Subplotting*

Suppose that we would like to have *sin (x)* be separately plotted from *cos (x)*. MATLAB allows such a luxury via holding on the plot mode and invoking a subplot function. Since we have two functions, we will have two separate plots at this stage. Now we can either sort the two subplots as 2 rows × 1 column, or 1 row × 2 columns. In the first case, the two subplots will be vertically aligned above one another; in the second case, the two subplots will be horizontally aligned beside one another.

Let us consider the first case, that is, 2 by 1 matrix of subplots.

First, define the interval for x to be from 0 to $2*\pi$:

```
>>N=200; h=2*pi/N; x= 0:h:2*pi;
```

Second, we will define both *sin(x)* and *cos(x)*:

```
>> y1=cos(x); y2=sin(x);
```

Third, invoke the subplot statement for each case:

```
>> subplot(211);plot(x,y1);xlabel('x');ylabel('cosine');
>> subplot(212);plot(x,y2);xlabel('x');ylabel('sine');
```

Figure 1.17 shows the MATLAB-generated figure that contains two subplots. The first subplot command picks the first row of this panel and plots the cosine function in it. The second picks the second row and plots the sine function in it. In this way, the two subplots are vertically aligned.

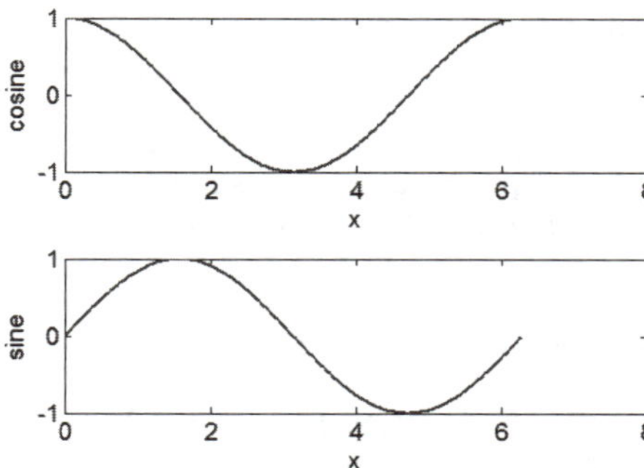

Figure 1.17 2 × 1 matrix of subplots that are vertically aligned.

Let us consider the second case; that is, the 1 × 2 matrix of subplots.

First, define the interval for x to be from 0 to $2*\pi$:

```
>>N=200; h=2*pi/N; x= 0:h:2*pi;
```

Second, we will define both sin(x) and cos(x):

```
>> y1=cos(x); y2=sin(x);
```

Third, invoke the subplot statement for each case:

```
>> subplot(121);plot(x,y1);xlabel('x');ylabel('cosine');
>> subplot(122);plot(x,y2);xlabel('x');ylabel('sine');
```

Figure 1.18 shows the MATLAB-generated figure that contains two subplots. The first subplot command picks the first column of this panel and plots the cosine function in it. The second picks the second column and plots the sine function in it. In this way, the two subplots are horizontally aligned.

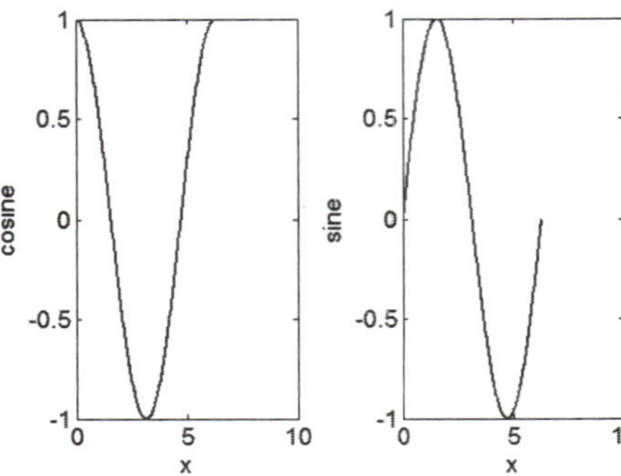

Figure 1.18 1 × 2 matrix of subplots that are horizontally aligned.

If three figures are to be presented, again, you may choose to show them in different ways. Consider the following example:

First, define the interval for x to be from 0 to $2*\pi$:

```
>>N=200; h=2*pi/N; x= 0:h:2*pi;
```

Second, we will define both *sin(x)*, *cos(x)*, and *tan(x)*:

```
>> y1=cos(x); y2=sin(x); y3=tan(x);
```

Presentation with vertical alignment:

Third, invoke the subplot statement for each case:

```
>> subplot(311);plot(x,y1);xlabel('x');ylabel('cosine');
>> subplot(312);plot(x,y2);xlabel('x');ylabel('sine');
>> subplot(313);plot(x,y3);xlabel('x');ylabel('tan');
```

Presentation with horizontal alignment:

Third, invoke the subplot statement for each case:

```
>> subplot(131);plot(x,y1);xlabel('x');ylabel('cosine');
>> subplot(132);plot(x,y2);xlabel('x');ylabel('sine');
>> subplot(133);plot(x,y3);xlabel('x');ylabel('tan');
```

Exercise 1.2 Try to find out how to present four figures as subplots under both vertical and horizontal alignment cases. Consider the fourth plot to be *y4=cot (x)*.

1.8.6 Axes Control

The axes of the graph can be controlled by the user with the help of the **axis** command that accepts a row vector composed of four components.

```
>>axis([xmin xmax ymin ymax]);
```

sets the limits for the x- and y-axis of the current axes.

The first two of these are the minimum and maximum limits for the x-axis and the last two are for y-axis.

Example 1.18

```
>> axis([0 10 0 20]);
```

scales the graph with the x-axis in [0, 10] range and y-axis in [0, 20] range.

In addition, MATLAB also allows users to set these axes with many options:

```
>> axis('auto');
```

scales the graph automatically for the user.

```
>>axis 'auto x'
```

computes only the x-axis limits automatically.

```
>>axis manual
```

or

```
>>axis(axis)
```

freezes the scaling at the current limits, so that if **hold is on (hold on** plot feature is a MATLAB command that retains the current plot and certain axis properties so that subsequent graphing commands add to the existing graph; **hold off** is the default), subsequent plots will use the same limits.

```
>>axis tight
```

sets the axis limits to the range of the data.

```
>>axis equal
```

sets the aspect ratio so that the data units are the same in every direction. The aspect ratio of the x-, y-, and z-axes is adjusted automatically according to the range of data units in the x, y, and z directions.

```
>>axis square
```

makes the current axes region square (or cubed when 3D). This option adjusts the x-axis, y-axis, and z-axis so that they have equal lengths and adjusts the increments between data units accordingly.

```
>>axis off
```

turns off all axis lines, tick marks, and labels.

```
>>axis on
```

turns on all axis lines, tick marks, and labels.

Exercise 1.3 Make a plot using the following command:

```
>> N=100; h=pi/N; x= -pi:h:pi; y1=cos(x); y2=sin(x); plot(x,y1,
'm-',x,y2,'b--')
```

For each of the following **axis** commands, watch out how the x- and y-axes change in terms of minimum, maximum, and location of ticks:

```
>> axis equal
>> axis square
>> axis normal
>> axis ([-3 3 -1 1])
```

1.8.7 Graphics of Functions of Two Variables

A MATLAB surface is defined by the z coordinates associated with a set of (x, y) coordinates. For example, let us create a set of (x, y) coordinates, using the **meshgrid** function, which generates the required x and y matrices:

```
>> [x,y] = meshgrid(-20:20);
```

Then, we define the surface z: $z = \sqrt{x^2 + y^2}$ which is the distance of each (x, y) point from the origin (0, 0).

Using MATLAB:

```
>> z = sqrt(x.^2 + y.^2);
```

We can plot the surface z as a function of x and y (Fig. 1.19):

```
>> mesh(x,y,z);
```

A clearer plot can be produced using a polar grid, instead of a rectilinear grid. First, we define a polar grid of points:

```
>>[r,theta] = meshgrid(0:.01:1, 0:pi/60:2*pi);
```

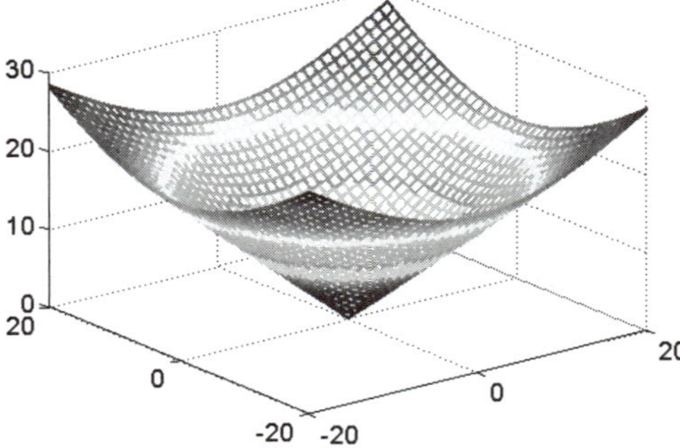

Figure 1.19 3D plot showing the distance, z (measured from the origin), as a function of x, y.

Then we define *x*, *y*, and *z* in terms of polar coordinates as follows:

```
>> x=r.*cos(theta);
>> y=r.*sin(theta);
>> z=r;
```

Use the mesh command to produce the 3D plot (Fig. 1.20):

```
>> mesh(x,y,z);
```

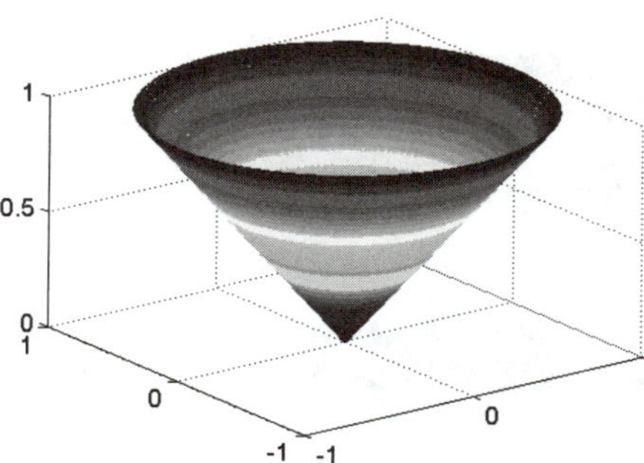

Figure 1.20 3D plot utilizing polar coordinates.

The following commands will produce a two-cone shape (Fig. 1.21):

```
>>[r,theta] = meshgrid(0:.01:1, 0:pi/60:2*pi);
>> x=r.*cos(theta);
>> y=r.*sin(theta);
>> z=r;
>> mesh(x,y,z);
>> hold on;
>> mesh(-x,-y,-z);
```

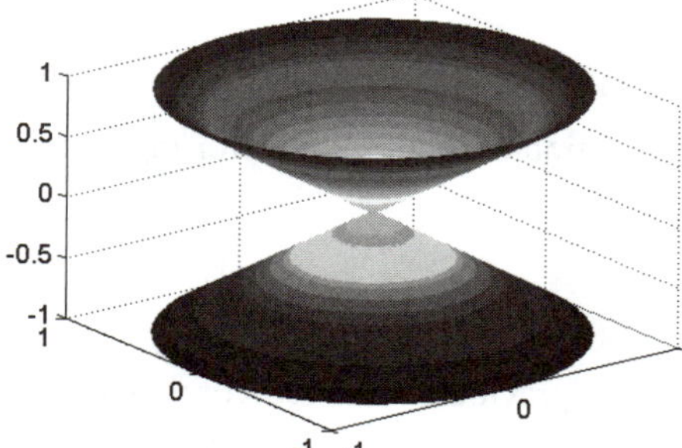

Figure 1.21 3D plot for a two-cone shape utilizing polar coordinates.

Different plots can be produced if the user manipulates values of *r*, *theta*, and the scales for *x*, *y*, and *z*. The following commands produce the multicone shape shown in Fig. 1.22.

```
>> [r2,theta] = meshgrid(0:-.005:-1, 0:pi/100:2*pi);
>> x2=r2.*cos(theta);
>> y2=r2.*sin(theta);
>> z2=r2;
>> mesh(x2,y2,z2);
hold on;
>> mesh(x2,y2,-z2);
>> mesh(2.*x2,2.*y2,4.*z2);
>> mesh(2.*x2,2.*y2,-4.*z2);
```

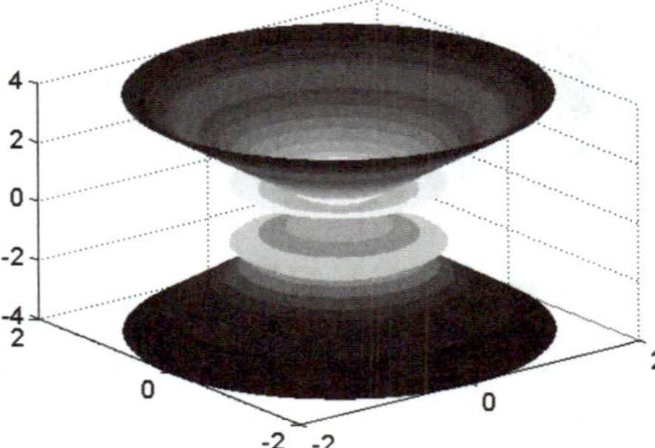

Figure 1.22 3D plot for a multicone shape utilizing polar coordinates.

1.9 Interrupting Calculations

If MATLAB is hung up in a calculation, or is just taking too long to perform an operation, you can usually abort it by typing CTRL+C (that is, hold down the key labeled CTRL, or CONTROL, and press C).

1.10 Syntax Errors

If you make an error in an input line, MATLAB will beep and print an error message. For example, here's what happens when you try to evaluate:

```
>> cos pi
```

MATLAB's output is:
??? Undefined function or method 'cos' for input arguments of type 'char'.

```
>> cos (pi)
ans =
  -1
```

To fix an error in a previously entered command, you can use the UP-ARROW key to redisplay the previous command; edit the command using the LEFT- and RIGHT-ARROW keys; fix the error; and then press RETURN or ENTER. The UP- and DOWN-ARROW keys allow you to scroll back and forth through all the commands you have typed in a MATLAB session and are very useful when you want to correct, modify, or reenter a previous command. In addition, as mentioned earlier in Sec. 1.1, you may use the History Command Window to highlight any previous command, copy it, and paste it into the Command Window where you can modify and finally re-evaluate the command.

See Sec. 3.5 for more information on types of errors involved in MATLAB scripting.

1.11 Suppressing Output

Typing a semicolon at the end of an input line suppresses printing of the output of the MATLAB command. The semicolon should generally be used when defining large vectors or matrices (such as **X = 1:0.01:3**;). It can also be used in any other situation where the MATLAB output need not be displayed.

1.12 Problems

1.1 Using the MATLAB Command Window and its built-in library functions (see Table 1.1), evaluate the following expressions:

a) $\ln(10)$, $\log(10)$, $\sqrt{7}$, $\sqrt{-7}$, $e^{3.5}$

b) Given that $A = 5 + \sqrt{-8}$, find the imaginary, real, and conjugate term of A

c) Given that $B = 1.5$, find the *cosine(B)*, *sine(B)*, *cosh(B)*, and *sinh(B)*

d) Given that $C = 2.345$, find ceil(C), fix(C), floor(C), and round(C)

e) Given that the angle is expressed in *degrees* as: deg_angle = 60° and in *radians* as rad_angle = pi/3, find the cosine, sine, and tangent value of the angle

1.2 Using MATLAB's Symbolic Math Toolbox (see Sec. 1.6), carry out the following algebraic steps:

a) Expand the term: (x-3)*(x+5)

b) Compact (put it into brackets) the term: $x^2 - 2x - 8$

c) Simplify the term: ((x^2-x-2)/(x-2))

d) Simplify the term: ((x^2-x+7)/(x+2))

1.3 If $D = \frac{1}{7}$, $E = \frac{1}{6}$, and $F = \frac{1}{5}$, using the symbolic notation, which avoids the floating point approximations inherent in quotients, carry out the following mathematical operations:

a) `>>2*D+3*F`

b) `>>E/F`

c) `>>(D+F)/E`

d) `>>E^F`

1.4 Given that $G = \begin{bmatrix} 1 & 2 & 7 \\ 3 & 4 & 8 \\ 5 & 6 & 9 \end{bmatrix}$ and $H = \begin{bmatrix} 9 & 8 & 7 \\ 6 & 5 & 4 \\ 3 & 2 & 1 \end{bmatrix}$, evaluate the following expressions:

Hint: To define **G** in MATLAB, at the command prompt: **>>G = [1, 2, 7;3, 4, 8;5, 6, 9]**

a) `>>I=2*G+3*H`

b) `>>J=4*G-2*H`

c) `>>K=[2*G,5*H]`

d) `>>L=[2*G;5*H]`

e) Assuming ***L*** represents a distribution of some discrete values, find the minimum, maximum, mean, median, standard deviation, and variance of ***L***.

1.5 Using the colon operator, **:** (Sec. 1.7.3), create a row vector called **row_vect** and column vector called **col_vect**, each with 100 different data points such that all lie within the range [0 1].

1.6 Produce the following plot using:

>> N=100; h=pi/N; x= -pi:h:pi; y=(cos(x)).^2; plot(x,y)

with a red color 16 font size for the *y*-axis; blue color with 16 font size for the *x*-axis; and with the title: "Graph of $\cos^2$ (x) function within $[-\pi, \pi]$ range" (Fig. 1.23).

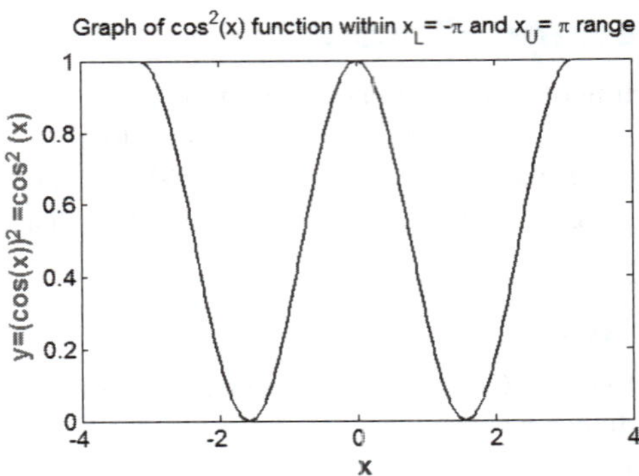

Figure 1.23

1.7 Run the following code to see what kind of 3D object you get:

```
>> [r2, theta] = meshgrid(0:-.01:-1, 0:pi/120:2*pi);
>> x2=r2.*cos(2*pi-theta);
>> y2=r2.*sin(theta);
>> z2=r2;
>> mesh(x2,y2,z2);
```

1.8 Run the following code to see what kind of 3D object you get:

```
>> [x,y] = meshgrid(-200:200);
>> z = x.^2+y.^2;
>> mesh(x,y,z);
```

1.9 Change **z** in Problem 1.8 to **z =x.^3+y.^2;** and see the new 3D object.

1.10 Repeat with **z =x.^4+y.^3;** and see the newest 3D object.

2 Living with Matrices

Chapter Outline

In Sec. 1.7, vectors, simplified versions of matrices, were introduced. This chapter focuses on matrices because they represent the symbolic language that can be easily understood and assimilated by MATLAB.

2.1 Defining Matrices

A matrix is essentially a 2D array of numbers composed of rows (row vectors) and columns (column vectors). A matrix can be entered in MATLAB in any of the following three ways:

1. Using the carriage return key:

```
>> A=[1 2 3
4 5 6
7 8 9 ];
```

2. Using semicolons to indicate the next line:

```
>> B=[3 4 5; 6 7 8; 9 10 11];
```

3. Using the colon (range) notation (see Sec. 1.7.3) with semicolon:

```
>> C=[2:4; 5:7; 8:10];
```

Notice how MATLAB saves variables such as **A**, **B**, and **C** (Fig. 2.1). On the other hand, some matrices can be defined simply by using functions. For example, the **zeros** function defines a matrix with all entries zeros; the function **ones** defines a matrix filled with ones; and **rand** defines a matrix with all entries random numbers in the [0, 1] range.

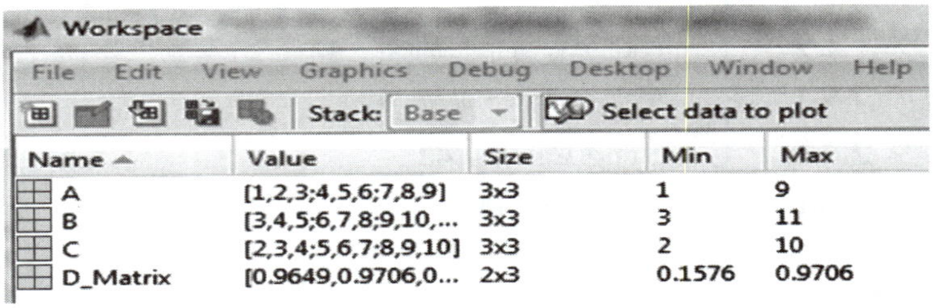

Name	Value	Size	Min	Max
A	[1,2,3;4,5,6;7,8,9]	3x3	1	9
B	[3,4,5;6,7,8;9,10,...	3x3	3	11
C	[2,3,4;5,6,7;8,9,10]	3x3	2	10
D_Matrix	[0.9649,0.9706,0...	2x3	0.1576	0.9706

Figure 2.1 The Workspace Window shows how MATLAB stores variables A, B, C, & D_Matrix. The first three are 3×3 matrices and the last is a 2×3 matrix.

Example 2.1

```
>> zeros(2)
ans =
     0     0
     0     0
>> ones(2)
ans =
     1     1
     1     1
```

```
>> rand (3)
ans =
        0.8147   0.9134   0.2785
        0.9058   0.6324   0.5469
        0.1270   0.0975   0.9575
```

The argument in each case is the size of the matrix. The same functions can also be used for defining some nonsquare matrix.

```
>> Any_Matrix = rand (r,c)
```

means that **Any_Matrix** will be randomly created with **r** rows and **c** columns.

```
>> D_Matrix=rand (2,3)
D_Matrix =
    0.9649   0.9706   0.4854
    0.1576   0.9572   0.8003
```

So, **D_Matrix** has two rows and three columns (see Fig 2.1).

2.1.1 *Referencing an Element in a Matrix*

Let us say that we would like to manipulate a given element in a matrix by some further arithmetic operations or introduce some definitions based on a particular combination of matrix elements. For example, if **A** is a 2×2 (square) matrix of the form:

$$A = \begin{bmatrix} a_{11} & a_{12} \\ a_{21} & a_{22} \end{bmatrix} \tag{2.1}$$

then each element is subscribed inside by two indices: the first index stands for the row location of the element and the second index stands for the column location. Thus, the element that lies top left will be located in the first row and first column; that is why it is subscribed by 11. The bottom left element lies in the second row and first column; that is why it is subscribed by 21. Consider the following example.

Example 2.2 Consider the matrix A:

$$A = \begin{bmatrix} 3 & 7 \\ 5 & 9 \end{bmatrix}$$

which can be defined in MATLAB (see Sec. 2.1) as follows:

```
>> A=[3 7; 5 9]
A =
     3   7
     5   9
```

For example, the determinant of a 2 × 2 (square) matrix, shown in Eq. (2.1), is defined as:

$$\det(A) = a_{11} \times a_{22} - a_{12} \times a_{21} \quad (2.2)$$

Application of Eq. (2.2) to the previous example will result in:

$$\det(A) = 3 \times 9 - 7 \times 5 = 27 - 35 = -8 \quad (2.3)$$

Let us apply these concepts using MATLAB:

```
>>Adet=det (A)
Adet=
    -8
```

which is the same as that of Eq. (2.3). Also, notice that you can recall the definition of determinant using Eq. (2.2). Thus:

```
>> det_of_A=A(1,1)*A(2,2)-A(1,2)*A(2,1)
det_of_A =
          -8
```

Notice how each element in **A** is referenced; the name of the matrix followed by two parentheses bracketing the two indices for that particular element.

2.2 Size of Matrices

The size function returns the size of any matrix. For example, if you invoke the size statement for matrices previously defined, you will see the following:

Example 2.3

```
>> size (A)
ans =
        3    3
>> size (D_Matrix)
ans =
        2    3
```

Notice that the size of a variable, like a matrix, is also shown in Fig. 2.1 via the Workspace Window.

2.3 The Identity Matrix

The identity matrix has the form:

$$I = \begin{bmatrix} 1 & \cdots & 0 \\ \vdots & \ddots & \vdots \\ 0 & \cdots & 1 \end{bmatrix}$$

This means that all off-diagonal elements are zero except those that lie on the diagonal that extends from top left to bottom right. For example, the 3 × 3 identity matrix will be:

$$I = \begin{bmatrix} 1 & 0 & 0 \\ 0 & 1 & 0 \\ 0 & 0 & 1 \end{bmatrix}$$

The identity matrix can be defined in MATLAB with the help of the 'eye' function with an argument representing the size of the matrix:

Example 2.4

```
>> eye(2)
ans =
     1     0
     0     1
>> eye(3)
ans =
     1     0     0
     0     1     0
     0     0     1
```

2.4 Diagonal Matrix

The diagonal matrix is similar to the identity matrix in that both have off-diagonal elements that are all zeros. For a square matrix of size **n** × **n**, the diagonal will be of size **n**. This diagonal, which is a row vector, is used as an argument to the **diag** function:

Example 2.5

```
>> diag ([1,2])
ans =
     1     0
     0     2
```

Example 2.6

```
>>kiko= diag ([1,2,4,5])
kiko =
     1     0     0     0
     0     2     0     0
     0     0     4     0
     0     0     0     5
>> sum(diag(kiko))
ans =
    12
```

2.5 Specialized Matrices

With Version 7.9.0 (R2009b), MATLAB has 12 built-in specialized matrices. Of course, you can search in the built-in Product Help using the keywords "specialized matrices." Examples of specialized matrices are:

1. Companion Matrix

Example 2.7 Find the roots of the polynomial $(x - 1)(x + 2)(x - 3) = (x - 1)(x^2 - x - 6) = x^3 - 2x^2 - 5x + 6$.

First, we define the coefficients of polynomial matrix as **u**, starting with the highest order in x:

```
>>u = [1  -2  -5  6]
```

Second, the companion of **u** is defined by:

```
>>A = compan(u)
A =
     2     5    -6
     1     0     0
     0     1     0
```

Third, the eigenvalues of **A** are basically the roots of the polynomial:

```
>> TheRoots=eig(A)
TheRoots =
   -2.0000
    3.0000
    1.0000
```

This answer also represents the roots of row vector **u**.

```
>> Theroots=roots(u)
Theroots =
   -2.0000
    3.0000
    1.0000
```

Example 2.8 Given the polynomial: $3x^3 - 5x^2 - 8x + 6$, find its roots. Form the row vector made up of the coefficients of x^3, x^2, x, and the constant.

```
>> Coeff_Vect=[3 -5 -8 6]
Coeff_Vect =
     3    -5    -8     6
>> Compan_Mat=compan(Coeff_Vect)
Compan_Mat =
    1.6667    2.6667   -2.0000
    1.0000    0         0
    0         1.0000    0
```

```
>> Roots=eig(Compan_Mat)
Roots =
    2.4260
   -1.3638
    0.6045
```

Alternatively,

```
>> The_Roots=roots(Coeff_Vect)
The_Roots =
    2.4260
   -1.3638
    0.6045
```

Both approaches yield the same result. The polynomial becomes:

$$3x^3 - 5x^2 - 8x + 6 = (x - 2.426)(x + 1.3638)(x - 0.6045)$$

Exercise 2.1

1. Given the polynomial: $-3x^3 + 5x^2 + 8x - 6$, find its roots.
2. Given the polynomial: $3x^3 - 5x^2 + 8x + 6$, find its roots.
3. Given the polynomial: $3x^3 + 5x^2 + 8x + 6$, find its roots.

What is the difference between one case and another, as far as the type and sign of the root are concerned?

2. Magic Matrix

A magic matrix: M = magic(n) returns an n-by-n matrix constructed from the integers 1 through n^2 with equal row and column sums. The order n must be a scalar greater than or equal to 3.

Example 2.9

```
>>M = magic(3)
M =
     8     1     6
     3     5     7
     4     9     2
```

This is called a magic square because the sum of the elements in each column is the same.

```
>> Column_sum=sum(M) [or, use sum(M,1))
Column_sum =
       15     15     15
```

This is also equivalent to:

```
>>Columnsum=sum(M,1)
Columnsum =
           15     15     15
```

And the sum of the elements in each row is the same.

```
>>Row_sum=sum(M,2)
Row_sum =
       15
       15
       15
```

This is also a special magic square because the diagonal elements have the same sum.

```
>>Diagsum=sum(diag(M))
Diagsum =
     15
```

3. Pascal Matrix
A = pascal(n) returns the Pascal matrix of order n: a *symmetric positive* definite matrix with integer entries taken from Pascal's triangle. It has eigenvalues ranging in magnitude from 10^{-4} to 10^{3}. The *inverse of A also has integer entries*. It has applications in electrical engineering (Psenicka, B., "The Bilinear Z Transform by Pascal Matrix and Its Application in the Design of Digital Filter" *IEEE Signal Proc Let*, 9: 368–370, 2002) and in applied mathematics (see Call G. S. and Vellman D. J., "Pascal's Matrices," *Am Math Mon*, 372–376, April 1993; Zobitz J. M., "Pascal Matrices and Particular Solutions to Differential Equations," *IIME Journal*, 11 (8): 437–444, 2003).

Example 2.10

```
>> My_Pascal=pascal(2)
My_Pascal =
     1     1
     1     2
```

A is a symmetric matrix if it fulfills the following condition: $a_{ij} = a_{ji}$ for all i & j.

```
>> Pas_inv=inv(My_Pascal)
Pas_inv =
   2    -1
  -1     1
>> eigenval=eig(My_Pascal)
eigenval =
    0.3820
    2.6180
```

2.6 The Colon Operator

The colon operator (see Sec. 1.7.3) is extremely powerful, and provides a very efficient way of handling matrices, e.g., if Z is the matrix defined by:

$$Z = \begin{bmatrix} 1 & 3 & 5 \\ 7 & 9 & 11 \\ 13 & 15 & 17 \end{bmatrix}$$

This is how it is introduced in MATLAB:

Example 2.11

```
>> Z=[1 3 5;7 9 11; 13 15 17]
Z=
     1     3     5
     7     9    11
    13    15    17
```

Let us evaluate some expressions containing colon (range) notation.

```
>>Z(:,1)
ans =
     1
     7
    13
```

`Z(:,1)` is equivalent to $\begin{bmatrix} Z_{11} \\ Z_{21} \\ Z_{31} \end{bmatrix} = \begin{bmatrix} 1 \\ 7 \\ 13 \end{bmatrix}$.

Here, we fix the column index to be 1 and permutate rows from 1 to 3.

```
>>Z(1,:)
ans =
1     3     5
```

`Z(1,:)` is equivalent to $\begin{bmatrix} Z_{11} & Z_{12} & Z_{13} \end{bmatrix} = \begin{bmatrix} 1 & 3 & 5 \end{bmatrix}$.

Here, we fix the row index to be 1 and permutate columns from 1 to 3.

```
>> Z(1:2,1:2)
ans =
     1     3
     7     9
```

`Z(1:2,1:2)` is equivalent to $\begin{bmatrix} Z_{11} & Z_{12} \\ Z_{21} & Z_{22} \end{bmatrix} = \begin{bmatrix} 1 & 3 \\ 7 & 9 \end{bmatrix}$.

Here, we permutate both columns and rows from 1 to 2.

```
>>Column_Vect=Z(:)

Column_Vect =
     1
     7
    13
     3
     9
    15
     5
    11
    17
```

NOTE This operation fills in, column-wise, the content of a matrix into a column vector. You can refill the content of a column vector into a new matrix.

Example 2.12 Create a new matrix, New_Matrix, from a 3 × 3 zero-based matrix.

```
>>New_Matrix=zeros(3)
New_Matrix =
      0      0      0
      0      0      0
      0      0      0
```

Create or fill in the new matrix, **New_Matrix**, from the previously defined (Example 2.11) column vector, **Column_Vect**.

```
>> New_Matrix (:)=Column_Vect(:)
New_Matrix =
      1      3      5
      7      9     11
     13     15     17
```

NOTE Use the colon operator and the empty array to delete entire rows or columns.

Example 2.13

```
>> New_Matrix(:,3) = [ ]
New_Matrix =
      1      3
      7      9
     13     15
```

deletes the third column of **New_Matrix**.

Example 2.14 Use the colon to extract a sub-matrix out of the given matrix

$$Z = \begin{bmatrix} 1 & 3 & 5 \\ 7 & 9 & 11 \\ 13 & 15 & 17 \end{bmatrix}.$$

```
>> Sub_Z=Z(:,[1,2])
Sub_Z =
      1      3
      7      9
     13     15
```

This operation extracts the first two columns out of **Z**. You can also get the same result using the **logical** operator on **Z**:

```
>> Same_Sub_Z=Z(:,logical([1 1 0]))
Same_Sub_Z =
      1      3
      7      9
     13     15
```

2.7 Manipulating Matrices

Here are some functions for manipulating matrices. See help for details.

Table 2.1 MATLAB built-in functions that are operable on matrices.

#	Function	Description
1	**det**	The determinant of a matrix
2	**eig**	Eigenvalue decomposition
3	**exp**	Exponential of a matrix; element-by-element
4	**fliplr**	Flips from left to right
5	**flipud**	Flips from up to down
6	**inv**	The inverse of a matrix
7	**lu**	***[L,U,P] = lu(A)*** returns unit lower triangular matrix ***L***, upper triangular matrix ***U***, and permutation matrix ***P*** so that *P*A = L*U*
8	**rot90(A)**	90-degree counterclockwise rotation of matrix A
9	**rot90(A,K)**	K*90-degree rotation of A, K = ±1, ±2, . . .
10	**tril**	Extracts the lower triangular part
11	**triu**	Extracts the upper triangular part

Example 2.15

```
>> A=[1 2 3; 4 5 6; 7 8 9]
A =
     1  2  3
     4  5  6
     7  8  9
>> Flipped_from_L_2_R=fliplr(A)
Flipped_from_L_2_R =
     3  2  1
     6  5  4
     9  8  7
>> Flipped_from_U_2_D=flipud(A)
Flipped_from_U_2_D =
     7  8  9
     4  5  6
     1  2  3
>> Counter_Clock_90_Rot= rot90(A)
Counter_Clock_90_Rot =
     3  6  9
     2  5  8
     1  4  7
>> Lower_Triangle=tril(A)
Lower_Triangle =
     1  0  0
     4  5  0
     7  8  9
```

Notice that the upper triangle of *A* is set to zero.

```
>> Upper_Triangle=triu(A)
Upper_Triangle =
     1  2  3
     0  5  6
     0  0  9
```

Here, the lower triangle of **A** is set to zero.

2.8 Array (Element-by-Element) Operations on Matrices

What is the difference between $A.*B$ vs. $A*B$? The difference between the two types of matrix multiplication will be realized toward the end of Sec. 2.9.

Example 2.16 Consider the following two matrices **A** & **B**:

$$A = \begin{bmatrix} 1 & 2 \\ 3 & 4 \end{bmatrix} \qquad B = \begin{bmatrix} 5 & 6 \\ 7 & 8 \end{bmatrix}$$

A.*B is an **element-wise** (i.e., element by element) multiplication of A & B. That means C elements are created as follows:

$$C = A.*B = \begin{bmatrix} a_{11} \times b_{11} & a_{12} \times b_{12} \\ a_{21} \times b_{21} & a_{22} \times b_{22} \end{bmatrix} \tag{2.4}$$

Applying Eq. (2.4) to the previous matrices:

$$C = A.*B = \begin{bmatrix} 1\times5 & 2\times6 \\ 3\times7 & 4\times8 \end{bmatrix} = \begin{bmatrix} 5 & 12 \\ 21 & 32 \end{bmatrix}$$

Using MATLAB

```
>> A=[1 2;3 4]
A =
     1     2
     3     4
>> B=[5 6;7 8]
B =
     5     6
     7     8
>> C=A.*B
C=
     5    12
    21    32
```

NOTE Because **A.*B** is carried out on elemental level, it requires that both **A** and **B** have the same size (see Sec. 2.2). **Matrix addition and subtraction** are defined based on an element-by-element array operation.

Example 2.17

$$A=\begin{bmatrix}1 & 2\\ 3 & 4\end{bmatrix} \qquad B=\begin{bmatrix}5 & 6\\ 7 & 8\end{bmatrix}$$

$$E=A+B=\begin{bmatrix}1+5 & 2+6\\ 3+7 & 4+8\end{bmatrix}=\begin{bmatrix}6 & 8\\ 10 & 12\end{bmatrix}$$

$$F=A-B=\begin{bmatrix}1-5 & 2-6\\ 3-7 & 4-8\end{bmatrix}=\begin{bmatrix}-4 & -4\\ -4 & -4\end{bmatrix}$$

2.9 Matrix Multiplication

The matrix multiplication for A and B requires that the *number of columns of A must equal the number of rows of B*; otherwise the matrix multiplication cannot be carried out.

$$\begin{aligned}D(m\times k)=A(m\times n)*B(n\times k)&=\begin{bmatrix}a_{11} & a_{12}\\ a_{21} & a_{22}\end{bmatrix}*\begin{bmatrix}b_{11} & b_{12}\\ b_{21} & b_{22}\end{bmatrix}\\ &=\begin{bmatrix}a_{11}\times b_{11}+a_{12}\times b_{21} & a_{11}\times b_{12}+a_{12}\times b_{22}\\ a_{21}\times b_{11}+a_{22}\times b_{21} & a_{21}\times b_{12}+a_{22}\times b_{22}\end{bmatrix}\end{aligned} \tag{2.5}$$

Example 2.18 Consider the previous two matrices ***A*** & ***B***:

$$A=\begin{bmatrix}1 & 2\\ 3 & 4\end{bmatrix} \qquad B=\begin{bmatrix}5 & 6\\ 7 & 8\end{bmatrix}$$

Application of Eq. (2.5) results in:

$$D=\begin{bmatrix}1 & 2\\ 3 & 4\end{bmatrix}*\begin{bmatrix}5 & 6\\ 7 & 8\end{bmatrix}=\begin{bmatrix}1\times5+2\times7 & 1\times6+2\times8\\ 3\times5+4\times7 & 3\times6+4\times8\end{bmatrix}=\begin{bmatrix}19 & 22\\ 43 & 50\end{bmatrix}$$

Using MATLAB:

```
>> D=A*B
D =
     19      22
     43      50
```

which is the same answer as that given by Eq. (2.5).

NOTE So far, you should realize the difference between A.^2 and A*A. The first accounts for an element by element self-multiplication for the matrix A; the second accounts for the typical matrix by matrix multiplication as defined by Eq. (2.5) for a 2×2 matrix.

Example 2.19 Consider the following two matrices A & B:

$$A=\begin{bmatrix}1 & 2\\3 & 4\end{bmatrix} \qquad B=\begin{bmatrix}2\\3\end{bmatrix}$$

$$D(2\times1)=A(2\times2)*B(2\times1)=\begin{bmatrix}a_{11} & a_{12}\\a_{21} & a_{22}\end{bmatrix}*\begin{bmatrix}b_{11}\\b_{21}\end{bmatrix}=\begin{bmatrix}a_{11}\times b_{11}+a_{12}\times b_{21}\\a_{21}\times b_{11}+a_{22}\times b_{21}\end{bmatrix}$$

$$D(2\times1)=\begin{bmatrix}1\times2+2\times3\\3\times2+4\times3\end{bmatrix}=\begin{bmatrix}8\\18\end{bmatrix}$$

NOTE The product matrix, **D**, has the number of rows equal to that of **A** and the number of columns equal to that of **B**. In general, $\mathbf{A}*\mathbf{B}\neq\mathbf{B}*\mathbf{A}$. In fact, in Example 2.19, **B*A** cannot be carried out.

2.9.1 *Matrix Exponentiation*

The matrix operation A^2 means $A\times A$, where A must be a square matrix. The operator ^ is used for matrix exponentiation.

Example 2.20

```
>> A=[1 2;3 4]
A =
     1     2
     3     4
>> A^2
ans =
     7    10
    15    22
>> A*A
ans =
     7    10
    15    22
```

2.10 String Arrays

2.10.1 *Assignment*

Strings may be assigned to variables by enclosing them in single quotes:

Example 2.21

```
>> s1= 'Hi Reader'
s1 =
Hi Reader
```

If an apostrophe is to be included in the string, it must be repeated:

Example 2.22

```
>> s2='Reader''s opinion'
s2 =
Reader's opinion
```

2.10.2 *Input*

Strings may be entered in response to the input statement in two ways:

1. Using the generic symbol '*s*' which relieves the user from entering the bracketing apostrophes:

   ```
   >> name = input( 'Please, Enter your Name: ', 's' );
   Please, Enter your Name: Kimo
   ```

   Once you run the first statement, MATLAB prompts you, asking you to enter your name at the prompt but without apostrophes.

2. Entering the string but with two bracketing apostrophes:

   ```
   >> name = input( 'Please, Enter your Name: ' );
   Please, Enter your Name: 'Kimo'
   ```

For the second case, if you try to enter your name without bracketing apostrophes, then you will get an error message from MATLAB as in the following example:

Example 2.23

```
>> name = input( 'Please, Enter your Name: ' );
Please, Enter your Name: Kimo
??? Error using ==> input
Undefined function or variable 'Kimo'.
Please, Enter your Name:
```

Notice that MATLAB did not accept Kimo as an input because it was not bracketed in two apostrophes and it prompts the user to enter the name a second time.

2.10.3 *Concatenation of Strings*

Strings are vectors; they may be concatenated (combined) with square brackets:

Example 2.24

```
>> My_Name='Kamal'
My_Name =
Kamal
>> My_Full_Name=[My_Name,' Al-Malah']
My_Full_Name =
Kamal Al-Malah
```

2.10.4 Interconversion between Double and Char

You can convert between double (numbers) and characters (strings) via the following commands:

Example 2.25

```
>> double('Kamal')
ans =
     75     97    109     97    108
```

Vice versa, you may convert from numbers to strings:

```
>> Num_2_Letter=char(65:90)
Num_2_Letter =
             ABCDEFGHIJKLMNOPQRSTUVWXYZ
```

That means 65 represents A and 90 represents Z; are orderly represented in ascending fashion.

Example 2.26 To see the equivalence between lowercase alphabets (i.e., a to z) and their numerical representation, apply the following command:

```
>> Numerical_Equivalence_alphabets = double('a'):double('z')
Numerical_Equivalence_alphabets =

  Columns 1 through 15
97     98     99    100    101    102    103    104    105    106    107
108    109    110    111

  Columns 16 through 26
112    113    114    115    116    117    118    119    120    121    122
```

You can use **char** and **double** to generate rows of identical characters.

Example 2.27

```
>> x = char(ones(3,10)*double('?'))
x =
??????????
??????????
??????????
```

Notice that **ones(3,10)** creates a numerical matrix with all entries set to one being multiplied by the equivalent numerical value of the question mark "?". This will not change the numerical value of "?" as it is multiplied by one. The **char** operator will convert back to "?".

Example 2.28

```
>> x = char(round(rand(10,10))*double('*'))
x =
* **  *  *
  * *** *
*  * *  *
 *  *  **
 **** *
 ** * * *
      ****
      ** *
   *   ***
 * * ***
```

NOTE If you rerun Example 2.28, you may end up with a different image as the function **rand** assigns different random numbers each time it is invoked.

If a character variable is involved in an arithmetic expression, MATLAB uses the ASCII code of the character in the calculation.

Example 2.29

```
>> name='Nimo'
name =
Nimo
>> double(name)
ans =
    78   105   109   111
>> name+1
ans =
    79   106   110   112
```

Notice that the variable **name** is assigned a string value equal to "Nimo"; once you add 1 to it MATLAB converts **Nimo** to its numerical equivalence and then adds 1 to the numerical value of each character.

Notice the numerical representation of ASCII characters, including the Arabic numbers 0, 1, 2, …, 8, and 9, as in the following example:

Example 2.30

```
>> double('0'):double('9')
ans =
    48    49    50    51    52    53    54    55    56    57
```

2.10.5 *Strings Comparison*

The alphabetical sorting of names, variables, or entries of a column in a spreadsheet is based on the numerical representation of ASCII characters. For example, let us sort the following names alphabetically based on the first character in each name:

`7.9Version, matlab, Matlab, R2009b`

The comparison is carried out based on the numerical equivalence of the first character appearing in each name. If the numerical representation for the first character is equal, then the comparison is carried out using the second character in each name. If required, you may use the trailing characters (i.e., third, fourth, etc.). The process of comparison continues until we sort all names.

To see the numerical equivalence of the *first letter*, because they are all different from one another, apply the double command for each:

Example 2.31 Sorting the previous words based on the first character of each

```
>> Char_2_Dbl=double(['7' 'm' 'M' 'R'])
Char_2_Dbl =
          55   109    77    82
```

Based on the numerical value assigned to the first character, the alphabetical *ascending sorting* is as follows:

7.9Version should come, first; **Matlab**, second; **R2009b**, third; and **matlab**, fourth.

You can define an array of characters or strings (MATLAB calls it cell array) as follows:

Example 2.32

```
>> String_Array={['7.9Version']; ['matlab']; ['Matlab'];
['R2009b']}
or,
>>String_Array2={'7.9Version'; 'matlab'; 'Matlab'; 'R2009b'}
String_Array2 =
     '7.9Version'
     'matlab'
     'Matlab'
     'R2009b'
>> Sorted_Names=sort(String_Array2)
Sorted_Names =
     '7.9Version'
     'Matlab'
     'R2009b'
     'matlab'
```

The ascending order (sorting) is similar to what has been explained earlier in Example 2.31.

NOTE Both **String_Array** and **String_Array2** are 4 × 1 cell arrays. Always refer to the MATLAB Workspace Browser window to see what kinds of matrices are created as a result of executing any assignment statement (or, command).

Exercise 2.2 Try the following assignment commands and see how MATLAB (via the Workspace Browser window) treats each case:

```
>> Char_Array=['This ', 'shows ', 'the difference', ' in
definition']
>> OneCell_Array={['This ', 'shows ', 'the difference', ' in
definition']}
>> MultiCell_Array={'This ', 'shows ', 'the difference', '
in definition'}
```

2.11 Printing Output

To demonstrate how to output some data, let us consider the following small piece of code. The code, in brief, gets a temperature value in Fahrenheit [°F] via the keyboard from the user, converts it into the equivalent temperature value in Centigrade [°C] and then prints the result, with some remarks, both onto the screen and into a data file named **F2C.dat**. We will explain each step so that the user can follow what is going on/off.

1. Entering the Data

We can enter the temperature value in degrees Fahrenheit as a number using one of the following two statements:

```
>> Deg_F = input('Input the temperature in Fahrenheit[F]:');
```

Alternatively, we can enter the temperature value in degrees Fahrenheit as a string using the following statement:

```
>> Deg_F = str2num(input('Input the temperature in
Fahrenheit[F]:', 's'));
```

Notice that the command **str2num** converts the **input** from a string to a number, although we literally enter the input as a number, but with such an input statement having **'s'** as an optional, or extra argument, whatever you input is treated as a string.

2. Conversion from Fahrenheit to Centigrade

```
>> Deg_C = 5/9*(Deg_F-32);
```

This commands converts from Fahrenheit to Centigrade.

3. Printing out the Result on Screen

```
>> fprintf('%5.2f(in Fahrenheit) is %5.2f(in Centigrade)
\n',Deg_F,Deg_C
```

This command prints on the screen, as one line and in an order fashion, the following four items:

First, the value of the variable: **Deg_F**;

Second, the phrase: "**(in Fahrenheit) is**";

Third, the value of the variable: **Deg_C**; and

Fourth, the phrase: "**(in Centigrade)**".

The backslash followed by n **'\n'** means that the next print statement will be executed on a new line. The result of applying this command is shown here:

```
>> fprintf('%5.2f(in Fahrenheit) is %5.2f (in Centigrade)
\n',Deg_F,Deg_C)
212.00(in Fahrenheit)is 100.00 (in Centigrade)
```

4. Printing out the Result to a File

```
>> fid=fopen('F2C.dat', 'w');
```

This command opens a file named **F2C.dat** for writing; that is why there is an optional switch 'w'.

```
>>fprintf(fid,'%5.2f(in Fahrenheit) is %5.2f (in Centigrade)
\n',Deg_F,Deg_C);
```

This command, similar to the previously discussed print on screen command, prints out one line into the opened file **F2C.dat** each time it is executed.

```
>> fclose(fid);
```

This closes the file that has already been opened for data input by MATLAB. So, the aforementioned commands can be put together here:

```
>>Deg_F = str2double(input('Input the temperature in
Fahrenheit[F]:', 's'));
>> Deg_C = 5/9*(Deg_F-32);
>> fprintf('%5.2f(in Fahrenheit) is %5.2f(in Centigrade)
\n',Deg_F,Deg_C)
>>fid=fopen('F2C.dat', 'w');
>>fprintf(fid,'%5.2f(in Fahrenheit) is %5.2f(in Centigrade)
\n',Deg_F,Deg_C);
>> fclose(fid);
```

Notice that we can save all of those aforementioned commands in one file called an M-file. For example, you may name the file as **F2C.m**, which can be recalled in the future by its name without the extension, *m*, and executed as many times as we wish without having to rewrite these commands. This is the advantage of making M-files that suit one's needs. See Chap. 3 for further details on creating M-files.

The term **'%5.2f'** means that the floating point number type will be shown with a minimum of five characters, whereas two of them lie just after the decimal point. See Table 2.2 for more specifiers to be used with **fprintf** command.

Table 2.2 Different specifiers and special characters that can be used with the **fprintf** statement.

Type	Printing Form:	Special	
Specifier	**fprintf**('**format string**', variables to be printed, ...)	Character Meaning	
%c	Single character	\n	New line
%d	Decimal integer number type	\t	Tab
%e	Decimal exponential type	\b	Backspace
%f	Floating point number type	\r	CR return
%g	The more compact of %e or %f, with no trailing zeros	\f	Form feed
%G	The more compact of %E or %f, with no trailing zeros	%%	%
%s	String of characters	''	'
%x	Hexadecimal integer number		

Compare the following formats just to see how numbers are presented in terms of precision (i.e., how many significant numbers after the decimal point):

```
>> fprintf('X is %5.2f meters or %7.2f mm\n', 99.67897, 99678.970)
X is 99.68 meters or 99678.97 mm
>> fprintf('X is %5.3f meters or %7.3f mm\n', 99.67897, 99678.970)
X is 99.679 meters or 99678.970 mm
```

The **%s** is used to display string characters as in the following example:

```
>> fprintf('My name is %s and my height is: %5.2f cm\n', 'Kamal', 170.00)
My name is Kamal and my height is: 170.00 cm
```

2.12 Problems

2.1 Create the following types of matrices:

a. $A = 3 \times 4$ matrix with all elements set to zero
b. $B = 3 \times 2$ matrix with all elements set to one
c. $C = 2 \times 3$ matrix with all elements randomly generated
d. $D = 3 \times 3$ identity matrix with diagonal elements =1 and off-diagonal = 0
e. $E = 4 \times 4$ diagonal matrix with diagonal elements 1, 2, 3, 4 and else 0

2.2 Given that $F = \begin{bmatrix} 2 & 7 \\ 8 & 1 \end{bmatrix}$, find the determinant of F using:

a. Eq. (2.2): $\det(F) = f_{11} \times f_{22} - f_{12} \times f_{21}$
b. MATLAB built-in determinant function

2.3 Solve (i.e., find the roots of) the following polynomials using: (1) the eigenvalues of the companion matrix of the vector of coefficients; and (2) MATLAB built-in **roots** function for the vector of coefficients: $ax^4 - bx^3 + cx^2 - dx + e = 0$.

a. $7x^4 - 3x^3 + 3x^2 - 8x + 4 = 0$
b. $-7x^4 + 3x^3 - 3x^2 + 8x - 4 = 0$
c. $7x^4 + 3x^3 - 3x^2 + 8x - 4 = 0$
d. $7x^4 + 3x^3 + 3x^2 - 8x - 4 = 0$

2.4 Consider the quartic $y = x^4 + x^2 + c$. For which values of c does the equation have two real roots?

2.5 Consider the following 4×4 matrix: $G = \begin{bmatrix} 16 & 2 & 3 & 13 \\ 5 & 11 & 10 & 8 \\ 9 & 7 & 6 & 12 \\ 4 & 14 & 15 & 1 \end{bmatrix}$.

a. Find the sum of each row.
b. Find the sum of each column.
c. Find the sum of the diagonal elements.
d. What kind of common matrix is G?

2.6 Given the matrix $H = \begin{bmatrix} 11 & 13 & 15 \\ 17 & 19 & 21 \\ 23 & 35 & 47 \end{bmatrix}$, use the colon ':' operation to generate the following matrices:

a. $I = \begin{bmatrix} 19 & 21 \\ 35 & 47 \end{bmatrix}$

b. $J = \begin{bmatrix} 11 & 13 \\ 17 & 19 \\ 23 & 35 \end{bmatrix}$

c. $K = \begin{bmatrix} 11 & 0 & 0 \\ 0 & 19 & 0 \\ 0 & 0 & 47 \end{bmatrix}$

d. $L = [11 \quad 17 \quad 23 \quad 13 \quad 19 \quad 35 \quad 15 \quad 21 \quad 47]$

2.7 Given $M = \begin{bmatrix} 4 & 6 & 8 \\ 3 & 5 & 7 \\ 10 & 12 & 14 \end{bmatrix}$, use MATLAB built-in matrix-related functions to obtain the following matrices:

a. $N = \begin{bmatrix} 8 & 6 & 4 \\ 7 & 5 & 3 \\ 14 & 12 & 10 \end{bmatrix}$

b. $O = \begin{bmatrix} 10 & 12 & 14 \\ 3 & 5 & 7 \\ 4 & 6 & 8 \end{bmatrix}$

c. $P = \begin{bmatrix} 4 & 0 & 0 \\ 3 & 5 & 0 \\ 10 & 12 & 14 \end{bmatrix}$

2.8 The following 3 × 3 matrix is given; define it in MATLAB.

$$Q = \begin{bmatrix} 2 & 7 & 3 \\ 8 & 1 & 1 \\ 6 & 5 & 8 \end{bmatrix}$$

Answer the following questions:

a. Minimum of Q?
b. Maximum of Q?
c. Mean of Q?
d. Size of Q?
e. Determinant of Q?
f. Median of Q?
g. Standard of Q?
h. Q*Q?
i. Q.*Q?
j. Inverse of Q?
k. Transpose of Q?

2.9 The following symmetric (i.e., R′=R, transpose(R)=R) 3 × 3 matrix is given; define it in MATLAB.

$$R = \begin{bmatrix} 1 & 2 & 3 \\ 2 & 4 & 5 \\ 3 & 5 & 6 \end{bmatrix}$$

Answer the following questions:

a. Determinant of R?
b. Inverse of R, R^{-1}; is R^{-1} symmetric?
c. S = R*R; is S symmetric?
d. T = R.*R; is T symmetric?
e. Try another symmetric matrix, *R2*, defined as: $R2 = \begin{bmatrix} 7 & 8 & 9 \\ 8 & 4 & 11 \\ 9 & 11 & 6 \end{bmatrix}$.

See if matrices in parts b, c, and d remain symmetric. What do you conclude?

NOTE You can also create a symmetric 3 × 3 matrix by invoking **A3=pascal(3)** at the command prompt of MATLAB.

2.10 Consider the following strings:
`'Kamal','Al-Malah','Mike','Penn','Amy','Stone-Braker','McGraw'`.

Create a cell (i.e., string) array made of those words and alphabetically sort them.

2.11 Make a small m-file (i.e., a set of MATLAB-understood commands) that will read the weight of a person in pounds (lb) and convert the weight into kilograms (kg). The conversion factor is: 1 lb = 0.45359 kg. Your program should print out the weight in both units on the screen and save to a file. See Sec. 2.11 for more information.

3 MATLAB Scripting Language: M-File

Chapter Outline

3.1 What Is an M-File?

MATLAB is a high-level technical computing language and interactive environment for algorithm development, data visualization, data analysis, and numerical computation. Using MATLAB, you can solve technical computing problems faster than with traditional programming languages, such as C, C++, and FORTRAN. With the MATLAB language, you can program and develop algorithms faster than with traditional languages because you do not need to perform low-level administrative tasks, such as declaring variables, specifying data types, and allocating memory. In many cases, MATLAB eliminates the need for 'for' loops. As a result, one line of MATLAB code can often replace several lines of C or C++ code. At the same time, MATLAB provides all the features of a traditional programming language, including arithmetic operators, flow control, data structures, data types, object-oriented programming (OOP), and debugging features.

MATLAB can execute a sequence of statements stored in files. These are called M-files because they must have the file type .m as the last part of their filename. Use the MATLAB Editor or another text editor to create a file containing the same statements you would type at the MATLAB command line (using Command Window). Save the file under a name that ends in .m. Much of your work with MATLAB will be in creating and refining M-files. M-files are usually created using your favorite text editor or with MATLAB's M-file Editor/Debugger. See also Help: MATLAB: Using MATLAB: Development Environment: Editing and Debugging M-Files.

There are two types of M-files: script files and function files. Script files basically summarize a series of MATLAB commands entered through the Command Window. For example, to complete a figure with a caption, you have to enter several commands one by one in the command window. Typing errors will be time consuming to fix because if you are working in the command window, you need to retype all or part of the program. To preserve large sets of commands, you can store them in a special type of file called an *M-file*. To run your *script M-file*, just type the filename (omitting the **.m** extension at its end) at the MATLAB prompt. However, let us first define the MATLAB search path.

3.2 MATLAB Search Path, Path Management, and Startup

MATLAB search path is an ordered list of directories that MATLAB searches to find script and function *M-files* stored on disk. The MATLAB's path is defined from the File menu via clicking on the Set Path sub-menu, as shown in Figs. 3.1 and 3.2*a*.

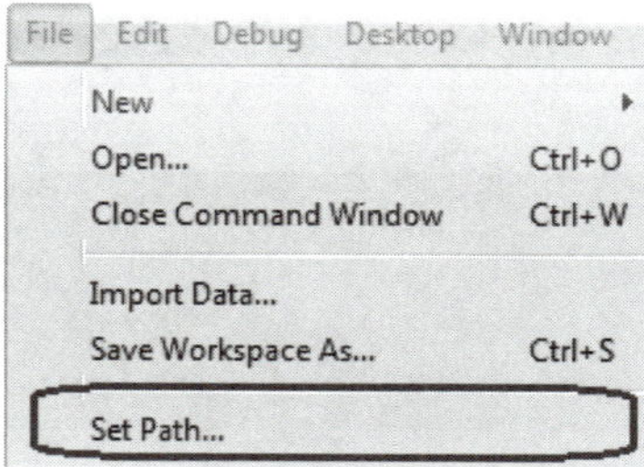

Figure 3.1 MATLAB's path is set via the File menu and then via the Set Path sub-menu.

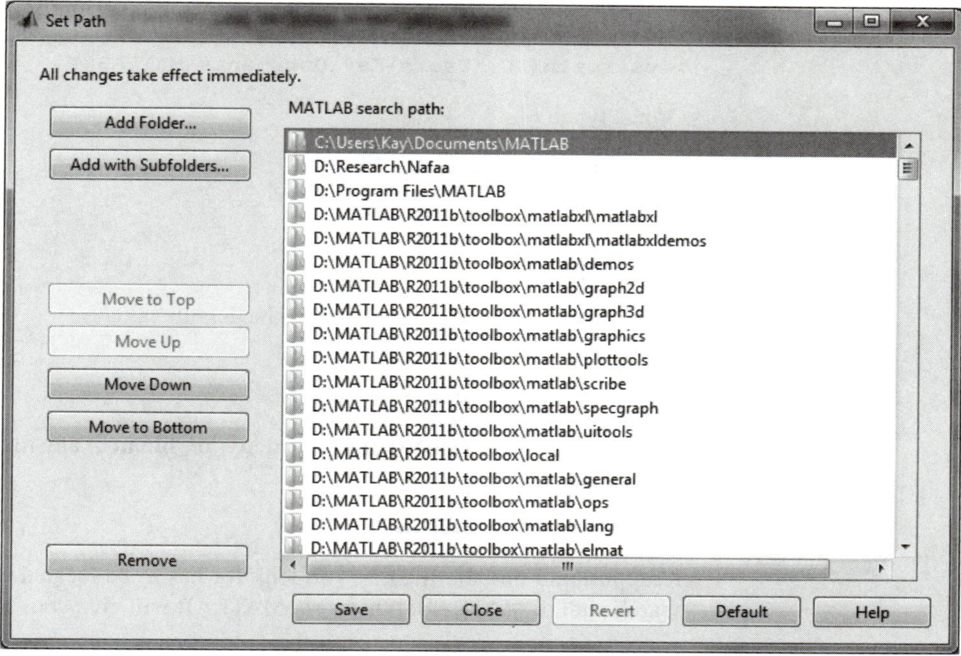

Figure 3.2(*a*) Upon clicking the Set Path sub-menu from the File menu, the Set Path main window pops up. This is where you can add a folder and/or its subfolders to MATLAB's search path.

In addition, you can directly invoke the Set Path sub-menu via the command prompt:

```
>> editpath
```

To see the *current search path*, enter **matlabpath** at the command prompt:

```
>>matlabpath
```

All defined MATLAB search path folders and subfolders will show up on the screen.

To see the startup user's path, type at the prompt command:

```
>> userpath
ans =
C:\Users\Kay\Documents\MATLAB;
```

To change the path to a new folder, type:

```
>>userpath('D:\Program Files\MATLAB')
```

This will change the default value to the new value; that is, 'D:\Program Files\MATLAB'. If you close your MATLAB session at this point and reopen MATLAB, you will notice that MATLAB will start with the user's current folder that was assigned earlier through the **userpath** command.

To go back to the default value either spell out the default value:

```
>>userpath('C:\Users\Kay\Documents\MATLAB')
```

Or just type:

```
>> userpath('reset')
```

Now, if you retype:

```
>> userpath
```

you will see that MATLAB reverts to the default value:

```
ans =
C:\Users\Kay\Documents\MATLAB;
```

To see the directory path to a file named **my_m_file.m** or any file, type at the command prompt:

```
>> which my_m_file.m
```

Keep in mind that the file that you look for has to be located within the current search path of MATLAB; otherwise, MATLAB will give you:

```
'my_m_file' not found.
```

The final word on this issue is that MATLAB comes with the built-in **pathdef.m** that can be updated via instructions explained here. It is preferred that the user add a new path via the Set Path main window (see Fig. 3.2*a*); however, an advanced user can edit this M-file and add any path after the following p= [... statement:

```
p = [...
%%% BEGIN ENTRIES %%%
   'D:\Research\Proteins;', ...
   'D:\Program Files\MATLAB;', ...
  matlabroot,'\toolbox\matlabxl\matlabxl;', ...
```

Figure 3.2(*b*) A portion of the built-in **pathdef.m** file in which the user may add a new string defining a new folder on disk.

The folder path should be inserted as a string, including the semicolon, between two apostrophes followed by a comma, and three dots (...) to indicate the continuation of the path definition to include down-coming statements.

Alternatively, in case you would like to replace this file by the original file that MATLAB comes with, you can click on the default button shown at the bottom of the Set Path window (see Fig. 3.2*a*), which will replace the content of `pathdef.m` by that of the file named **restoredefaultpath.m**. Both files are found under the directory: **"X:\Program Files\MATLAB\R2011a\toolbox\local"**.

NOTE The default directory may change from the aforementioned, based on where the user installs MATLAB and its version.

3.3 Script M-File

Example 3.1 Let us recall the set of commands needed to convert from degrees Fahrenheit to Centigrade (Celsius), introduced in Sec. 2.11.

```
>>Deg_F = str2double(input('Input the temperature in
Fahrenheit[F]:', 's'));
>> Deg_C = 5/9*(Deg_F-32);
>> fprintf('%5.2f(in Fahrenheit) is %5.2f(in Centigrade)
\n',Deg_F,Deg_C)
>>fid=fopen('F2C.dat', 'w');
>>fprintf(fid,'%5.2f(in Fahrenheit) is %5.2f(in Centigrade)
\n',Deg_F,Deg_C);
>> fclose(fid);
```

To create an *M-file* (script type) using the built-in MATLAB m-editor/debugger, go to **File** menu as shown in Fig. 3.3, click on **New** (sub-menu), and choose the **Script** sub-submenu. Alternatively, while activating the MATLAB main window, you may click CTRL+N to open up the *m-editor*.

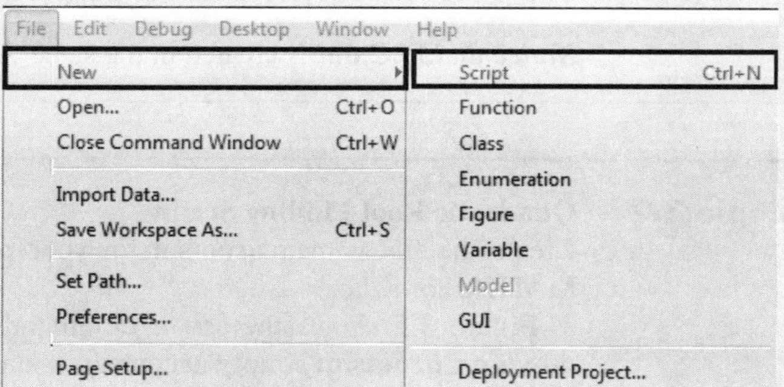

Figure 3.3 MATLAB's *m-editor* is opened via the **File** menu, followed by the **New** sub-menu, and finally clicking the **Script** sub-submenu. Alternatively, while keeping the main MATLAB window active, you may click CTRL+N to open up the *m-editor*.

The built-in MATLAB *m-editor* will open up. Here you can paste the set of previous commands as shown in Fig. 3.4.

After that, save the file as ***F2C.m***. Of course, the ***F2C.m*** file has to be saved in a directory that is part of MATLAB's search path. This search path is defined in Sec. 3.2.

After saving the M-file in MATLAB's search path, *type the name of the M-file without extension*. Notice that naming is *case sensitive*. You must be very aware of this issue or you will experience problems with MATLAB.

```
>> F2C
Input the temperature in Fahrenheit[F]:366
366.00(in Fahrenheit) is 185.56(in Centigrade)
```

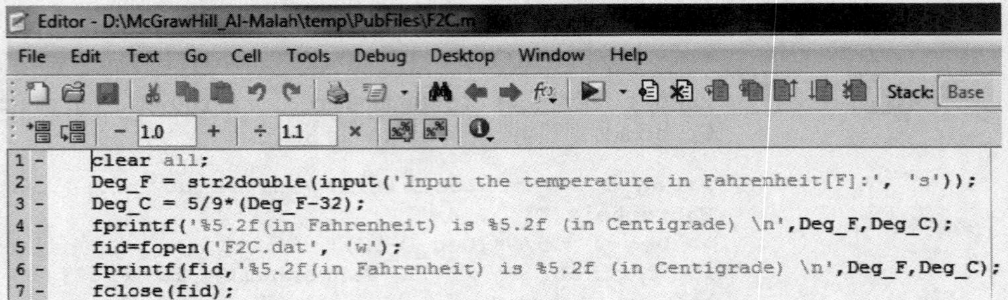

Figure 3.4 MATLAB's built-in *m-editor*, in which commands can be entered line by line or pasted from another application.

If you attempt to run:

```
>> f2c
??? Undefined function or variable 'f2c'.

>> F2c
??? Undefined function or variable 'F2c'.
```

Notice that **F2C.dat** is created in the same folder in which **F2C.m** resides.

Example 3.2 **Quadratic Root Finding Script**

Create the file named **qroots.m** in your present working directory using the MATLAB *m-editor.*

Figure 3.5 shows the set of commands as they appear in the MATLAB *m-editor*. **qroots.m** simply accepts **a**, **b**, and **c** coefficient values from the user

Editor - D:\McGrawHill_Al-Malah\public-files\PubFiles\qroots.m

```
% qroots: Quadratic root finding script for a quadratic equation in the
% form of a*x^2+b*x+c
format compact;
a_term= str2double(input('Input the coefficient of x^2:', 's'));
b_term= str2double(input('Input the coefficient of x:', 's'));
c_term= str2double(input('Input the constant:', 's'));
term1= -b_term/(2*a_term);
term2 = sqrt(b_term^2-4*a_term*c_term)/(2*a_term);
s1 = term1+term2;
s2 = term1-term2;
disp(['The quadratic equation is of the form: ', num2str(a_term),'*x^2+', ...
'(',num2str(b_term),')*x+','(',num2str(c_term),')'])
fprintf('The first root is %s \n', num2str(s1))
fprintf('The second root is %s \n', num2str(s2))
```

Figure 3.5 MATLAB *m-editor*, in which the content of the **qroots.m** file is shown.

as input parameters; calculates two terms: **term1** and **term2** that will be utilized in finding the two quadratic roots: **s1** and **s2**; and finally prints out the two roots on the screen. Again, the **str2double** function converts from string to double and the **num2str** function converts back to string. The **disp**(argument) function outputs the argument to the screen. The argument can be a single or concatenated string phrase.

To execute the script M-file, simply type the name of the script file, **qroots**, at the MATLAB prompt:

```
>> qroots
Input the coefficient of x^2:1
Input the coefficient of x:-1
Input the constant:-6
The quadratic equation is of the form: 1*x^2+(-1)*x+(-6)
The first root is 3
The second root is -2
>> qroots
Input the coefficient of x^2:1
Input the coefficient of x:2
Input the constant:3
The quadratic equation is of the form: 1*x^2+(2)*x+(3)
The first root is -1+1.4142i
The second root is -1-1.4142i
```

Notice that in the last case we ended up with two complex conjugate, not real, roots.

If you add the following command at the start of the M-file, **qroots.m**, then you will be able to see a line-by-line command within the M-file itself upon executing the M-file:

```
>> echo on;
```

This echo statement or command controls the display (or echoing) of statements in a function during their execution. Normally, statements in a function file are not displayed on the screen during execution. Command echoing is useful for debugging or for demonstrations, allowing the commands to be viewed as they execute.

This is how MATLAB behaves when you run **qroots** with echo on statement:

```
>> qroots
echo on;
format compact;
a_term= str2double(input('Input the coefficient of x^2:', 's'));
Input the coefficient of x^2:2
b_term= str2double(input('Input the coefficient of x:', 's'));
Input the coefficient of x:2
c_term= str2double(input('Input the constant:', 's'));
Input the constant:2
term1= -b_term/(2*a_term);
term2 = sqrt(b_term^2-4*a_term*c_term)/(2*a_term);
```

```
s1 = term1+term2;
s2 = term1-term2;
disp(['The quadratic equation is of the form: ',
num2str(a_term),'*x^2+', ...
'(',num2str(b_term),')*x+','(',num2str(c_term),')'])
The quadratic equation is of the form: 2*x^2+(2)*x+(2)
fprintf('The first root is %s \n', num2str(s1))
The first root is -0.5+0.86603i
fprintf('The second root is %s \n', num2str(s2))
The second root is -0.5-0.86603i
```

NOTE Notice here how MATLAB shows each command line within the M-file itself. Use **echo off** to suppress printing out each step, except for statements that do not end up with the suppressing symbol (;), or for non-suppressible data and graph display (e.g., **disp**, **plot**, **fplot**, **fprintf**, etc.).

3.3.1 *Effective Naming and Referencing of M-Files*

The following suggestions are either necessary or highly recommended for the effective use of MATLAB script and function M-files:

1. The name of a script file must follow the MATLAB convention for naming variables; that is, the *name must begin with a letter* and may include digits and the underscore character.
2. *Do not give a script file the same name as a variable it computes*, because MATLAB will not be able to execute that script file more than once unless the variable is cleared. Recall that typing a variable name at the command prompt causes MATLAB to display the value of that variable. If there is no variable by that name, then MATLAB searches for a script file having that name. For example, if the variable **qroots** is created in a script file having the name **qroots.m**, then after the script is executed the first time, the variable **qroots** will exist in the MATLAB workspace. If the script file is modified and an attempt is made to run it a second time, MATLAB will display the value of **qroots** and will not execute the script file.
3. *Do not give a script file the same name as a MATLAB command or function.* You can check to see whether a function already exists by using the **'which'** command. For example, to see whether **qroots** already exists, type

   ```
   >>which qroots
   ```

   If it doesn't exist, MATLAB will display **qroots** not found. If it does exist, MATLAB will display the full path to the function. For more details as to the existence of a variable, script, or function having the name **qroots**, type

   ```
   >>exist('qroots')
   ```

This command returns one of the following values:

0 if qroots does not exist

1 if qroots is a variable in the workspace

2 if qroots is an M-file or a file of unknown type in the MATLAB search path

3 if qroots is a MEX-file in the MATLAB search path

4 if qroots is a MDL-file in the MATLAB search path

5 if qroots is a built-in MATLAB function

6 if qroots is a P-file in the MATLAB search path

7 if qroots is a directory

4. As in interactive mode, *all variables created by a script file are defined as variables in the workspace.* After script execution, you can type **who** or **whos** to display information about the names, data types, and sizes of these variables.

5. You can use the type command to display an M-file without opening it with a text editor. For example, to view the file **qroots.m**, the command is

```
>>type qroots
```

3.4 Function M-File

A function file is a user-defined function using MATLAB. We use function files for repetitive tasks. The first line of a function file must contain the word function, followed by the output argument, the equal sign (=), function name, and the input argument enclosed in parentheses. *The function name and file name must be the same,* but the file name must have the extension .m. For example, the function file consisting of the following two lines

```
function y = myfunction(x)
y=x .^ 2 + cos(2 .* x)
```

is a function file and must be saved by its name followed by the .m extension. To create it, from the **File** menu of the main window, we choose **New** and then click on ***Function***. This takes us to the Editor window where we type these two lines and we save it as **myfunction.m**.

Example 3.3 Suppose we want to construct the squares and cubes of the elements of a vector. We can use the code:

```
function [sq,cube] = xpowers(input)
% xpowers calculates both the square and cubic values of an
input quantity.
sq = input.^2;
cube = input.^3;
```

The first line defines the function **xpowers**, which has two outputs **sq** and **cube** and one input. The second line is a comment statement that is used to tell what **xpowers** is all about. The third and fourth lines calculate the values of input squared and cubed, respectively. While the main MATLAB window is active, press CTRL+N to activate the MATLAB *m-file editor* and paste the previous four lines. Save the file as **xpowers.m**.

Let us define **x** as the input argument to be passed to the **xpowers** function. Notice that **x** is a row vector, not necessarily a scalar quantity.

```
>> x=1:2:12
x =
   1   3   5   7   9   11
```

To output the square and cubic values for each entry in the row vector **x**, enter the following command which calls the *user-defined* **xpowers** function:

```
>> [xsq,xcube] = xpowers(x)
xsq =
   1   9   25   49   81  121

xcube =
        1    27    125    343    729    1331
```

The output is two row vectors, the first containing the square and the second the cubic values of the elements of **x**. Notice that when the function is called we must know what form of output we expect, whether it be a scalar, a vector, or a matrix. The expected outputs should be placed within square brackets.

Notice that if you run the following command

```
>> xpowers(x)
ans =
   1   9   25   49   81  121
```

this means that without using the output within square brackets we obtain only the first output from left. This will be the case if you use an array with one argument:

```
>> [output]=xpowers(x)
output =
   1   9   25   49   81  121
```

Example 3.4 Find the zeros, minima, and maxima of the function

$$f(x)=\frac{1}{(x-1.0)^2+0.1}-\frac{1}{(x-1.5)^2+0.4}-4.0 \tag{3.1}$$

in the interval $-1.0 \le x \le 3.0$.

Solution The *following MATLAB built-in functions* will be used in this example:

- **fplot(fcn,lims)**—plots the function specified by the string **fcn** between the *x*-axis limits specified by **lims = [xmin xmax]**. Using **lims = [xmin xmax ymin ymax]** also controls the *y*-axis limits. The string **fcn** must be the name of an *M-file* function or a string with a variable.
- **fzero(f,x)**—tries to find a zero of a function of one variable, in which **f** is a string containing the name of a real-valued function of a single real variable. MATLAB searches for a value near a point where the function **f** changes sign, and returns that value, or it returns NaN if the search fails.

NOTE We must use **roots(p)** to find the roots of polynomials only, such as those in Examples 2.7 and 2.8 (see Sec. 2.5).

- **fminunc(fun,x0,options)**—tries to find a minimum of unconstrained multivariable function. **fun** is passed to **fminunc** as a handle to a function M-file; **x0** is the initial guess; and **options** is a struct (a structure array) that gives more specifications on how to handle the optimization algorithm and output the results.

We plot this function to observe the approximate zeros, minima, and maxima using the following script:

```
>>x=-1.0:0.01:3.0;
>>y=1./((x-1.0).^2 + 0.1)-1./((x-1.5).^2+0.4)-4.0;
```

NOTE Here we used ./ and .^, instead of / and ^, respectively, to tell MATLAB that the y vector is a function of the x vector mapping each point of x by the corresponding point of y. This kind of dot division, dot multiplication, and dot exponentiation should be applied for matrix (or vector as a subset of matrix) arguments.

```
>>plot(x,y);grid
```

The plot is shown in Fig. 3.6.

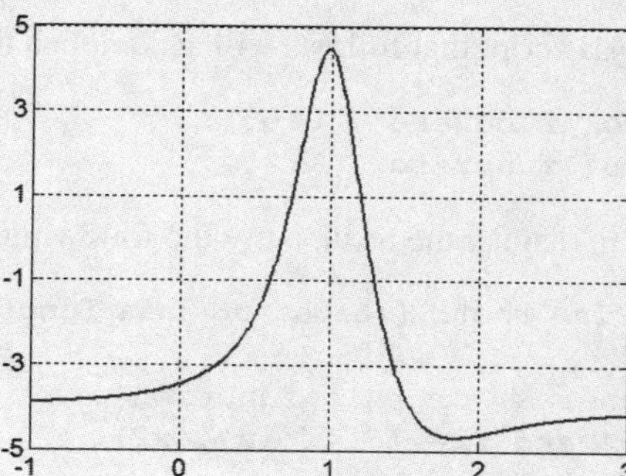

Figure 3.6 The plot of f(x) using the MATLAB **plot** command.

The roots (zeros) of this function appear to be in the neighborhood of $x = 0.65$ and $= 1.25$. The maximum occurs at approximately $x = 1.0$, where approximately $y_{max} = 4.5$, and the minimum occurs at approximately $x = 1.75$, where approximately $y_{min} = -4.7$. Next, we define and save *f(x)* as **funczero.m** (function M-file) with the following script:

```
function y=funczero(x)
% Defining the function for which the zeros to be found.
y=1./((x-1.0).^2 + 0.1)-1./((x-1.5).^2+0.4)-4.0;
```

To create and save such a file, while the main MATLAB window is active, press CTRL+N to activate the MATLAB *m-file editor*, paste the previous three lines, and then save it as **funczero.** MATLAB appends the extension **.m** to it.

Now, we can use the **fplot(fcn,lims)** command to plot as follows:

```
>>fplot('funczero', [-1.0 3.5]); grid
```

This plot is shown in Fig. 3.7. As expected, this plot is identical to the plot of Fig. 3.6, which was obtained with the **plot(x,y)** command as shown in Fig. 3.6.

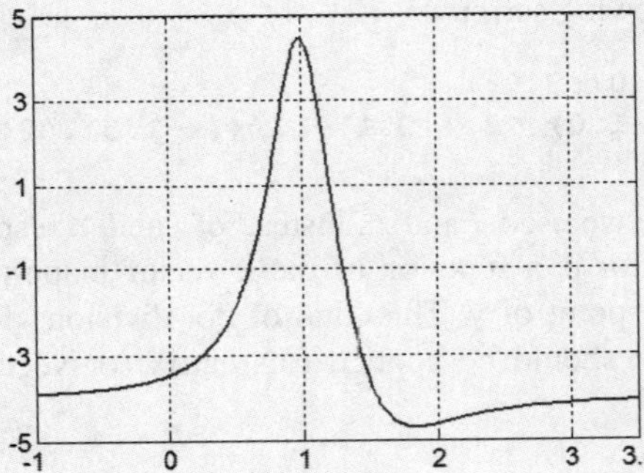

Figure 3.7 The plot of f(x) using the MATLAB **fplot** command.

We will use the **fzero(f,x)** function to compute the roots of Eq. (3.1) more precisely.

The MATLAB script that follows will accomplish this.

```
>> x1= fzero('funczero', 0.65);
>> x2= fzero('funczero', 1.25);
```

For MATLAB to display the results, use the following:

```
>> fprintf('The roots (zeros) of this function are r1= %3.4f\n',
x1);
The roots (zeros) of this function are r1= 0.6794
>> fprintf(' and r2= %3.4f \n', x2)
 and r2= 1.2496
```

The earlier MATLAB versions included the function **fmin(f,x1,x2)** and with this function we could compute both a minimum of some function $f(x)$ or a maximum of $f(x)$ because a maximum of $f(x)$ is equal to a minimum of $-f(x)$. This can be visualized by flipping the plot of a function upside down. This function is no longer used in MATLAB so we will compute the minima and maxima using the MATLAB **fminunc** function. Before doing that we have to define some terms related to **options** struct, which should be part of the minimization process.

```
>>options = optimset('LargeScale','off','Display','final-detailed');
```

Here, the flagger '**LargeScale**' tells MATLAB whether or not to use the user-defined gradient that can be used in the algorithm of converging toward the minimum point. The value assigned to '**LargeScale**' can be `'on'` or `'off'`. Hence, we use the large-scale algorithm if possible when set to the default `'on'`, and we use the medium-scale algorithm when set to `'off'`. However, the **LargeScale** algorithm requires the user to provide the gradient, otherwise `fminunc` uses the medium-scale algorithm.

The flagger **'Display'** has different levels of displaying the output (i.e., the results). Levels of display are:

- `'off'` displays no output.
- `'iter'` displays output at each iteration, and gives the default exit message.
- `'iter-detailed'` displays output at each iteration, and gives the technical exit message.
- `'notify'` displays output only if the function does not converge, and gives the default exit message.
- `'notify-detailed'` displays output only if the function does not converge, and gives the technical exit message.
- `'final'` (default) displays just the final output, and gives the default exit message.
- `'final-detailed'` displays just the final output, and gives the technical exit message.

```
>> x = fminunc(@funczero,1.6,options)
Optimization completed: The first-order optimality measure,
1.828415e-007, is less
than options.TolFun = 1.000000e-006.
Optimization Metric                                  Options
relative norm(gradient) = 1.83e-007                    TolFun = 1e-006
(default)
x =
   1.8437
```

Here, y is the minimum at $x = 1.8437$ and the optimization process was successfully completed with the optimality measure (i.e., relative norm gradient) being less than the specified function tolerance. This practically says that the gradient is normally zero (or very close to that) at the optimum (minimum or maximum) point.

To find the maximum point we, again, use **fminunc** but this time with a minus sign. Notice from Fig. 3.7 that the maximum lies at around $x = 1.0$. However, you have to add a minus sign to the function statement within the M-file itself. Thus, the **funczero** *m-file* becomes:

```
function y=funczero(x)
% Defining the function for which the zeros to be found.
y=-(1./((x-1.0).^2 + 0.1)-1./((x-1.5).^2+0.4)-4.0);
```

NOTE Do not forget to save the M-file so any changes or updates you make will take effect whenever MATLAB calls **funczero**, as in the next line command.

```
>> x = fminunc(@funczero,1.0,options)
Optimization completed: The first-order optimality measure,
1.062198e-007, is less
than options.TolFun = 1.000000e-006.
Optimization Metric                              Options
relative norm(gradient) = 1.06e-007                 TolFun = 1e-006
(default)
x =
  0.9883
```

Notice here that $x = 0.9883$ represents the point at which the minimum of $-f(x)$ occurs, which at the same time represents the maximum of $f(x)$. Look at Fig. 3.7 to see where the maximum of $f(x)$ occurs.

3.5 Errors and Debugging

You will find that you will seldom get your scripts to run correctly the first time. However, you should not panic, as this is also the case for experienced programmers. You will need to learn to identify and correct the errors, known in computer jargon as *bugs*. Before we move to the types of errors involved in MATLAB scripting and programming, let us introduce the concept of representing number, in particular for non-integer numbers.

3.5.1 Scientific Notation, Significant Figures, and Precision

Both very large and very small numbers are commonly encountered in process calculations. A convenient way to represent such numbers is to use scientific notation, in which a number is expressed as the product of another number (usually between 0.1 and 10) and a power of 10.

Examples are:

$$123{,}000{,}000 = 1.23 \times 10^8 \ (or\ 0.123 \times 10^9)$$

$$0.000028 = 2.8 \times 10^{-5} \ (or\ 0.28 \times 10^{-4})$$

NOTE The **significant figures of a number are** the digits from the first nonzero digit on the left to either the last digit (zero or nonzero) on the right if there is a decimal point, or the last nonzero digit of the number if there is no decimal point.

Example 3.5 Count how many significant figures are found in each of the following numbers.

Solution

$2300 = 2.3 \times 10^3$	It has two significant figures.
$2300. = 2.300 \times 10^3$	It has four significant figures.
$2300.0 = 2.3000 \times 10^3$	It has five significant figures.
$23{,}040 = 2.304 \times 10^4$	It has four significant figures.
$0.035 = 3.5 \times 10^{-2}$	It has two significant figures.
$0.03500 = 3.500 \times 10^{-2}$	It has four significant figures.

3.5.2 Errors Related to Floating-Point Arithmetic

These errors occur when your script is run on a particular set of data. They typically occur when the result of some operation leads to *NaN (not a number), Inf (infinity), or an empty array*. For example, the fraction 1/3 is a non-terminating decimal number; to exactly represent this fraction, the trailing number three (3) should indefinitely follow the decimal point (0.333···333···333). However, depending on the number of *significant figures* (Sec. 3.5.1) to retain after the decimal point, the error, resulting from the *truncation and/or round-off step*, may affect the following computational steps that will utilize this cosmetically modified number and may result in some sort of numerical instability.

Almost all operations in MATLAB are performed in double-precision arithmetic conforming to IEEE Standard 754 (double precision calls for 52 mantissa bits). This represents the highest degree of resolution by which MATLAB can see two very close numbers as two different entities. The following examples illustrate the concept of floating point-related computational problems.

Example 3.6 The decimal number 4/3 is not exactly represented as a binary fraction. For this reason, the following calculation does not give zero, but rather reveals the quantity **eps**.

```
>>e = 1 - 3*(4/3 - 1)
e =
 2.2204e-16
```

eps represents the floating-point relative accuracy, which is defined as the distance from 1.0 to the next largest double-precision number, that is **eps** = `2^(-52)`= 2.2204e^{-16}. This means that from the *MATLAB degree of resolution standpoint*, any number that lies within the range 1.0± `2.2204e`$^{-16}$ will be seen (treated) as 1.0.

Example 3.7

```
>> one_plus=1.00+0.9*eps

One_plus =
  1.0000

>> one_minus=1.00-0.9*eps

one_minus =
  1.0000
```

Obviously, one, one$^+$, and one$^-$ are all seen by MATLAB as a single "sharp" one; there is no noticeable difference among them.

The following is an example of a *cancellation* caused by swamping (loss of precision that makes the addition insignificant).

Example 3.8

```
>> ignored_number=sqrt(1+1e-16)-1.0
ignored_number =
  0
```

Consequently, more often than not, the results of an arithmetic operation cannot be represented exactly as a valid floating-point number. Suppose the "correct answer" falls between adjacent normalized floating-point numbers. The computer must round to one of the numbers, and this produces "round off" error. Such types of errors can produce serious problems in algorithms that involve a large number of repetitive calculations.

3.5.3 *Numerical Stability*

Numerical methods for solving differential equations are important tools for analyzing engineering systems. Although valuable algorithms have been developed that facilitate construction of approximate solutions, all available methods are vulnerable to limitations inherent in the underlying approximation processes. The numerical instability in this case may stem from two sources. The first is the onset of an initial error caused by either a relatively large integration step size, or initial value (initial/boundary condition). This type of error will grow rapidly and swamp out the result without necessarily having round-off errors. The second has to deal with function evaluations that are carried out at each step, involving *round-off and arithmetic truncation.* Ultimately, the final answer (solution) is vulnerable to accumulated errors issuing from *the implemented approximation scheme*. These errors sometimes lead to an accumulative effect that grows exponentially (or blows up), which eventually destroys solution validity.

3.5.4 Syntax Errors

The most common type of error is the syntax error, a typing error in a MATLAB command (for example, **srqt** instead of **sqrt**). These errors are said to be *fatal*, as they cause MATLAB to stop execution and display an error message. Unfortunately, because the MATLAB command interpreter can easily be confused by these errors, the displayed error message may not be very helpful. It is your job as a programmer to make sense of the message and correct your script, a process known as *debugging*. An incorrect use of variable name, resulting from a typographical error, is also classified under this category.

Here are some examples of syntax errors; the associated messages are included.

Example 3.9

```
>> 2(4+5)
??? 2(4+5)
     |
Error: Unbalanced or unexpected parenthesis or bracket.
>> 2*((4+5)
??? 2*((4+5)
       |
Error: Expression or statement is incorrect--possibly
unbalanced (, {, or [.
>> sine(60)
??? Undefined function or method 'sine' for input arguments
of type 'double'.
```

Missing multiplication operators and parentheses are among the most common errors. There are, however, a large number of possible syntax errors, many of which you will discover in writing scripts for this course. As time goes on, you will gradually become more adept at understanding and correcting your mistakes.

3.5.5 Run-Time Logic Errors

After correcting syntax errors, your script will execute, but it still may not produce the results you desire. This has to do with logical errors. These are errors in your programming logic, when your script executes properly, but does not produce the intended result. When this occurs, you need to re-think the development and implementation of your problem-solving algorithm.

3.5.6 Remedies to Minimize Errors

The following are suggestions to help with the debugging of MATLAB scripts or M-files:

1. Execute the script on a simple version of the problem, using test data for which the results are known, to determine which displayed results are incorrect.
2. Remove semicolons from selected commands in the script so that *intermediate* results are displayed.

3. Add statements that display variables of interest within the script.
4. Place the keyboard command at selected places in the script to give temporary control to the keyboard. By doing so, the workspace can be interrogated and values changed as necessary.

3.6 Problems

3.1 What is the intrinsic difference between a scrip and function M-file?

3.2 To see the *current search path*, at the MATLAB command prompt you enter:

a) userpath
b) currentpath
c) searchpath
d) matlabpath
e) editpath

3.3 To see the directory path to a file named **myf.m**, at the MATLAB command prompt you enter:

a) what myf
b) who myf
c) where myf
d) which myf
e) whos myf

3.4 To see MATLAB-related files in a directory named **mydir**, at the MATLAB command prompt you enter:

a) what mydir
b) who mydir
c) where mydir
d) which mydir
e) whos mydir

3.5 What is the difference between **who** and **whos** command if you search for a user-defined variable in MATLAB's environment (i.e., workspace)?

NOTE For a function M-file, choose a file name that you can easily remember, which should reflect what it does. For example, you may name the file in Problem 3.6 **circle_area**.

3.6 Write a *script M-file* that takes the radius **r** and returns the area of a circle (πr^2). Make sure that your code works for variables that are scalars, vectors, and matrices. Use the proper input statement to enter the radius and output statement to print out the area of a circle.

3.7 Write a *function M-file* that takes the radius **r** as an input argument for your function and returns both the area of a circle (πr^2) and the circumference ($2\pi r$) as two output arguments for the given function input. Make sure that your code works for variables that are scalars, vectors, and matrices. You may use the proper output statement to print out the area and circumference of a circle.

3.8 Write a *function M-file* that takes the radius **r** as an input argument for your function and returns the area of a sphere ($4\pi r^2$) and its volume ($(4/3)\pi r^3$) as two output arguments for the given function input. Make sure that your code works for variables that are scalars, vectors, and matrices. You may use the proper output statement to print out the area and volume of a sphere.

3.9 Write a *function M-file* that takes the radius **r** and height **h** as two input arguments for your function and returns both the total surface area of a cylinder ($2\pi r^2+2\pi rh$) and its volume ($\pi r^2 h$)

as two output arguments for the given function input. Make sure your code works for variables that are scalars, vectors, and matrices. You may use the proper output statement to print out the area and volume of a cylinder.

3.10 Write a *function M-file* that takes the temperature **T**, pressure **P**, and number of moles **n**, as three input arguments for your function and returns the volume **V** of the ideal gas (**nRT/P**) as one output argument for the given function input. Do not forget to define the ideal gas constant **R**. You should also tell the user what the units are for pressure and temperature so you will code the equation of state with the proper units for **R**. Make sure that your code works for variables that are scalars, vectors, and matrices. You may use the proper output statement to print out the volume of an ideal gas.

3.11 Write a code that takes a variable x and returns the value of natural log (x^2) and $\log_{10}$ (x^2). Make sure that your code works for variables that are scalars, vectors, and matrices. Try x = [1,2,3].

3.12 The code:

```
clear x y
x = -2:0.2:2;
y = 10-x.^2;
plot(x,y);
```

plots the function $y = 10 - x^2$ for $x \in [-2, 2]$ in steps of *2/10*. Modify the code so the function $y = x^3 + 4x$ is plotted between the same limits and then for $x \in [-4, 6]$ in steps of *1/5*. This code can be run from the prompt or can be entered using the m-editor and then saved as a script file.

3.13 Find the zeros, minima, and maxima of the function

$$f(x) = \frac{1}{(x+3.0)^2 + 0.1} - \frac{1}{(x-3.5)^2 + 0.4} - 3.0$$

in the interval $-5.0 \leq x \leq 5.0$

Use `plot, fplot, fzero,` and `fminunc` to help find the required features of f(x).

3.14 Find the zeros, minima, and maxima of the function

$$f(x) = \frac{1}{(x-1.0)^2 + 0.1} + \frac{1}{(x-1.5)^2 + 0.4} - 4.0$$

in the interval $-10.0 \leq x \leq 10.0$.

Use `plot, fplot, fzero,` and `fminunc` to help find the required features of f(x).

3.15 This task contains *codes that are written with deliberate mistakes*. You should try to debug the codes so that they actually perform the calculations they are supposed to.

1. Given:

$x = 5$

$x + 3 = y$

$z = \frac{1}{y^3 \pi}$

perform the calculations using the following MATLAB commands:

x=5;

x+3=y;

z=1/y^3Pi

3.16 Calculate the sum: $\sum_{j=1}^{N} \frac{1}{j} + \frac{1}{(j+3)(j+4)}$, *where the user inputs N.*

```
clear all;
sum=0.0;
N=str2double(input('Enter N: ','s' ));
for j=1:N
sum = 1/j + 1/(j+3)*(j+4);
end
disp( ['The answer is: ' num2str(sum)]);
```

- Make sure the code gives the correct answer. Try it with *N=1*. What is wrong with this code?
- Try to change the sum statement to: `sum = 1/j + 1/((j+3)*(j+4));`
- See how the result will change.
- Make a second trial but this time with *N=2*. Again there is something wrong with the code.
- Try to change the sum statement to: `sum = sum+ 1/j + 1/((j+3)*(j+4));`
- Finally, see how you manage to obtain the correct answer.

4 Image and Image Analysis

Chapter Outline

Digital cameras, mobiles equipped with cameras, scanners, and laptop cameras are now available at low prices; such devices can be used to produce images that can be transferred to laptops and PCs. Images can convey a great deal of useful information. As they say, "One picture is worth a thousand words."

Image data files fall into three general formats. The first is a *noncompressed* type, the second is a *compressed* type, and the third is *vector* graphics. Noncompressed types, like early versions of Microsoft Windows Bit Map Picture (BMP) and Tagged Image File Format (TIFF) formats, are image data files that simply store each pixel of image data in an array. Compressed bitmaps do the same thing, except some mathematical treatment is applied to the data to attempt to reduce the size of the file. Compressed bitmaps can be further divided into *with-loss* and *lossless* formats as well. With-loss formats actually lose some of the original data during compression and it is usually a decision as to how much of the data can be lost before it is no longer suitable for its intended use. The Joint Photographic Experts Group (JPEG) format is a with-loss type with varying degrees depending on the compression ratio. You will always get the image file size reduced with JPEG, but you might do so with a certain degree of degradation of the data. In other words, compression introduces artifacts or aberration in the image.

Compression with-loss algorithms work on reducing what is called data redundancy. The redundancy can be of different types. The first is interpixel redundancy, whereby the difference in brightness value between two or more neighboring pixels is negligibly small such that a neighborhood can be treated as one brightness value and there is no need to use all of the data in patches with a similar brightness. The second is coding redundancy, which deals with saving information about every bit (or pixel) of an image with equal rights (i.e., byte). However, it turns out that this approach is not the most efficient way of representing such microinformation; a more efficient way is to reserve more memory (bytes) for less frequently than for more frequently repeating bits in an image. The third is psycho-visual redundancy, which deals with the capability of the human eye to detect the difference in brightness value between identical objects or patches. In this regard, the human visual system is not well suited to pick up on details. For example, if the number of digital counts is reduced from 256 to 64 using straightforward gray-level quantization, then quite a few people will be able to notice the difference.

On the other hand, lossless formats like Portable Network Graphics (PNG), Graphics Interchange Format (GIF), and compressed TIF images offer varying degrees of compression based on the data but do not compress to the point of losing data.

In general, compression with loss can be used for general purpose photographic images, whereas compression without loss must be used when dealing with line art, technical drawings, cartoons, or images of important tiny details (e.g., space and medical images).

At last, the vector graphics is one in which the shape of a curve is defined by a math function. Vector graphics are basically points connected by curves of various shapes, filled with solid or gradient colors.

4.1 Reading a Graphics Image

You can use **imfinfo** to retrieve information about the image file. It has the form

imfinfo(filename, fmt)

where both input variables are strings, the first being the name of the file and the second being the image file format. For example, MATLAB includes a JPEG image of the football:

Example 4.1

```
>>imfinfo('football','jpg')
ans =

        Filename:
'D:\MATLAB\R2011a\toolbox\images\imdemos\football.jpg'
     FileModDate: '01-Mar-2001 09:52:38'
        FileSize: 27130
          Format: 'jpg'
   FormatVersion: ''
           Width: 320
          Height: 256
        BitDepth: 24
       ColorType: 'truecolor'
 FormatSignature: ''
 NumberOfSamples: 3
    CodingMethod: 'Huffman'
   CodingProcess: 'Sequential'
         Comment: {}
```

From this, we can see that **football.jpg** is a 256 × 320 truecolor image.

NOTE Notice that you do not have to specify a file extension in filename provided the extension is the same as **fmt**. Moreover, the **ColorType** shown in the above file information will tell us if the image is of type truecolor (i.e., RGB); indexed with a colormap; grayscale; or binary. A truecolor image can be uint8, uint16, single, or double. An indexed image can be logical (contains 0 or 1 pixel value), uint8, single, or double. A grayscale image can be logical, uint8, int16, uint16, single, or double. Finally, a binary image must be of class logical.

The **imread** function has the form:

```
>>[X,C]=imread(filename,fmt
```

which reads the indexed image in filename into **X** and its associated colormap into **C**. The colormap values are automatically rescaled into the range [0, 1]. Alternatively, if **C** is dropped as an output, then

```
>>[X]=imread(filename,fmt)
```

will return **X** as an array containing the image data. If the file contains a grayscale image, **X** will be an M-by-N array. If the file contains a truecolor image, **X** will be an M-by-N-by-3 array. For TIFF files containing color images that use the CMYK (cyan, magenta, yellow, and black) color space, **X** is an M-by-N-by-4 array.

Example 4.2

```
>> [Xforest_withC,Cforest]=imread('forest','tif');
>> [Xforest_noC]=imread('forest','tif');
>> [Xftbl_withC,Cftbl]=imread('football','jpg');
>> [Xftbl_noC]=imread('football','jpg');
```

Figure 4.1 shows properties of the **Xforest_withC** and **Xforest_noC** arrays. Notice that they are both identical. The image file **'forest.tif'** is resolved into an indexed image array accompanied by the colormap **Cforest** that is rescaled into the range [0, 1]. Both **Xforest_withC** and **Xforest_noC** are uint8 (unsigned 8-bit integer from 0 to 255), 2D, grayscale arrays. On the other hand, **Xftbl_withC** and **Xftbl_noC** are 3D truecolor RGB arrays. The image file **'football.jpg'** is resolved into a 3D truecolor image array and the accompanied colormap **C** is thus blank (i.e., not referenced). The coding of the "colored" image array **Xftbl** is a MATLAB RGB coding scheme in which three different but associated arrays are used to indicate the intensity of the three-color components of the image: red (R), green (G), or blue (B). This coding scheme produces what is known as a *truecolor* image.

Name ▲	Value	Size	Min	Class	Max
Cforest	<256x3 double>	256x3	0	double	1
Cftbl	[ ]	0x0		double	
Xforest_noC	<301x447 uint8>	301x447	6	uint8	251
Xforest_withC	<301x447 uint8>	301x447	6	uint8	251
Xftbl_noC	<256x320x3 uint8>	256x320x3	6	uint8	255
Xftbl_withC	<256x320x3 uint8>	256x320x3	6	uint8	255

Figure 4.1 MATLAB Workspace shows that the creation of image arrays mainly depends on the image color type embedded in the image file itself.

NOTES

1. As with the encoding used in an indexed image, the larger the pixel value, the brighter is the respective color.
2. As with **imfinfo**, it is not required to specify an extension with the file name if the extension correctly corresponds to the specified file format **fmt**.
3. **Bitdepth** is the number of bits used to represent each image pixel. **Bitdepth** is calculated by multiplying the bits-per-sample with the samples-per-pixel. Thus, a format that uses 8 bits for each color component (or sample) and 3 samples per pixel has a bitdepth of 24.

4.2 Graphics Image Display

Displaying the image we have just read is achieved by simply typing:

```
>> imshow(Xftbl_noC);
```

This will open a figure window and display the image within a certain box area, as shown in Fig. 4.2.

Figure 4.2 Visualization of 3D array image array, **Xftbl_noC**, using the **imshow** command.

Notice that **imshow** can be used for grayscale, indexed, and truecolor image. For the indexed type, the associated colormap can be input as a second argument:

```
>>imshow(X,map)
```

displays the indexed image **X** with the colormap **map**. A color map matrix may have any number of rows, but it must have exactly 3 columns. Moreover,

```
>>imshow(I,[low high])
```

displays the grayscale image **I**, specifying the display range for **I** in [low high]. In addition,

```
>>imshow('filename')
```

displays the image stored in the graphics file filename. Besides **imshow**, the user may also use any of the following commands:

```
>> figure;image(Xrgb);
>> figure;imagesc(Xrgb);
```

Both are best used for the RGB type of images. For an indexed image, the user cannot insert the colormap as a second input argument for either **image** or **imagesc** command, as is the case with the **imshow** command. For other types of images, both can be used. The difference between **image** and **imagesc** is simply that **imagesc** scales the image data to use the full color map. See Exercise 4.1.

At last, the command:

```
>> imtool(X);
```

opens the image under MATLAB image tool, wherein some useful features can be utilized to handle or deal with images. For example, pixel information in terms of X and Y coordinates and RGB values can be seen via hovering the mouse over any location within the image. Moreover, the **imtool** ruler can be used to measure a distance between two points, expressed in pixel values.

Example 4.3 Let us show images created in Example 4.2:

```
>> figure; image (Xforest_withC);
```

Figure 4.3 (*top*) is the result. If the image you read is an indexed image, i.e., a 2D index array with a corresponding N × 3 colormap, you will need to use the **colormap** function to get the correct coloring of the plot.

```
>> colormap(Cforest);
```

Now, the beauty shows up as in Fig. 4.3 (*bottom*). You may not notice the change if the figure lags behind the MATLAB main window; recheck the previous figure if it is not active (always search for beauty).

The MATLAB command:

```
>>imshow(Xforest_withC,Cforest)
```

will also produce Fig. 4.3 (*bottom*).

NOTE The array **Xforest_withC** is an array of indices (indexes) specifying which row from Cforest (256×3 array) the pixel color is to be taken from. **Cforest** is a 256-by-3 matrix of real numbers between 0.0 and 1.0. Each row is an RGB vector that defines one out of 256 colors. The *kth* row of the colormap, **Cforest**, defines the kth color, where Cforest(k,:) = [r(k) g(k) b(k)] specifies the intensity of red, green, and blue. So, in indexed images, the pixel values are themselves indices (indexes) to a table that maps the index value to a color value.

Figure 4.3 The '**image (Xforest_withC)**' command visualization (*top*) and invoking the associated colormap '**Cforest**' afterward will produce the beauty (*bottom*).

Figure 4.3 (*Continued*)

Exercise 4.1 Save the following set of commands to an M-file and then run the M-file itself. It will create different indexed 400 × 400 matrices.

```
clear all;
ImMat=rand(400,400);image(ImMat);
ImMat10=10*ImMat;figure; image(ImMat10);
ImMat100=100*ImMat; figure; image(ImMat100);
ImMat1000=1000*ImMat;figure; image (ImMat1000);
figure; imagesc(ImMat);figure; imagesc(ImMat10);
figure; imagesc(ImMat100);figure; imagesc (ImMat1000);
```

See the difference between using **image** and **imagesc**. **imagesc(X)** is the same as **image(X)**, except the data is scaled to use the full colormap.

Exercise 4.2 Save the following set of commands to an M-file and then run the M-file itself. It will create different indexed 256 × 256 matrices.

```
clear all;
rows = 256; columns = 256;
img = zeros(rows, columns); imshow(img);
```

```
img2=ones(rows,columns); figure; imshow(img2);
img(120:136, :) = 0.8;
img(:, 120:136) = 0.8; figure; imshow(img);
img2(120:136, :) = 0.2;
img2(:, 120:136) = 0.2; figure; imshow(img2);
```

NOTE Notice that grayscale images are stored as intensity class images wherein the pixel value represents the brightness or grayscale value of the image at that point.

4.3 MATLAB Image-Modifying Functions

Below is a summary of some MATLAB built-in image-modifying functions:

1. **gray2ind**

   ```
   >>[X, map] = gray2ind(I,n)
   ```

   This function converts the grayscale image **I** to an indexed image **X**, associated with its colormap **map**. **n** specifies the size of the colormap, **map(n)**. **n** must be an integer between 1 and 65536. If **n** is omitted, it will default to 64.

2. **ind2gray**

   ```
   >>I = ind2gray(X,map)
   ```

   This function converts the indexed image **X** with its associated colormap **map** to a grayscale (or, gray-level) image **I**. **ind2gray** removes the hue and saturation (chromatic) information from the input image while retaining the luminance (intensity or grayscale) information. Notice that **X** can be of class uint8, uint16, single, or double. **map** is double. **I** is of the same class as **X**.

3. **rgb2ind**

   ```
   >>[X,map] = rgb2ind(RGB, n,'dither_option')
   ```

   This function converts the **RGB** image to an indexed image **X** with its associated colormap **map**. **n** determines the length (number of colors) of **map**. **map** contains at most **n** colors. **n** must be less than or equal to 65,536. In general, the function **rgb2ind** is used to reduce the number of colors presented in an image.

NOTE Notice '**dither_option**' can have one of two values: '**dither**,' the default value, which dithers (scatters different colored pixels in an image to make it appear as though there are intermediate colors in images with a limited color palette; dithering is often used to reduce banding in GIFs with graduated color transitions), if necessary, to achieve better resolution at the expense of spatial resolution; on the contrary, **'nodither'** maps each color in the original image to the closest color in the new map (depending on the value of **n**).

Notice that the input **RGB** image can be of class uint8 (values between 0 and 255); uint16 (values between 0 and 65535); or double. The output array, **X**, is of class uint8 if **n** is less than or equal to 256; otherwise it is of class uint16.

4. **ind2rgb**

```
>>RGB = ind2rgb(X,map)
```

This function converts the indexed image **X** and its associated colormap **map** to **RGB** (truecolor) format. **X** can be of class uint8, uint16, or double. **RGB** is an m-by-n-by-3 array of class double.

5. **rgb2ntsc**

```
>>Xntsc = rgb2ntsc(RGB);
```

This function converts the truecolor RGB image to NTSC (television) image. In the NTSC color space, the luminance (intensity) is the grayscale signal used to display pictures on monochrome (black and white) televisions. The other components carry the hue and saturation (chromatic) information.

RGB can be of class uint8, uint16, int16, single, or double. The output is double.

6. **ntsc2rgb**

```
>>Xrgb=ntsc2rgb(Xntsc);
```

This function converts NTSC to RGB image. The input image *must be of class double*. The output is of class double.

7. **imadjust**

```
>>J = imadjust(I)
```

This function maps the intensity values in grayscale image **I** to new values in **J** such that 1% of data is saturated at low and high intensities of **I**. This increases contrast of the output image **J**.

8. **im2bw**

```
>>BW = im2bw(I, level)
```

This function converts the grayscale image **I** to a binary image. The output image **BW** replaces all pixels in the input image with luminance greater than **level** with the value 1 (white) and replaces all other pixels with the value 0 (black). Specify **level** in the range [0, 1]. This range is relative to the signal levels possible for the image's class. Therefore, a level value of 0.5 is midway between black and white, regardless of class. To compute the level argument, you can use the function **graythresh**. If you do not specify **level, im2bw** uses the value 0.5.

9. **im2double**

```
>>Idouble = im2double(I)
```

This function converts the intensity image **I** to double precision **Idouble**, rescaling the data if necessary. If the input image is of class double, the output image will be identical.

```
>>RGBdouble = im2double(RGB)
```

This converts the truecolor image **RGB** to double precision **RGBdouble**, rescaling the data if necessary.

```
>>Idouble= im2double(BW)
```

This converts the binary image **BW** to a double-precision intensity image **Idouble**.

```
Xdouble = im2double(X,'indexed')
```

This converts the indexed image **X** to double precision **Xdouble**, offsetting the data if necessary.

10. **im2uint8**

```
>>Iuint8 = im2uint8(Igray)
```

This converts the grayscale image **Igray** to uint8 **Iunit8**, rescaling the data if necessary.

```
>>RGBuint8 = im2uint8(RGB)
```

This converts the truecolor image **RGB** to uint8 **RGBuint8**, rescaling the data if necessary.

```
>>Iuint8 = im2uint8(BW)
```

This converts the binary image **BW** to uint8 grayscale image **Iuint8**, changing 1-valued elements of **BW** to 255 in **Iuint8**.

```
>>Xuint8 = im2uint8(X1,'indexed')
```

This converts the indexed image **X1** to uint8 **Xuint8**, offsetting the data if necessary.

NOTE Notice that it is not always possible to convert an indexed image to uint8. If **X1** is of class double, the maximum value of **X1** must be 256 or less; if **X1** is of class uint16, the maximum value of **X1** must be 255 or less.

11. **im2uint16**

    The same task as that of **im2uint8**, except it converts into uint16 instead of uint8.

12. **mat2gray**

    ```
    >>Igray = mat2gray(A)
    ```

    This function converts the matrix **A** to the intensity image **Igray**. The returned matrix **Igray** contains values in the range 0.0 (black) to 1.0 (full intensity or white). The values of minimum and maximum intensity match with the minimum and maximum values found in **A**.

4.4 Image Saving

We may at some point in the process of image acquisition want to save the changes and/or corrections we have made. This is done by the MATLAB **imwrite** function. Refer to the MATLAB built-in Product Help for more information on **imwrite** options.

```
>> imwrite(im, map, 'filename', 'fmt');
```

'**fmt**' is one of the above-mentioned types of choice, like: '**png**', '**jpg**', '**tif**', and '**gif**'. '**BitDepth**' parameter can be further specified for some types.

```
>> imwrite(im, map, 'filename', 'fmt', 'BitDepth',#);
```

where **#** depends on the type of image; for grayscale images this can be 8, 12, or 16; and for color images this can be 8 or 12.

You can also add the following arguments:

```
..., 'Compression',Val
```

where **Val** can be '**none**' (the default); '**jpeg**'(valid only for grayscale and RGB images); or `'rle'` (valid only for grayscale and indexed images).

The write mode is done via:

```
...,'WriteMode', Val2
```

where `Val2` can be: `'overwrite'` (the default), or `'append'`.

Example 4.4 Suppose that we would like to carry out the following image-based treatment step:

```
>> Xfabric=imread('fabric','png');
>> Xntsc = rgb2ntsc(Xfabric);
```

To save **Xntsc** to a file named **myphoto.png**, then we may use:

```
>> imwrite(Xntsc, 'myphoto.png', 'png','BitDepth',8, ...
'Compression','jpeg','WriteMode','append');
```

4.5 Image Resizing

Image resizing can be done via:

```
>>B = imresize(A, scale)
```

This returns image **B** that is scale times the size of **A**. The input image **A** can be a grayscale, RGB, or binary image. If scale is between 0 and 1.0, B is smaller than **A**. If scale is greater than 1.0, **B** is larger than **A**.

```
>>B = imresize(A, [numrows numcols])
```

This returns image **B** that has the number of rows and columns specified by **[numrows numcols]**. Either **numrows** or **numcols** may be NaN, in which case **imresize** computes the number of rows or columns automatically to preserve the image aspect ratio.

```
>>[Y newmap] = imresize(X, map, scale)
```

This returns image **Y** via resizing the indexed image **X**. **scale** can either be a numeric scale factor or a vector that specifies the size of the output image (**[numrows numcols]**). By default, **imresize** returns a new, optimized colormap **newmap** with the resized image.

Example 4.5

```
>> [Xonion]=imread('onion','png');
```

To reduce the dimensions of the image array by a factor of one half, use:

```
>> Xonion_reduced=imresize(Xonion,0.5);
```

On the other hand, to increase the dimensions of the image array by a factor of 2, use:

```
>> Xonion_enlarged=imresize(Xonion,2.0);
```

4.6 Mathematical Manipulation of Image Arrays

Images are simply 2D or 3D arrays. Consequently, all the array manipulations and operations are applicable. Let us say that a file is read via the **imread** statement and is stored to an image array. We can carry out some mathematical operations on the image array, which will result in alteration or modification to RGB values of the associated image.

Example 4.6 Let's read the image file of '**onion.png**' and store as a true-color RGB image array, **Xonion**.

```
>> [Xonion]=imread('onion','png');
```

We can carry out the following arithmetic operations on the matrix **Xonion**:

```
>> Xonion2=1.5*Xonion+50.0;
>> imshow([Xonion, Xonion2]);
```

You will notice that **Xonion2** image becomes brighter than the original, unmodified image.

On the other hand,

```
>> Xonion3=Xonion.^2;
>> imshow([Xonion, Xonion3]);
```

Xonion3 image fades out compared with the original, unmodified image.

Example 4.7 Let us stamp a box with different colors on the previous image.

```
>> WBox=uint8(255+zeros(30,30));
>> Xonion4=Xonion;
>> Xonion4(50:79,120:149,1)=WBox;
>> imshow(Xonion4);
```

Notice that a red box is stamped on the image. This is simply because we have set the color value to 255 for the first layer (i.e., index 1 in the third dimension of the image array); this means a red (R) color will fill in the box area.

```
>> Xonion4(50:79,120:149,2)=WBox;
>> imshow(Xonion4);
```

Notice that a yellow box is stamped on the image. Here it is a bit different; the second layer (i.e., index 2) is set to 255, which translates into fully green (G) color. What is wrong? Well, the blending (addition) of red (R) and green (G) will give the yellow color as a transition color between red (R) and green (G) zones in the three primary color (R, G, B) bandwidth of light.

```
>> Xonion4(50:79,120:149,3)=WBox;
```

Notice that a white box is stamped on the image. Here the third layer (i.e., index 3) is set to 255, which translates into fully blue (B) color. What is wrong? Well, the blending (addition) of red (R), green (G), and blue (B) will result in the white color as the overlap region among the three primary colors of light.

Example 4.8 Consider the following M-file (**Example4_8.m**) code:

```
clear all;
Im=zeros(400,400); % creates 400x400 zero-valued array
mycol=1:1:160000; % creates a column to fill in the zero-valued
% array
Im(:)=mycol(:); % fill in the zero-value array
ImMatpoly1=2*Im+7;% different arrays are created as a function
% of input Im
```

```
figure; imagesc(ImMatpoly1);
ImMatpoly2=(Im).^2+2*Im+7;
figure; imagesc(ImMatpoly2);
ImMatpoly3=(Im).^3+(Im).^2+2*Im+7;
figure; imagesc(ImMatpoly3);
ImMatpoly4=(Im).^4+(Im).^2+2*Im+7;
figure; imagesc(ImMatpoly4);
ImMatp_1=(Im).^0.1+7;
figure; imagesc(ImMatp_1);
```

This will create different images with rainbow colors (i.e., the prism-resolved white light spectra).

Example 4.9 Consider the following M-file (**Example4_9.m**) code:

```
clear all;
Im=zeros(400,400); % creates 400x400 zero-valued array
mycol=1:1:160000; % creates a column to fill in the zero-valued
% array
Im(:)=mycol(:); % fill in the zero-value array

ImMatsin=sin(Im).^3+sin(Im).^2+sin(2*Im)+7;
figure; imagesc(ImMatsin);
ImMatcos=cos(Im).^3+cos(Im).^2+cos(2*Im)+7;
figure; imagesc(ImMatcos);

ImMatpolysin=((Im).^2+2*Im+7)*sin(Im)+cos(Im);
Im2=ImMatcos./ImMatpolysin;
figure; imagesc(Im2);
```

This will create different images similar to the decoration imprinted on carpets.

4.7 Image Acquisition/Quality Control

This section demonstrates MATLAB's capability of data acquisition and analysis. This will be demonstrated via running the M-file named **Im_Process.m** that is subdivided into sections. MATLAB has the option to convert the output of an M-file into a full-fledged report (e.g., HTML format) that includes both the MATLAB scripting language and the results of executing them. It is worth mentioning here that the output file format can be **html, xml, latex, doc, ppt, or pdf**. To configure publishing **Im_Process.m**, as an example:

- Open **Im_Process.m** using the built-in MATLAB M-file editor.
- Go to **File** menu; under **File** menu, choose **Publish Configuration for Im_Process.m** followed by **Edit Publish Configuration for Im_Process.m**.

- **The Edit Configurations** window pops up where you pick up your choice in terms of Output file format and Output Folder (i.e., location within the hard disk).
- The main item (i.e., the bulleted list) within the published content can be created via using %% in the M-file. For example, the first item shown below is generated via using the following line in the **Im_Process.m** file:

```
%% 0. CLOSE ALL FIGURES AND CLEAR VARIABLES BEFORE
YOU START
```

The explanation for each step is provided in advance so that the user can follow up and grab the idea lying behind using image acquisition for quality control purposes.

0. Close all figures and clear variables before you start

```
>>close all; clear all;
```

1. Read the image file

```
>>FileName='D:\MATLAB\R2011b\toolbox\images\imdemos\candy';
>>candy=imread(FileName,'jpg');
```

2. Conversion from truecolor to grayscale image

   I=rgb2gray(candy) converts the truecolor image candy to the grayscale intensity image, **I**. It eliminates the hue and saturation (chromatic) information but retains the luminance (intensity).

```
>>I=rgb2gray(candy);
>>figure(1); imshow(I);
```

Figure 4.4 The transformed grayscale intensity image of the original truecolor candy image.

3. First morphological image treatment: close operation (object interconnection)

   imclose performs morphological closing on the grayscale or binary image **I**, returning the closed image, **bckgrnd**. The structuring element, **strel**, must be a single structuring element object, as opposed to an array of objects. Create a disk-shaped structuring element. Use a disk-structuring element to preserve the circular nature of the object (i.e., the Mars pellet). Specify a radius of 12 pixels so that the largest gap gets filled. Perform a morphological close operation on the image.

NOTE A 12-pixel radius is smaller than the radius of the smallest Mars pellet found in the picture. You can invoke the image tool (**>>imtool**) and then open the candy image file. After that, use the ruler shown at the top of the image tool, see Fig. 4.12, to decide on the proper size of the pellet object, expressed in pixels.

The morphological close operation (**imclose**) is essentially a combined process of a dilation (see **imdilate** for help) followed by an erosion (see **imerode** for help), using the same structuring element for both operations.

```
>>bckgrnd=imclose(I,strel('disk',12));
>>figure(2); imshow(bckgrnd);
```

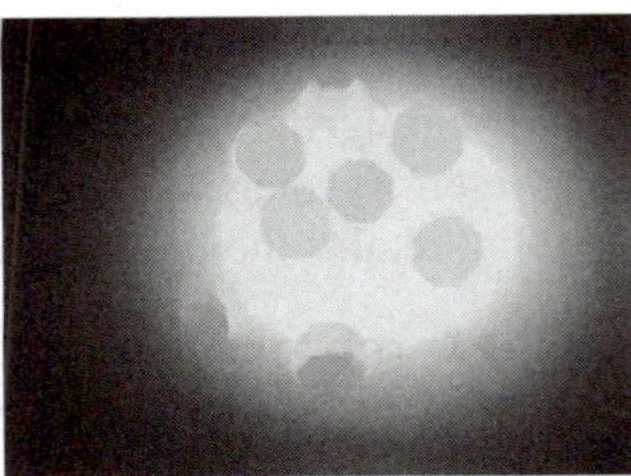

Figure 4.5 The output of the **imclose** operation on the original image using a disk-structuring element with radius = 12 pixels. Objects (pellets) are interconnected and gaps among them are filled.

4. Second morphological treatment: removal of background luminances

   I2 = imsubtract(bckgrnd,I) subtracts each element in array **I** from the corresponding element in array **bckgrnd** and returns the difference in the corresponding element of the output array **I2.**

```
>>I2=imsubtract(bckgrnd,I); % Equivalent to I2=bckgrnd-I;
>>figure(3);imshow(I2,[]);
```

   See Fig. 4.6.

5. Setting a threshold for light intensity (luminance)

   level=graythresh(I2) computes a global threshold (level between 0 and 1) that can be used to convert an intensity image to a binary image with **im2bw**.

```
>>level=graythresh(I2);
```

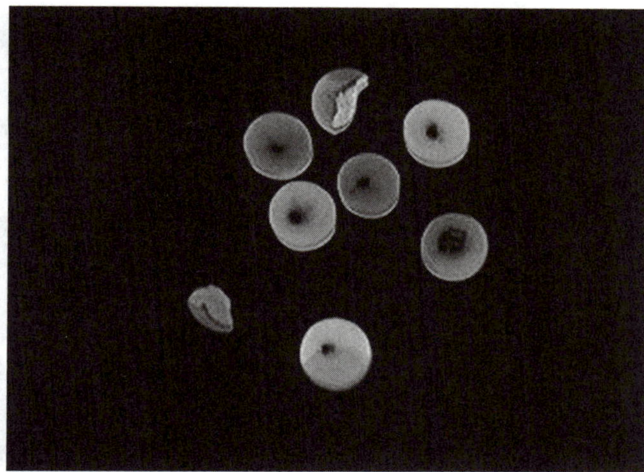

Figure 4.6 Background luminance (intensity) is removed via subtracting the array **I** from the array **bckgrnd**, resulting in array **I2**.

6. Conversion of gray image into binary image while utilizing the threshold

   BW = im2bw(I2, level) converts the grayscale image **I2** to a binary image, **BW**. The output image **BW** replaces all pixels in the input image with luminance greater than **level** with the value 1 (white) and replaces all other pixels with the value 0 (black). Notice that **I**, **I2**, and **bckgrnd** are grayscale arrays of uint8 type, which have values that range from 0 to 255. On the other hand, **BW** is logical that has 0 or 1 pixel value.

```
>>BW=im2bw(I2,level);
```

7. Third morphological treatment: removal of artifacts (small objects less than 20 pixels in size)

   BW2=bwareaopen(BW,20) removes from a binary image, **BW**, all connected components (objects) that are fewer than 20 pixels (object area), producing another binary image, **BW2**. The default connectivity is 8 for two dimensions. 8 means 8-connected neighborhood.

```
>>BW2=bwareaopen(BW,20);
>>figure(4); imshow(BW2);
```

   See Fig. 4.7.

8. Fourth morphological treatment: removal of border artifacts (connected to the image border)

   BW3 = imclearborder(BW2) suppresses structures that are both lighter than their surroundings and are connected to the image border. (In other words, use this function to clear the image border.) **BW2** can be a grayscale or binary image. The output image, **BW3**, is grayscale or binary, respectively. The default

Figure 4.7 Removal of artifacts (i.e., connected objects less than 20 pixels in area).

connectivity is 8 for two dimensions. Since we do not have border artifacts, the image shown in Fig. 4.7 and that in Fig. 4.8 are essentially identical.

```
>>BW3=imclearborder(BW2);
>>figure(5); imshow(BW3);
```

See Fig. 4.8.

9. Fifth morphological treatment: filling in the gaps within objects

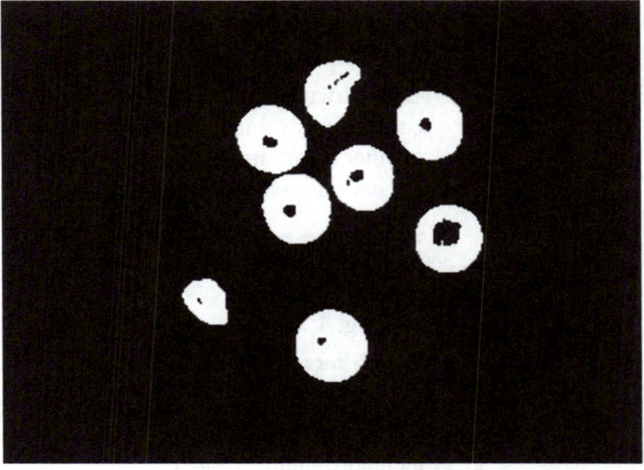

Figure 4.8 Removal of border artifacts (if any exists), connected to the image border.

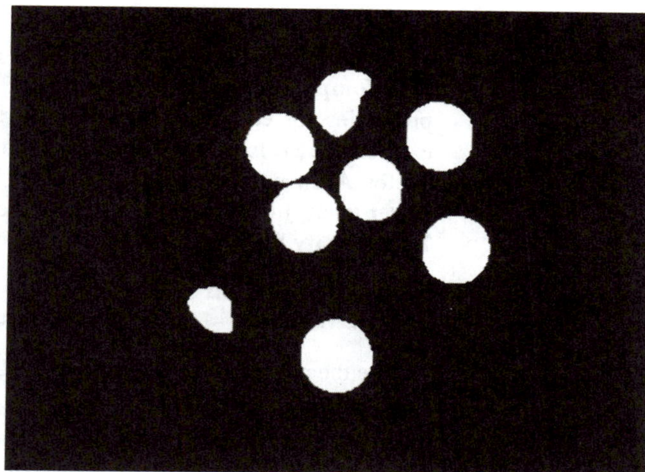

Figure 4.9 Filling in the gaps (i.e., holes) within objects.

fill=imfill(BW3,'holes') fills holes in the binary image **BW3**. A hole is a set of background pixels that cannot be reached by filling in the background from the edge of the image.

```
>>fill=imfill(BW3,'holes');
>>figure(6); imshow(fill);
```

See Fig. 4.9.

10. Find connected components in a binary image

    ConComp = bwconncomp(fill) returns the struct, **ConComp**, of the connected components that are found in the final, finished binary image, **fill**. **ConComp** is a struct with four fields. One important field is number of objects, **NumObjects**. A struct is a structure array, which has fields. The most common way to access the data in a struct is by specifying the name of the field that we want to reference. Use the name of the struct followed by the dot '.' and immediately by the name of the field.

```
>>[ConComp]=bwconncomp(fill);
```

11. Determine an object's geometrical and shape properties

    Stats = regionprops(ConComp, properties) measures a set of properties for each connected component (object) in **ConComp**, which is a struct returned by the previous **bwconncomp** command. Examples of properties are **Area, BoundingBox, Centroid, Eccentricity, EquivDiameter, Extent, MajorAxisLength, MinorAxisLength, and Perimeter**. *See* **regionprops** *using Product Help*. Here, the chosen properties are **Eccentricity**, **Area**, and **BoundingBox**.

```
>>Stats=regionprops(ConComp,'Eccentricity','Area','Bound
ingBox')
```

12. Redefining variables for simplicity

    Stats is a structure array with length equal to the number of objects as **ConComp.NumObjects**. The fields of the structure array denote different properties for each region, as specified by properties. **Area=[Stats.Area]** extracts the **Stats** field, that is, **Area**, into a separate vector, called **Area**. The vector **Area** has one value for each object (pellet). **Eccentricity=[Stats. Eccentricity]** facilitates the usage of the property: **Eccentricity**. The vector **Eccentricity** has one value for each object.

```
>>Area=[Stats.Area];
>>Eccentricity=[Stats.Eccentricity];
```

13. Define the minimum area below which the pellet fails the quality control test

    A restriction to accept a candy pellet as undamaged is to have its actual area be greater than a minimum area named **MinArea**. **MinArea**, a scalar quantity, is defined as the arithmetic mean minus one fourth times the standard deviation of the pellets (object's) area. Keep in mind that **Area** is a 1×8 row vector, where 8 is the number of found objects using **bwconncomp** command in Step 10.

```
>>MinArea=mean(Area)-0.25*std(Area);
```

14. Use minarea and eccentricity values to identify broken objects (candy pellets)

    If the area of the pellet (object) is below the threshold value, **MinArea**, and if the eccentricity is above 0.5, then the object is said to have defects. The eccentricity is the ratio of the distance between the foci of the ellipse and its major axis length. The value is between 0 and 1. A zero value means a full circle and a value of one means a line segment, in 2D shapes. **IndexOfDefect** is the index of the matrix element returned by the **find** statement once the condition for failure is met, or fulfilled. **StatDefect=Stats(IndexOfDefect)** picks up only the defected objects. **StatDefect** is also a struct, similar to **Stats**; however, it contains only objects that fail the quality control test. It is worth mentioning here that for each entry found in either struct, it has three associated attributes: **Area**, **BoundingBox**, and **Eccentricity**.

```
>>IndexOfDefect=find(Area<MinArea & Eccentricity> 0.5);
>>StatDefect=Stats(IndexOfDefect);
```

15. Labeling of broken objects

    Labeling, in the form of drawing a bounding box, is now carried out on the original, unmodified image. The command: **hold on** retains the current plot and certain axes properties so that subsequent graphing commands add to the existing graph. The "for-loop" is carried out all over defected objects; where a rectangle is drawn around each defected object. Keep in mind that a defected object is the object that has area that is below the threshold value, **MinArea**, and its eccentricity value lies above 0.5. Notice that the command: **rectangle('Position', [x y w h])** will add a rectangle at the specified positions. Here, the matrix **[x y w h]** is replaced by **StatDefect(index).BoundingBox**, where the latter has four values equivalent to [x y w h].

Figure 4.10 Labeling of defected objects. A defected object is an object that has **Area** below the threshold value, **MinArea**, and its **Eccentricity** value lies above 0.5.

```
>>figure(7); imshow(candy);
>>hold on;
>>for index=1:length(StatDefect)
>>h=rectangle('Position',StatDefect(index).
BoundingBox,'LineWidth',3;
>>  set(h,'EdgeColor',[0.75 0 0]);
>>end
>>hold off;
```

4.8 Image Acquisition/Bacterial Colony Counting

The following example demonstrates how to carry out bacterial colony counting, which is a significant technical hurdle for vaccine studies as well as various microbiological studies. MATLAB can be used to process images obtained by a digital camera or document scanner. The image shows a square Petri dish with pneumococcal colonies. The colonies are in 24 spots, arranged in 3 columns and 8 rows. Each spot has a varying number of colonies. The image was obtained with a document scanner as described in the following reference. Figure 4.11 shows a snapshot taken by Screen_Capture software (https://sites.google.com/site/almalahweb/software), developed by the author.

As indicated by Putman et al. (see the reference below), the colonies are in 24 spots, arranged in 3 columns and 8 rows. Table 4.1 shows the number of colonies as reported by Putman et al.

As Table 4.1 shows, the total number of colonies is 1038.

The following contents are obtained via MATLAB Microsoft Word interface to produce polished, integrated, interactive documents for reports, presentations, or online publishing. A detailed explanation is given at the beginning of Sec. 4.7.

Figure 4.11 A snapshot of pneumococcal bacterial colony forming units (CFUs) or colonies in a square Petri dish.

Table 4.1 Number of pneumococcal bacerial colonies formed in a square Petri dish.

141	96	109
66	69	89
33	54	59
13	30	53
4	5	46
1	4	43
2	4	29
3	11	74
Sum = 263	Sum = 273	Sum = 502
Total = 263 + 273 + 502 = 1038 colonies		

Putman, M, Burton, R, and Nahm, MH, "Simplified method to automatically count bacterial colony forming unit," *Journal of Immunological Methods, 302*: 99–102, 2005.

NOTE Regarding the definition of each MATLAB built-in function used in this section, please refer, first, to Sec. 4.7 and, second, to MATLAB Product Help. In this section, the differences from the previous section will be pointed out.

0. Close all figures and clear variables before you start

```
>>close all; clear all;
```

1. Read the image file

```
>>cfu=imread('D:\Program Files\MATLAB\CFU','bmp');
```

2. Conversion from truecolor to grayscale image

   See Sec. 4.7 for more information on **rgb2gray** function.

```
>>I=rgb2gray(cfu);
```

3. First morphological image treatment: close operation (object interconnection)!

   SE = strel('disk', R, N) creates a flat, disk-shaped structuring element, where **R** specifies the radius. **R** must be a non-negative integer. **N** must be 0, 4, 6, or 8. When **N** is greater than 0, the disk-shaped structuring element is approximated by a sequence of N periodic-line structuring elements. When **N** equals 0, no approximation is used, and the structuring element members consist of all pixels whose centers are no greater than **R** away from the origin. If **N** is not specified, the default value will be 4. Specify a *radius of one pixel* so that objects with radius 1 pixel or larger will be accounted for. See Sec. 4.7 for further information on the **imclose** function.

NOTE You can invoke the image tool (**imtool** at the command prompt) and then open the **cfu** image file. After that, use the ruler shown at the top of the image tool to decide on the proper size of the CFU, expressed in pixels. Figure 4.12 shows that the size of one CFU is, at least, 2 pixels in width for a square (equivalent to 1.13 pixel in radius for a circle). The image is magnified to 3200% resolution.

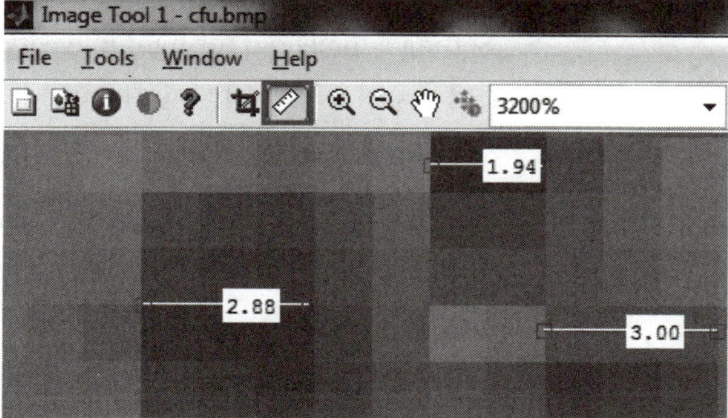

Figure 4.12 Using the '**imtool**' ruler, the size of an object, expressed in pixels, can be estimated by clicking on one spot and moving the dragged mouse to another spot. Notice that you can magnify (or, zoom in) the image as large as you like (here, 3200%).

```
>> SE = strel('disk', 1, 4);
>>bckgrnd=imclose(I,SE);
>>figure(1); imshow(bckgrnd);
```

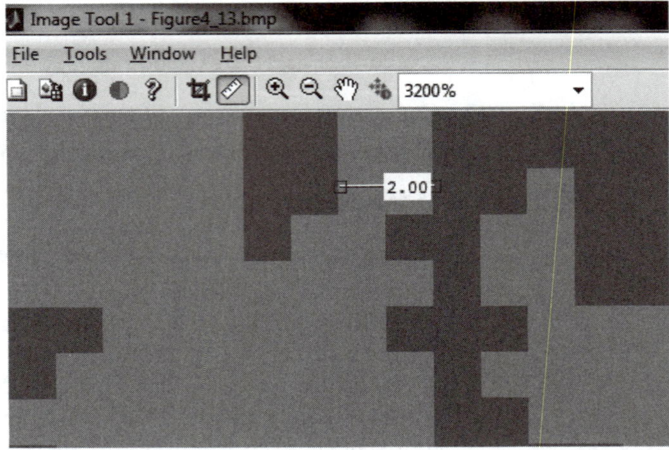

Figure 4.13 The output of the **imclose** operation on the original image using a disk-structuring element with a radius of 1 pixel. CFU objects are interconnected and gaps are filled. Interconnection of objects is seen here as the image is magnified to 3200% resolution. This figure resembles Fig. 4.5 in Sec. 4.7.

4. Second morphological treatment: removal of background luminance

   See Sec. 4.7 for more information on **imsubtract** function.

```
>>I2=imsubtract(bckgrnd,I);
% figure(2);imshow(I2,[])%no need to show the figure;
```

5. Setting a threshold for light intensity (luminance)

   See Sec. 4.7 for more information on **graythresh(I2)** function.

```
>>level=graythresh(I2);
```

6. Conversion of gray image into binary image while utilizing the threshold

   See Sec. 4.7 for more information on the **im2bw** function.

```
>>BW=im2bw(I2,level);
```

7. Third morphological treatment: removal of artifacts [small objects less than 2 pixels (1.4 pixel width for a square object)]

   See Sec. 4.7 for more information on the **bwareaopen** function.

```
>>BW2=bwareaopen(BW,2);
>>figure(3); imshow(BW2);
```

   See Fig. 4.14.

8. Fourth morphological treatment: removal of border artifacts (connected to the image border)

Figure 4.14 Removal of artifacts (i.e., connected objects less than 2 pixels in size).

See Sec. 4.7 for more information on the **imclearborder** function.

```
>>BW3=imclearborder(BW2);
>>figure(4); imshow(BW3);
```

See Fig. 4.15.

9. Fifth morphological treatment: filling in the gaps within objects

   See Sec. 4.7 for more information on the **imfill** function.

```
>>fill=imfill(BW3,'holes');
% figure(5); imshow(fill); % no need to show this figure.
```

10. Count the number of connected objects

    See Sec. 4.7 for more information on the **bwconncomp** function. **ConComp = bwconncomp(fill)** returns the connected components, **ConComp**, found in the final, finished binary image, **fill**.

```
>>[ConComp]=bwconncomp(fill);
```

11. Determine object's geometrical and shape properties

    See Sec. 4.7 for more information on the **regionprops** function. Because a typical CFU or colony is almost circular, the equivalent diameter will be used as the criterion upon which the decision is to be made for either accepting or rejecting an object.

```
>>Stats=regionprops(ConComp,'EquivDiameter');
```

Figure 4.15 Removal of border artifacts connected to the image border; objects on the outskirts were removed.

12. Redefining variables for simplicity

    See Sec. 4.7 for more information on **Stats** as a struct (i.e., structure array).

```
>>EqDiam=[Stats.EquivDiameter];
```

13. Define the mean equivalent diameter and its standard deviation.

    A restriction to accept the CFU object is to have its equivalent diameter fall within a certain range, as shown in the next step.

```
>>MEqDiam=mean(EqDiam);% mean diameter of CFU objects.
>>Stdev=std(EqDiam);% the standard deviation of CFU objects.
```

14. CFU with either a too small or too large equivalent diameter will be rejected.

    If the equivalent diameter (EqDiam) of the CFU object is found to be too large or too small, compared with the mean equivalent diameter of all found objects, then this CFU will be rejected (i.e., considered an artifact). **IndexOfReject** is the index of the matrix element returned by the find statement once the condition is met or fulfilled. **StatReject=Stats(IndexOfReject)** picks up only the rejected objects.

```
IndexOfReject=find(EqDiam>(MEqDiam+2.0*Stdev)    |
EqDiam<(MEqDiam-2.0*Stdev));
StatReject=Stats(IndexOfReject);
```

15. Counting # of CFUs (colonies)

Counting colony forming unit (CFU) objects is done via subtracting the number of rejected objects from the total count of candidate objects. Keep in mind that an artifact is either too small or too large, compared with the mean value. A 95% confidence interval (i.e., $\pm 2\sigma$) is taken around the mean value of the equivalent diameter of all found CFU objects.

```
TotalNum=ConComp.NumObjects;
display ([‘The total number of candidate CFUs:
’,num2str(TotalNum)]);
CFUReject=length(StatReject);
display ([‘The number of rejected (too small or too large)
CFUs is: ’, ...
  num2str(CFUReject)]);
CFUNUM=ConComp.NumObjects-length(StatReject);
display ([‘The number of accepted CFUs is:
’,num2str(CFUNUM)]);
```

Here are the final results:

The total number of candidate CFU is 1108.

The number of rejected (too small or too large) CFUs is 21.

The number of accepted CFUs is 1087.

The manual count, as given by Putman et al. (2005), is 1038 (Table 4.1). Thus, the percent relative error (PRE) is:

$$PRE = \frac{|1038 - 1087|}{1038} \times 100\% = 4.7\%.$$

NOTE Finally, it is worth noting that *parameters used for pixel size* in steps 3 and 7 and the range of CFU acceptance, defined in step 14, all can be adjusted or *tuned-up* for a given image that represents a physical (real) system such that the estimated count, carried out by MATLAB, is very close to the manual count. In essence, this prerequisite step will *tune up* parameters associated with MATLAB image-related functions. After the completion of this tune-up step, the previous algorithm of MATLAB image acquisition can then be used for handling other images (e.g., counting CFUs in an image) created with the same accuracy and precision.

4.9 MATLAB Image Acquisition Tool (imaqtool)

At the end of this chapter, we will show how to interface the laptop or PC hardware and its image devices to the MATLAB environment. At the command prompt, run the following command:

```
>>imaqtool
```

Figure 4.16 shows the MATLAB image acquisition tool window, which is subdivided into panes or windows.

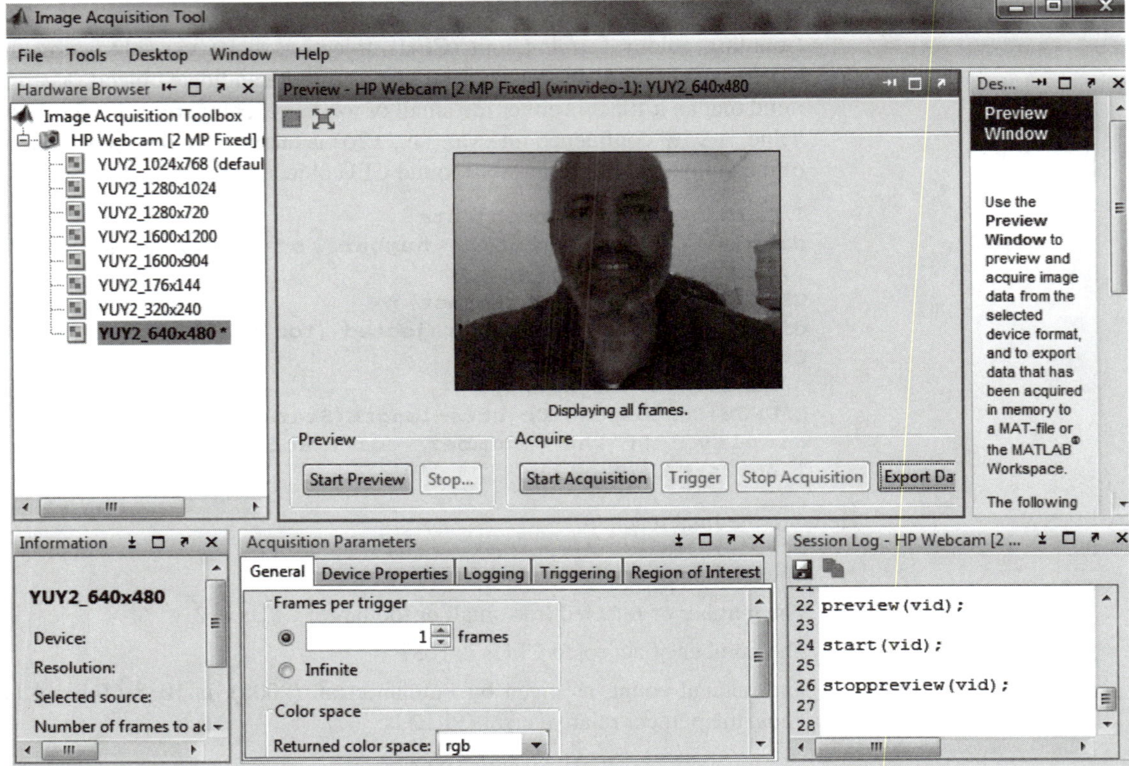

Figure 4.16 The MATLAB image acquisition tool window, wherein the user has the opportunity to customize the process of interfacing between MATLAB and the image device installed on a given hardware.

At the top left, the type of image device attached to the laptop/PC will be shown. For example, with my HP laptop, it shows that it is connected to HP Webcam [2MP fixed]. If I click on the submenu YUY2_640×480, as an example, then MATLAB will be ready to receive or acquire the input signal from such a device. Moreover, it also gives the user the chance to first preview a single image or multi images (i.e., frames of images, or movie) before he/she launches the process of data acquisition. To preview, click the "**Start Preview**" button in the middle pane. If you are happy with what you see, then stop the preview process (click on the "**Stop Preview**" button) and be ready for the next step, that is, the acquisition step. Before you click on the "**Start Acquisition**" button, however, you ought to customize some settings pertaining to receiving the image data.

Down in the middle pane, the user can change what kind of input signal he/she wishes to receive; for example, is it a single snapshot or a series of snapshots? The returned color space: is it grayscale, RGB, or YCbCr (YCbCr: luminance (Y) and chrominance (Cb and Cr) color values)?

The "**Device Properties**" tab is used to tune up light-related properties, such as brightness, contrast, sharpness, hue, saturation, and so on. The "**Logging**" tab pertains to saving the data to memory, disk, or both. The "**Triggering**" tab is for choosing between automatic or manual mode, and finally the "**Region of Interest**" tab is for specifying the offset borders and dimensions of the region of interest. Once you are done with the customization step, click the "**Start Acquisition**" button to capture the image data. Great, the snapshot is captured, what to do next?

Well, the user will tell MATLAB what to do with the acquired data (i.e., the captured image or images). To export the data to a place where you can handle the data later using MATLAB different image-based tools and functions, click on the "**Export Data**" button (top middle pane). A window pops up as shown in Fig. 4.17.

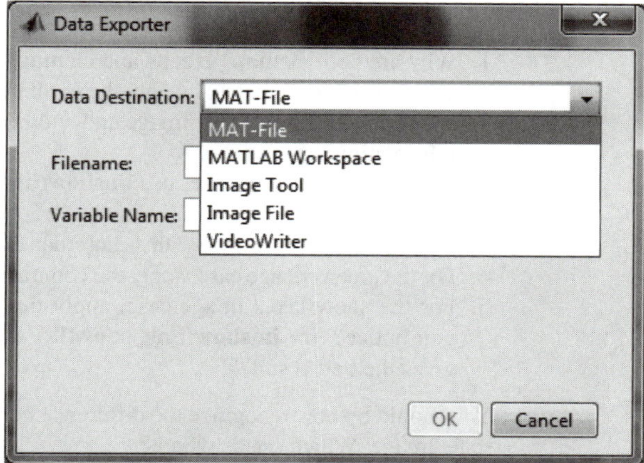

Figure 4.17 Clicking on the Export Data button will bring up a pop-up window wherein the user has the chance to tell MATLAB where to keep the image data and in what format.

The user now has the choice to save the acquired image in different formats, such as to MAT-File, to MATLAB Workspace, to Image Tool, to Image File (you have to specify a name for the file), or to VideoWriter. Once the image is defined in the MATLAB environment, all MATLAB different image-based tools and functions, explained earlier, can now be implemented.

4.10 Problems

NOTE Different image files are installed in the imdemos folder that is located where MATLAB is installed by the user (say, C:\Program Files\MATLAB\R2011b\toolbox\images\imdemos)

4.1 Read the following image files and assign an image and colormap array for each via the following command:

```
>> [img_canoe, clrmap_canoe]= imread('canoe.tif');
>> [img_greens, clrmap_greens]= imread('greens.jpg');
>> [img_snowflk, clrmap_snowflk]= imread('snowflakes.png');
```

You should have in the workspace the following defined image arrays as shown in the figure below:

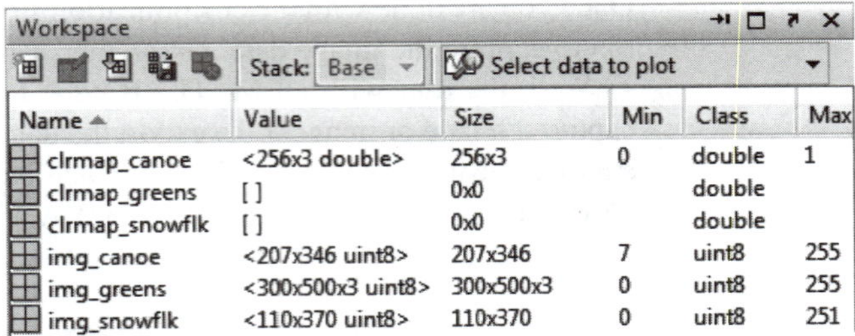

Name	Value	Size	Min	Class	Max
clrmap_canoe	<256x3 double>	256x3	0	double	1
clrmap_greens	[]	0x0		double	
clrmap_snowflk	[]	0x0		double	
img_canoe	<207x346 uint8>	207x346	7	uint8	255
img_greens	<300x500x3 uint8>	300x500x3	0	uint8	255
img_snowflk	<110x370 uint8>	110x370	0	uint8	251

1. Why are both **clrmap_greens** and **clrmap_snowflk** empty matrices whereas **clrmap_canoe** is a 256×3 matrix? Apply the command **imfinfo** for each image to see the difference between one type of image and another; look at the entry: ColorType in the image information data.
2. For the canoe image case, use **imshow(img_canoe)** and see how the produced image looks.
3. Repeat part 2) but this time with the insertion of the associated colormap array **clrmap_canoe**.
4. For the greens image case, apply the command **image(img_greens)**. What do you notice?
5. For the snowflakes image case, apply the command **image(img_snowflk)**. What do you notice? Try **imshow(img_snowflk)** and **imagesc(img_snoflk)**. Which command gives the best result?

You should by now recognize the difference between an indexed, RGB truecolor, and grayscale image. Which one is which?

4.2 Convert the grayscale **img_snowflk** (Problem 4.1) into an indexed-image **indx_snowflk** using different **n** values, where $1 \le n \le 65536$.

```
>> [indx_snowflk, snow_map] = gray2ind(img_snowflk, n);
>> figure,imshow(indx_snowflk, snow_map);
```

4.3 Convert back the indexed image **indx_snowflk** (Problem 4.2) into a grayscale image **grscale_snowflk** using the colormap **snow_map** provided.

```
>> [grscale_snowflk] = ind2gray(indx_snowflk, snow_map);
>> figure,imshow(grscale_snowflk);
```

4.4 Convert the RGB image **img_greens** (Problem 4.1) into an indexed image **indx_greens** and colormap array **greens_map** with **n=16, 64, 128,** and **256** (n must be ≤ 65,536).

```
>>[indx_greens,greens_map] = rgb2ind(img_greens, 16,'dither');
>>figure,imshow(img_greens);figure,imshow(indx_greens,greens_
map);
```

Repeat the same task but, with `'nodither'` instead of `'dither'`.

```
>>[indx_greens,greens_map] = rgb2ind(img_greens, 16,'nodither');
>>figure,imshow(img_greens);figure,imshow(indx_greens,greens_
map);
```

Create a logical image array **BW_greens** (i.e., only black and white colors) with '**nodither**' option.

```
>>BW_greens=rgb2ind(indx_greens,greens_map,'nodither');
```

You will notice that **BW_greens** will be affected by the **n** value.

NOTE Carry out all steps in Problem 4.4 for a given **n** and then repeat them for a new **n**. You will notice that **BW_greens** will be affected by the **n** value.

4.5 Read the **cameraman.tif** that is found in the MATLAB installed directory and assign it to an image array, named **CamMan**. Your task is to create a white box (30 × 30 pixels) anywhere within the image and show the modified image, where the white box appears embossed within the image. Repeat the task and create a black box with a size of 30 × 30 pixels.

4.6 Read the **greens.jpg** that is found in the MATLAB-installed directory and assign it to an image array named **ImgGreens**. Your task is to create a black box (30 × 30 pixels) anywhere within the image and show the modified image, wherein the black box appears embossed within the image. Repeat the task three times, and each time create a box with a size of 30 × 30 pixels for the three RGB colors: red, green, and blue.

4.7 For the **CamMan** image matrix (Problem 4.5), carry out the following arithmetic operations on the image array and see how the brightness is affected, accordingly:

```
>> CamManDrk=CamMan-100;
>> CamManLght=CamMan+100;
>> CamManSqrd=CamMan.^2;
>> CamManThrsh30=image(CamMan>30);
```

1. Show the original (**CamMan**) and modified images on the screen.
2. Save the modified images each in a separate image file with a common extension.

4.8 Given the figure below, your job is to write an M-file that will enable the user to perform the following tasks:

1. Determine the number of objects found in each of the three categories (i.e., rectangle, circle, and triangle).
2. Determine the minimum and maximum area for each of the three categories.
3. Use the **BoundingBox** property to draw a box around both the minimum and maximum area for each of the three categories.

Hint 1: After transforming the truecolor image to the grayscale intensity image, and carrying out the required MORPHOLOGICAL TREATMENT steps (i.e., see Secs. 4.7 and 4.8), define Stats as a structure array with length equal to the number of objects: using the following command:

```
Stats=regionprops(ConComp,'Extent','Area','BoundingBox');
```

where **ConComp** is a struct with four fields; one important field is number of objects, **NumObjects**.

Hint 2: Use the following **imclose** as the FIRST MORPHOLOGICAL IMAGE TREATMENT: CLOSE OPERATION (OBJECT INTERCONNECTION).

```
bckgrnd=imclose(I,strel('disk',20));
```

You should be able to obtain a figure that is similar to the following figure:

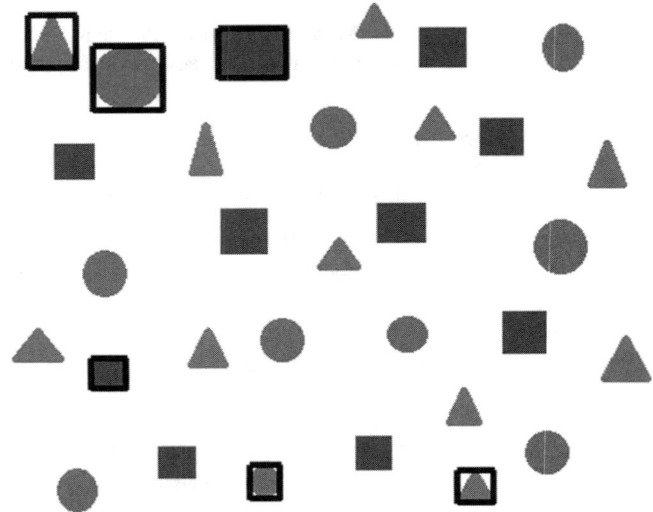

Moreover, an output shows the number of objects found in each category:

```
>> RegShape
The number of Rectangular objects is equal to: 10
The number of Triangular objects is equal to: 11
The number of Circular objects is equal to: 10
```

5 Curve-Fitting

Chapter Outline

Researchers in different fields of study often need to fit an equation or model to experimental data. This can be difficult because many curve-fitting tools require the user to have advanced mathematical or programming experience. In addition, it is common for researchers to perform many fitting trials with multiple data sets, which can be time consuming.

In this chapter, you will learn how to perform curve fitting with MATLAB in a manner suitable for those with little or no programming experience. You will also see how to automate your entire curve-fitting workflow, including:

- Importing data
- Visualizing data
- Fitting a built-in or customized equation to your data. The user will be introduced to the Curve/Surface Fitting Toolbox.

NOTE We will show how to proceed with curve-fitting process showing both the old and new look of the MATLAB Curve Fitting Toolbox. In general, Fig. 5.2*a* will be reserved for the old look and Fig. 5.2*b* for the new look whenever there is a difference worth mentioning.

5.1 (Two-Parameter) Linear Regression

Given a set of experimental data in the form of (x_i, Y_i), we will propose a model in the form of:

$$Y_{model} = \beta_o + \beta_1 x_i \tag{5.1}$$

where Y_{model} should *numerically approach* Y_i, which represents the experimental value of the *dependent (monitored or measured) variable*, and x_i is the experimental value of the *independent (manipulated)* variable.

The objective of curve-fitting or nonlinear regression, in general, is to *minimize the difference* between the measured value Y_i and the predicted value given by the proposed model; in this case, $Y_{model} = \beta_o + \beta_1 x_i$. Consequently, the objective function to be minimized can be represented by π:

$$\pi = \sum_{i=1}^{n} [Y_i - Y_{model}]^2 = \sum_{i=1}^{n} [Y_i - (\beta_o + \beta_1 x_i)]^2 \tag{5.2}$$

Taking the first derivative of π with respect to β_o and equating to zero, we get:

$$\frac{\partial \pi}{\partial \beta_o} = -2\sum_{i=1}^{n} [Y_i - (\beta_o + \beta_1 x_i)] = 0 \tag{5.3}$$

$$\sum_{i=1}^{n} \beta_o + \sum_{i=1}^{n} \beta_1 x_i = \sum_{i=1}^{n} Y_i \tag{5.4}$$

Taking the first derivative of π with respect to β_1 and equating to zero, we get:

$$\frac{\partial \pi}{\partial \beta_1} = -2\sum_{i=1}^{n} x_i[Y_i - (\beta_o + \beta_1 x_i)] = 0 \tag{5.5}$$

$$\sum_{i=1}^{n} \beta_o x_i + \sum_{i=1}^{n} \beta_1 x_i x_i = \sum_{i=1}^{n} x_i Y_i \tag{5.6}$$

Equations (5.4) and (5.6) are simultaneously solved to find the curve-fitted (or point estimate) parameters β_o and β_1. Both equations can be combined in the matrix form as:

$$\begin{bmatrix} \sum_{i=1}^{n} 1 & \sum_{i=1}^{n} x_i \\ \sum_{i=1}^{n} x_i & \sum_{i=1}^{n} x_i \end{bmatrix} \begin{bmatrix} \beta_o \\ \beta_1 \end{bmatrix} = \begin{bmatrix} \sum_{i=1}^{n} Y_i \\ \sum_{i=1}^{n} x_i Y_i \end{bmatrix} \tag{5.7}$$

In compact (i.e., matrix) notation:

$$[X_{mat}] \times [\beta] = [XY_{mat}] \tag{5.8}$$

where

$$[X_{mat}] = \begin{bmatrix} \sum_{i=1}^{n} 1 & \sum_{i=1}^{n} x_i \\ \sum_{i=1}^{n} x_i & \sum_{i=1}^{n} x_i^2 \end{bmatrix} \tag{5.9}$$

$$[\beta] = \begin{bmatrix} \beta_o \\ \beta_1 \end{bmatrix} \tag{5.10}$$

$$[XY_{mat}] = \begin{bmatrix} \sum_{i=1}^{n} Y_i \\ \sum_{i=1}^{n} x_i Y_i \end{bmatrix} \tag{5.11}$$

Let us multiply both sides of Eq. (5.8) by $[X_{mat}]^{-1}$, which then becomes:

$$[X_{mat}]^{-1}[X_{mat}] \times [\beta] = [X_{mat}]^{-1}[XY_{mat}] \tag{5.12}$$

Equation (5.12) becomes:

$$[I] \times [\beta] = [X_{mat}]^{-1}[XY_{mat}] \tag{5.13}$$

The identity matrix, $[I]$, multiplied by any matrix, will be the matrix itself. Thus, Eq. (5.13):

$$[\beta] = [X_{mat}]^{-1}[XY_{mat}] \tag{5.14}$$

Equation (5.14) also gives the solution of finding the curve-fitted parameters: β_o and β_1.

NOTE The phrase *point-estimate* means that we have a distribution of values for both β_o and β_1. The regression process attempts to find the point estimate value (i.e., the mean value of the distribution for each parameter) plus its standard deviation around that estimated mean, assuming that the errors are randomly and normally distributed around their mean value and so do both distributions.

Example 5.1 Consider the following set of experimental data:

Table 5.1 Experimental data in the form of x_i and y_i.

x_i	Y_i	x_i^2	x_iY_i	$Y_{model} = \beta_o + \beta_1 x_i = -1.57143 + 4.32143x$
1	3	1	3	2.7500
2	7	4	14	7.0714
3	11	9	33	11.3928
4	17	16	68	15.7142
5	19	25	95	20.0356
6	23	36	138	24.3570
7	30	49	210	28.6784
$\sum_{i=1}^{n} x_i = 28$	$\sum_{i=1}^{n} Y_i = 110$	$\sum_{i=1}^{n} x_i^2 = 140$	$\sum_{i=1}^{n} x_iY_i = 561$	

Find the curve-fitted (point estimate) parameters using Eqs. (5.7) and (5.14).

Solution

Call Eq. (5.7): $$\begin{bmatrix} \sum_{i=1}^{n} 1 & \sum_{i=1}^{n} x_i \\ \sum_{i=1}^{n} x_i & \sum_{i=1}^{n} x_i^2 \end{bmatrix} \begin{bmatrix} \beta_o \\ \beta_1 \end{bmatrix} = \begin{bmatrix} \sum_{i=1}^{n} Y_i \\ \sum_{i=1}^{n} x_iY_i \end{bmatrix}$$

$$\begin{bmatrix} 7 & 28 \\ 28 & 140 \end{bmatrix} \begin{bmatrix} \beta_o \\ \beta_1 \end{bmatrix} = \begin{bmatrix} 110 \\ 561 \end{bmatrix} \qquad \begin{pmatrix} 5.15a \\ 5.15b \end{pmatrix}$$

The set of equations (5.15*a*, *b*) has two ***unknown terms*** (β_o and β_1) with two ***independent equations***.

$$7\times\beta_o + 28\beta_1 = 110 \tag{5.15a}$$

$$28\times\beta_o + 140\beta_1 = 561 \tag{5.15b}$$

From Eq. (5.15*a*), we have:

$$\beta_o = \frac{110 - 28\beta_1}{7} = \frac{110}{7} - 4\beta_1 \tag{5.16}$$

Substitute in Eq. (5.15*b*):

$$28\times\left(\frac{110}{7} - 4\beta_1\right) + 140\beta_1 = 561 \rightarrow\rightarrow\rightarrow 440 - 112\beta_1 + 140\beta_1 = 561 \tag{5.17}$$

$$28\beta_1 = 121 \rightarrow\rightarrow\rightarrow \beta_1 = \frac{121}{28} = 4.32143 \tag{5.18}$$

Substitute the value of β_1 in Eq. (5.16) to get the value of β_o:

$$\beta_o = \frac{110}{7} - 4\beta_1 = \frac{110}{7} - 4\times 4.32143 = -1.57143 \tag{5.19}$$

$$Y_{model} = \beta_o + \beta_1 x_i = -1.57143 + 4.32143x \tag{5.20}$$

Let us use MATLAB to calculate the parameters (β_o and β_1) based on Eq. (5.14):

$$[\beta] = [X_{mat}]^{-1}[XY_{mat}]$$

Substitute the numerical value of each sum that appears in Eqs. (5.9) and (5.11) by its equivalence from the last row in the table, where the sum of each column is shown.

$$[X_{mat}] = \begin{bmatrix} 7 & 28 \\ 28 & 140 \end{bmatrix} \qquad [XY_{mat}] = \begin{bmatrix} 110 \\ 561 \end{bmatrix}$$

$$[\beta] = inv(Xmat) * XYmat$$

```
>>   Xmat=[7 28;28 140]

Xmat =
     7     28
    28    140
```

```
>>     XYmat=[110;561]

XYmat =
   110
   561

>>     beta=inv(Xmat)*XYmat

beta =

   -1.5714
    4.3214
```

Comparing the results, we see that they are identical. Moreover, the predicted values of y can be given by MATLAB as:

```
>>     x=[1,2,3,4,5,6,7];
>>     y=-1.5714+(4.3214)*x
y =
2.7500    7.0714   11.3928   15.7142   20.0356   24.3570 28.6784
```

5.2 Importing Data

To demonstrate how to import data from another application into the MATLAB environment, let us assume that the experimental data exist in a file sheet such as Excel. The data represent the yearly incoming surface erythemal solar UV radiation dosage (J per cm^2) for a clear sky, after modulation due to average cloud cover, as a function of earth latitude. The data were excerpted from M. Ilyas [Climate augmentation of erythemal UV-B radiation dose damage in the tropics and global change, *Curr. Sci.* 93(11):1604–1608, 2007]. For simplicity, the file was copied to, or moved into, one of the MATLAB search path directories so that it will appear in the Current Folder Window of MATLAB (Fig. 5.1).

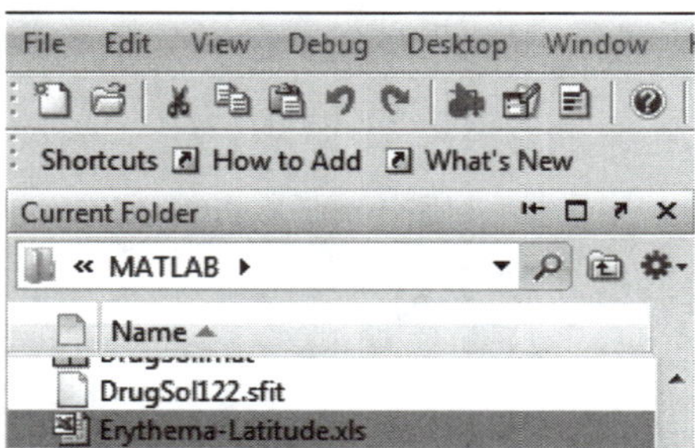

Figure 5.1 Current Folder Window that shows the Excel file named Erythema-Latitude.xls. Right click on this file and pick up Import Data from the pop-up context menu.

Now, right click on the file and pick up import data from the short pop-up menu. The MATLAB Import Data wizard will be invoked as shown in Fig. 5.2.

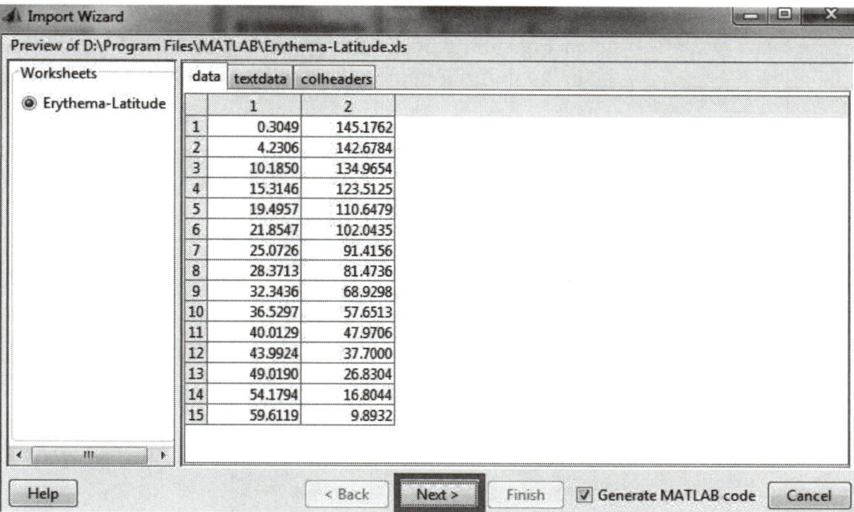

Figure 5.2(a) The first screen of MATLAB's Import Data Wizard (old look).

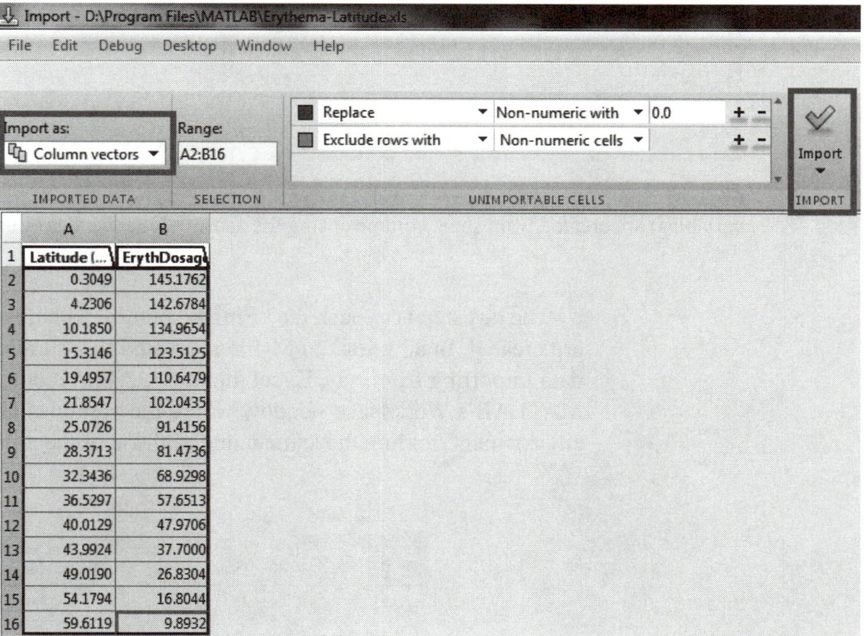

Figure 5.2(b) The first screen of MATLAB's Import Data wizard (new look). Here choose import as Column vectors and click the **Import** button.

Push the **Next** button (shown in Fig. 5.2*a*) to go to the next screen that is shown in Fig. 5.3. Choose the second option; that is, create vectors from each column using column names. Then click the **Generate MATLAB code** box.

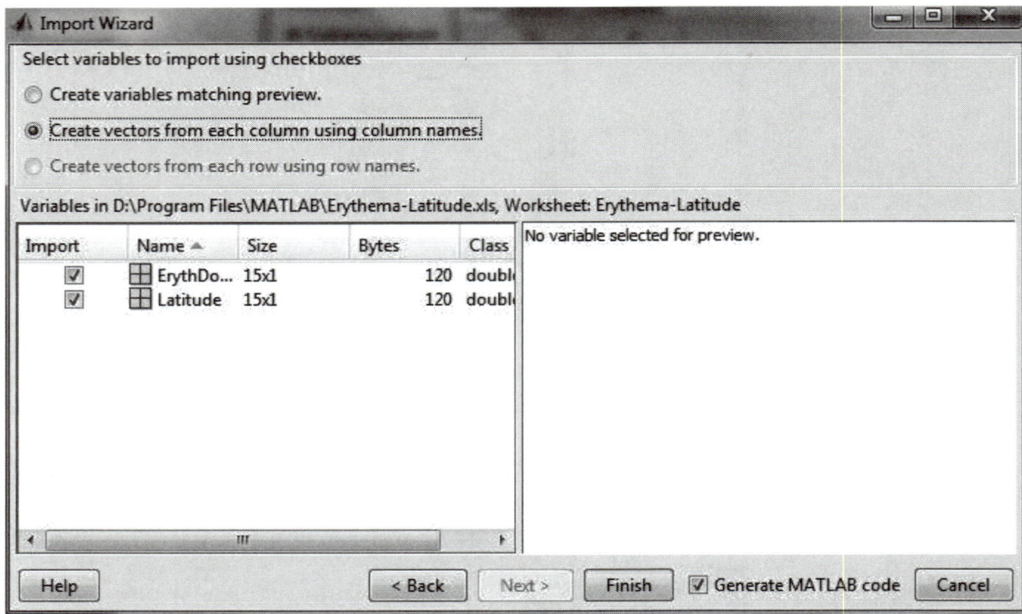

Figure 5.3 The second screen of the MATLAB Import Data wizard, in which vectors (i.e., Workspace variables) are created from sheet columns using the same column header or name.

The last step is to push the "**Finish**" button, whereby the variables (i.e., vectors) are created. In addition, an M-file is created that summarizes the entire process of data importing from, say, Excel into the MATLAB environment. Figure 5.4 shows MATLAB's Workspace window where the variables are created into MATLAB's environment holding the same name as those of the columns.

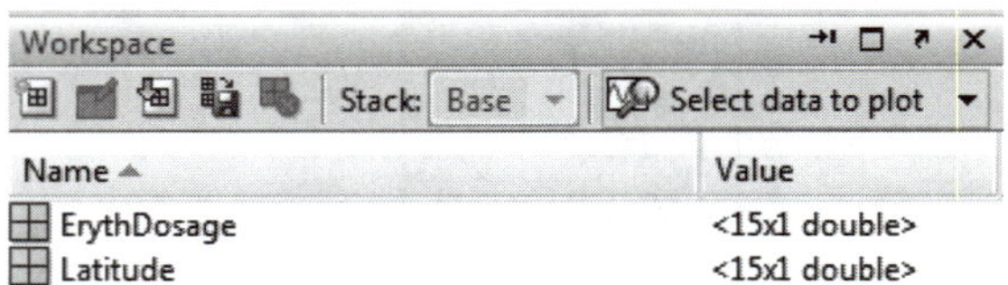

Figure 5.4 MATLAB's Workspace window showing the vectors (or variables) created via the import process.

Finally, Fig. 5.5 shows the M-file created by MATLAB, as a result of asking MATLAB to generate a code for the Data Import process.

```
function importfile(fileToRead)
%IMPORTFILE1(FILETOREAD1)
%  Imports data from the specified file
%  FILETOREAD1:  file to read
sheetName='Erythema-Latitude';
[numbers, strings, raw] = xlsread(fileToRead, sheetName);
if ~isempty(numbers)
    newData1.data =  numbers;
end

if ~isempty(strings) && ~isempty(numbers)
    [strRows, strCols] = size(strings);
    [numRows, numCols] = size(numbers);
    likelyRow = size(raw,1) - numRows;
    % Break the data up into a new structure with one field per column.
    if strCols == numCols && likelyRow > 0 && strRows >= likelyRow
        newData1.colheaders = strings(likelyRow, :);
    end
end

% Create new variables in the base workspace from those fields.
for i = 1:size(newData1.colheaders, 2)
     assignin('base', genvarname(newData1.colheaders{i}), newData1.
data(:,i));
end
```

Figure 5.5 The creation of the function M-file (**importfile.m**) by MATLAB, which can be exploited later to do the same procedure, but perhaps this time with another file name.

You can make use of the M-file by calling the M-file itself at the command prompt while supplying the name of the data file as an input argument. Here is the general statement:

```
>>    importfile('X:\MyFolder\myfilename')
```

Specifically:

```
>>    importfile('Erythema-Latitude.xls')
```

5.3 Data Visualization

Once you have the data in the form of vectors, you can start the process of data visualization. The idea here is to have an initial glimpse of the data before you attempt to fit them to any built-in or customized model. Highlight both variables and you will notice that MATLAB presents you with the plot options, at the right top of the Workspace window, as shown in Fig. 5.6.

For example, you can pick up your choice to be **plot(ErythDosage,Latitude)** as a 2D line graph using linear axes. To switch the order of variables, choose the

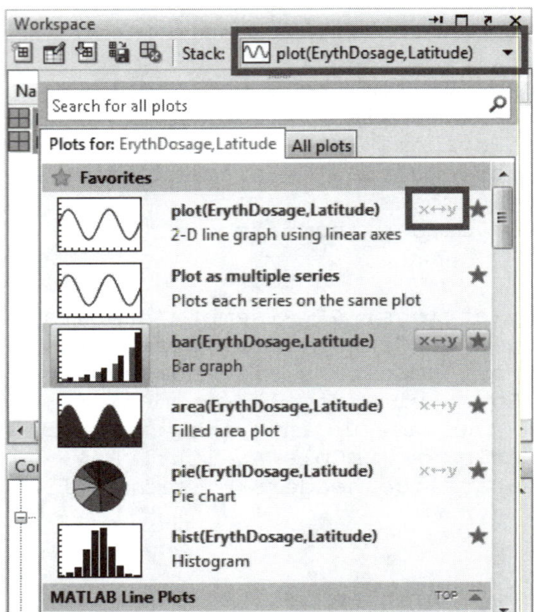

Figure 5.6 Highlighting both vectors in the Workspace will activate the plot options menu, where you can draw different plots.

$x \leftrightarrow y$ symbol (shown as the framed $x \leftrightarrow y$ symbol in Fig. 5.6), and the new plot becomes: plot(Latitude,ErythDosage), as shown in Fig. 5.7.

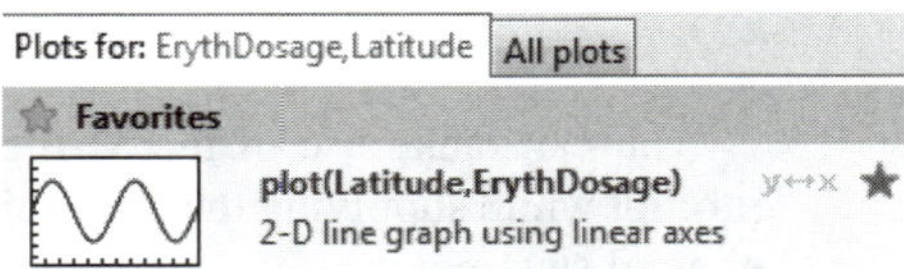

Figure 5.7 Switching the order of variables where the first variable is considered x, and the second, y.

The output of clicking this choice of plots is shown in Fig. 5.8.

This is how it appears in MATLAB's Command:

```
>>     plot(Latitude,ErythDosage,'DisplayName','ErythDosage vs.
Latitude','XDataSource','Latitude','YDataSource','ErythDosage')
```

Using the **Tools** menu (shown as the framed tool item in Fig. 5.8) in the same figure, you may choose the **Basic Fitting** sub-menu, in which you can make further testing on the relationship between the dependent and independent variable using built-in common relationships or models (Fig. 5.9).

Figure 5.10 shows the experimental data fitted to a quadratic equation (polynomial of degree two) and Fig. 5.11 shows the residuals (see Sec. 5.3) for such a fitting model.

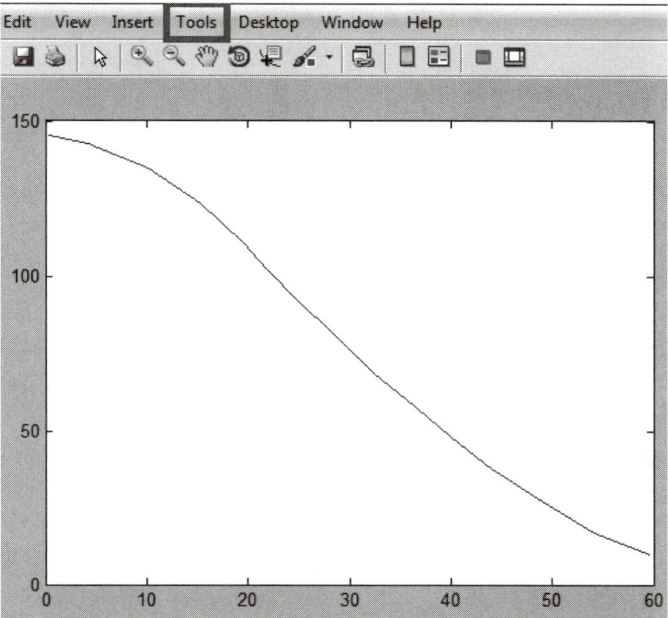

Figure 5.8 Plot of erythemal solar UV radiation dosage (J per cm^2), y, as a function of the earth's latitude, x.

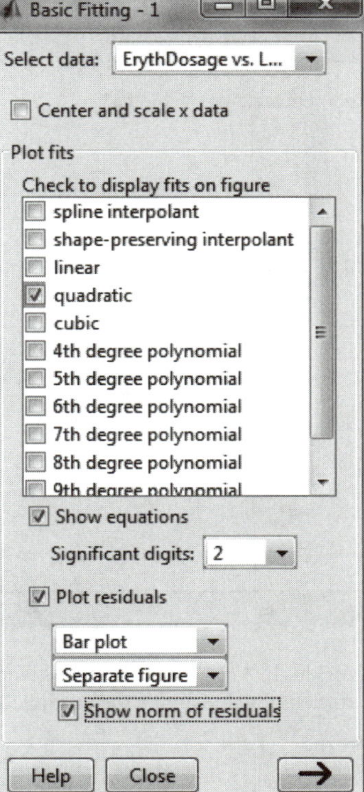

Figure 5.9 Using built-in common models to describe the relationship between y and x.

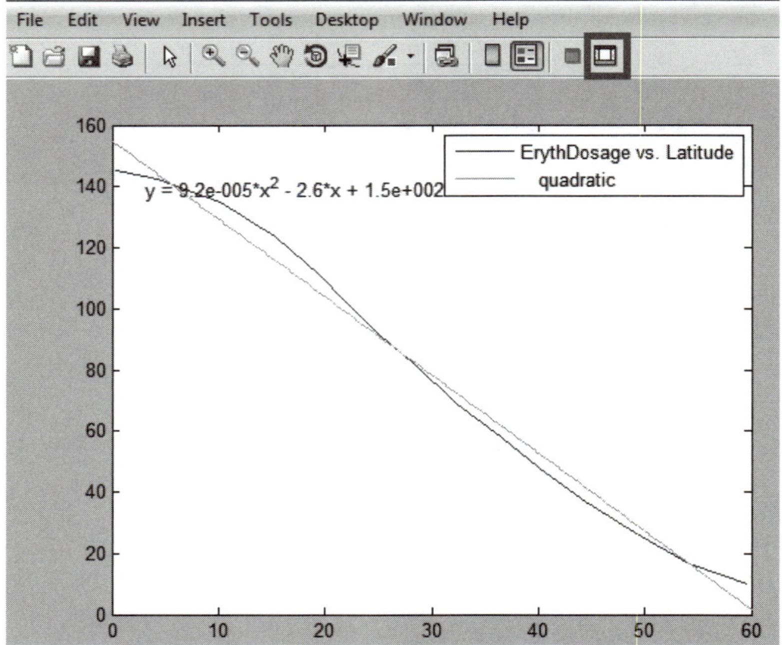

Figure 5.10 Curve-fitting y vs. x using a quadratic (of degree two) polynomial.

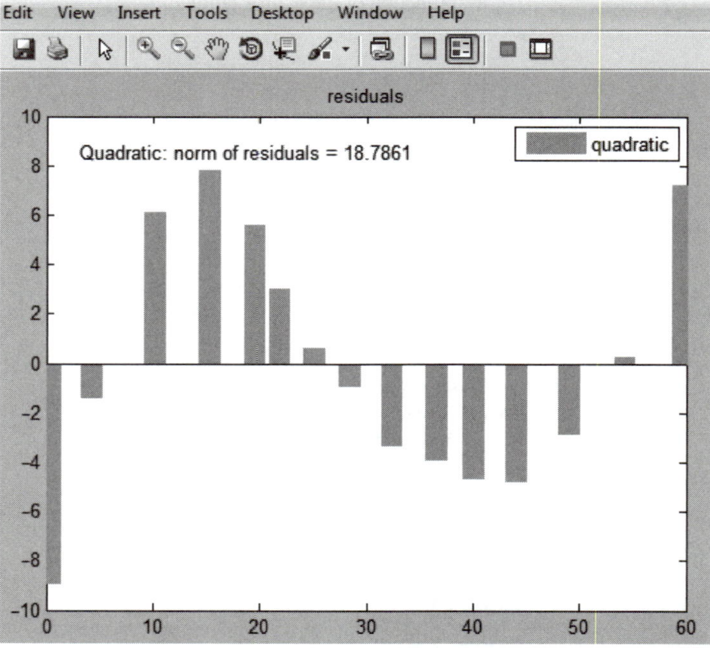

Figure 5.11 Residual $(y_i - \hat{y}_i)$ vs. x_i for the given data, where y_i represents the experimental value and $\hat{y}_i$ the predicted value for a given x_i, $Norm\ of\ Residuals = \sqrt{\sum_{i=1}^{N}\left(y_i - \hat{y}_i\right)^2}$.

You can make some cosmetic changes for the given figure using the **Plot Tool and Dock Figure** menu (shown as a framed menu icon in Fig. 5.10) to edit the figure and modify the content and location of the title, axis title, font size and color, and Min and Max for each axis. Clicking anywhere inside the plot area will furnish the plot with the relevant palette, where you can modify the title, axis, and font properties. You may left click on the figure's caption or text and drag it to another location within the plot area itself.

After you are done with the cosmetic changes, you may hide the Plot Tool by pushing the **Hide Plot Tool** (shown as framed icon in Fig. 5.12). Figure 5.13 shows the new appearance after applying some cosmetic changes.

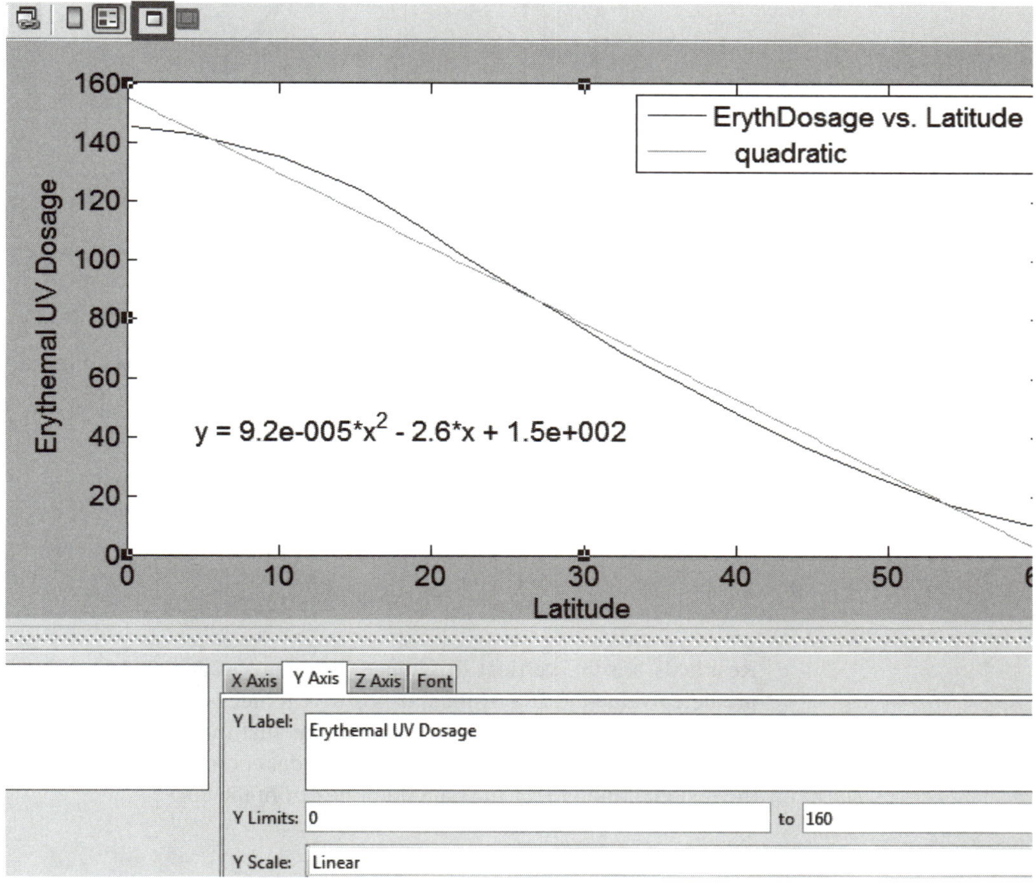

Figure 5.12 Using the **Plot Tool and Dock Figure** menu, one may modify many properties related to how the figure should look.

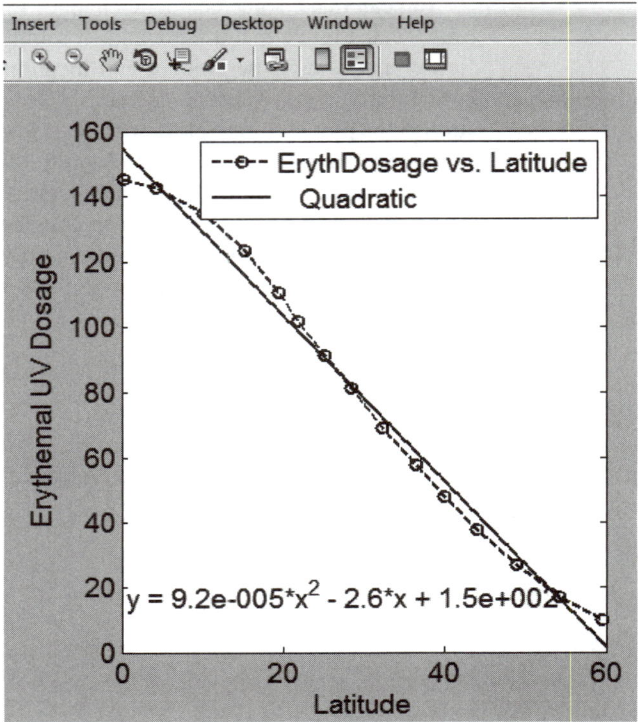

Figure 5.13 Hiding/closing the **Plot Tool and Dock Figure**, after carrying out some cosmetic changes pertaining to the appearance of the figure.

5.4 Statistical Definitions

5.4.1 Residuals

$$jth\ Residual = (y_i - \hat{y}_i)$$

Residuals are the vertical difference between the actual data points y_i and the predicted values $\hat{y}_i$. If a residual is positive, it means that the actual data point lies above the curve. A negative residual means that the actual data point lies below the curve. If the residual is zero, the actual data point lies on the curve. The larger the residual, the farther the data point lies from the curve.

5.4.2 Sum of Residuals

$$Sum\ of\ Residuals = \sum_{i=1}^{N}(y_i - \hat{y}_i)$$

where N represents the total number of data points, or observations. This is the total sum of the residuals for all data points. If the curve passed through each data point, the sum of residuals would be zero. Keep in mind though that a regression model can have large positive and negative residuals and still sum to a small number.

5.4.3 Residual Sum of Squares

The absolute residual sum of squares is the sum of the squares of the differences between the actual data points and the predicted values:

$$SSE = Residual\ or\ Error\ Sum\ of\ Squares\ (Absolute) = \sum_{i=1}^{N}(y_i - \hat{y}_i)^2$$

The principle behind nonlinear regression is to minimize the residual sum of squares (SSE) by adjusting the parameters in the regression model to bring the curve closer and closer to the data points. This parameter is also referred to as the error sum of squares, or SSE. Ideally, if the residual sum of squares is equal to 0.0, the curve passes through every data point.

5.4.4 Error Variance

$$\hat{\sigma}^2 = S^2 = \frac{\sum_{i=1}^{N}(y_i - \hat{y}_i)^2}{N-P} = \frac{SSE}{N-P}$$

where P is the number of parameters or coefficients in the regression model and N is the total number of data points, or observations. This is an estimate of the error variance. The standard error will be $S = \sqrt{\dfrac{SSE}{N-P}}$.

5.4.5 R^2, or the Coefficient of Multiple Determination

$$R^2 = \frac{SSR}{SST} = \frac{SST - SSE}{SST} = 1.0 - \frac{SSE}{SST} \qquad 0 \le R^2 \le 1.0$$

where:

$$SSR = Regression\ Sum\ of\ Squares = \sum_{i=1}^{N}(\hat{y}_i - \overline{y})^2$$

$$SSE = Rsidual\ or\ Error\ Sum\ of\ Squares = \sum_{i=1}^{N}(y_i - \hat{y}_i)^2$$

$$SST = Total\ Sum\ of\ Squares = \sum_{i=1}^{N}(y_i - \overline{y})^2$$

$$SST = SSE + SSR$$

Therefore, R^2 measures the proportion of variation in the data points, y_i, which is explained by the regression model. For example, if $R^2 = 0.95$, the 95% of the variation in the dependent, or response variable y is explained by the regression model.

A value of $R^2 = 1.0$ means that the curve passes through every data point. A value of $R^2 = 0.0$ means that the regression model does not describe the data any better than a horizontal line (i.e., the average $\bar{y}$) passing through the data points.

NOTE For nonlinear regression, one might end up with a negative R^2 value; an indication that *SSE is greater than SST*, which is not possible in the case of linear regression. The reason for this *abnormality* is simply in the nonlinear regression case, *the addition of SSE and SSR may not sum to SST*. So, keep in mind that the notion of describing linear regression model fitness (or, goodness) is being extended to the nonlinear regression case. This case, nevertheless, may happen if a set of data does not really follow the pattern of the model being applied.

5.4.6 AdjR² (R_a^2), Adjusted Coefficient of Multiple Determination

$$AdjR^2 = R_a^2 = 1.0 - \left[(1 - R^2) \times \left(\frac{N-1}{N-P}\right)\right] \qquad R_a^2 < R^2$$

$AdjR^2$ (R_a^2) is used to balance out the cost of using a model with more parameters against the increase in R^2. $AdjR^2(R_a^2)$ can be derived as follows:

$$MSE = SSE/(N - P)$$

$$MST = SST/(N - 1)$$

$$SSR = SST - SSE$$

By definition: $AdjR^2 = R_a^2 = 1.0 - \left(\frac{MSE}{MST}\right)$

$$R^2 = \frac{SSR}{SST} = \frac{SST - SSE}{SST} = 1.0 - \frac{SSE}{SST} \rightarrow\rightarrow\rightarrow \frac{SSE}{SST} = (1.0 - R^2)$$

$$AdjR^2 = R_a^2 = 1.0 - \left(\frac{MSE}{MST}\right) = 1.0 - \frac{SSE/(N-P)}{SST/(N-1)}$$

$$AdjR^2 = R_a^2 = 1.0 - \left[\left(\frac{SSE}{SST}\right) \times \left(\frac{N-1}{N-P}\right)\right] = 1.0 - \left[(1 - R^2) \times \left(\frac{N-1}{N-P}\right)\right]$$

5.5 The Model Goodness

5.5.1 By-the-Naked-Eye Test (No Statistics)

The first thing to check for model goodness is via looking, by the naked eye, at the curve or line generated by plotting the response variable, *y*, versus the predictor variable, *x*, in light of the deciphered relationship between *y* and *x*. However, the naked eye test is a bit subjective and is difficult to quantify in terms of numbers. Of course, if the experimental data dramatically depart from the regressed curve or line, then one can easily say that the model being examined is not applicable for the given set of data. On the other hand, if the model is pretty close to the experimental data, then the question arises, *How good is the model goodness*?

Here comes the role of statistics and statisticians. I recall here what a professor in statistics at Oregon Statistics (State) University (OSU) said: "Figures lie and only liars figure out." Do not panic; statistics is a powerful tool, if correctly implemented, which can be used to help a researcher, scientist, analyst, and even politician, to judge the goodness of a model that describes the behavior of a physical, chemical, or societal system or process performance. Even highly nonlinear dirty politics can be modeled but with an extra precaution when it comes to interpreting the results and reflecting back upon a world characterized by a folded, hideous intricate state in which you can barely see a small part of that world (the iceberg model).

5.5.2 The Model Goodness (with Statistics)

The model goodness or fitness may be examined by looking at the following points:

1. Check to see how well the regression model describes the actual data. This information can be obtained by looking at the following calculated parameters: R^2, $AdjR^2$, SSE, S, and 95% confidence interval for each estimated (regressed) parameter. For example, an R^2 value closer to one is a good indication of model goodness. If the confidence is very wide, the fit is not unique, meaning that different values chosen for the variables would result in nearly as good a result. Data containing a lot of scattering or collecting an insufficient amount of data would cause the confidence intervals to be excessive; however, the most common reason for an excessive confidence interval is fitting the data to a model with *variable redundancy*. In the equation, $\boldsymbol{y = exp(b_1 * b_2 * x_1)}$, the variables $\boldsymbol{b_1}$ and $\boldsymbol{b_2}$ are indistinguishable. There is no way for the algorithm to determine how to distribute values (the product of $\boldsymbol{b_1}$ and $\boldsymbol{b_2}$) between these two variables.
2. The *residuals should be randomly scattered around zero* with no discernible pattern, i.e., they should have no relationship to the value of the independent variable, x_i. *If there are groups of residuals with like signs, or the residuals increase or decrease as a function of the independent variable, it is probable that another functional approximation exists that will better describe the data.*
3. Check to see if the results are scientifically or statistically meaningful. Does the fitted value of any of the variables violate a possible physical reality? For example, if Arrhenius kinetics (***k=ko*exp(-Ea/RT)***) is used to describe the temperature dependence of the rate constant, ***k***, then parameters like $\boldsymbol{k_o}$ and ***Ea*** should be positive, nonzero values. If the regressed values are negative, then you conclude that the model is not adequate to describe the kinetic data; you have either made a syntax error in the function formula, or entered some wrong data points. A data point will be wrong if it is not correctly entered as a magnitude or not expressed in the proper physical units.
4. It is possible to converge on a false minimum in the objective function. This is a problem inherent in any iterative optimization procedure. Nonlinear regression ensures that once a solution is obtained, small changes in the parameters will worsen the fit. It is rare but possible, however, that some very large change may

actually converge to a better fit. This problem is rare, except in cases in which the data are widely scattered, there are too few collected data points, or the chosen model is completely wrong for the data. Determining good initial estimates is important for highly nonlinear models.

Here are the estimated parameters and their associated standard error for the second-order polynomial: $y = a*x^2 + b*x + c$, using Curve_It software developed by Dr. Al-Malah (https://sites.google.com/site/almalahweb/software):

$a = 9.11663696964859 \times 10^{-07}$, Standard Error = 0.00474977841168609

$b = -2.55726579140641$, Standard Error = 0.294466359539368

$c = 155.015944828759$, Standard Error = 3.95196401232341

Residual Standard Error, S = 5.5440646001718

R^2 Value = 0.987666603046649

Regression Sum of Squares, SSR = 29536.9378658428

Residual Sum of Squares, SSE = 368.839827490537

Looking only at R^2 value might mislead the investigator into believing that his or her model is the right model; nevertheless, if you look at the standard error associated with parameter ***a***, it will tell you that ***a*** is accompanied by a relatively large standard error compared with the magnitude of ***a*** itself. In addition, ***a*** itself is very close to zero (0.0000009 ± 0.0047), which then reduces the model to a straight line relationship.

This leads us to contemplate the distribution of errors; are they randomly distributed or scattered around their mean? Looking back at Fig. 5.11, the residuals are not randomly distributed around their zero mean.

The residuals are also shown in Fig. 5.14 using Curve_It software. Obviously, the points (i.e., errors) are not randomly scattered around their mean and there exists such a pattern, say, tilde or S shape with x_i.

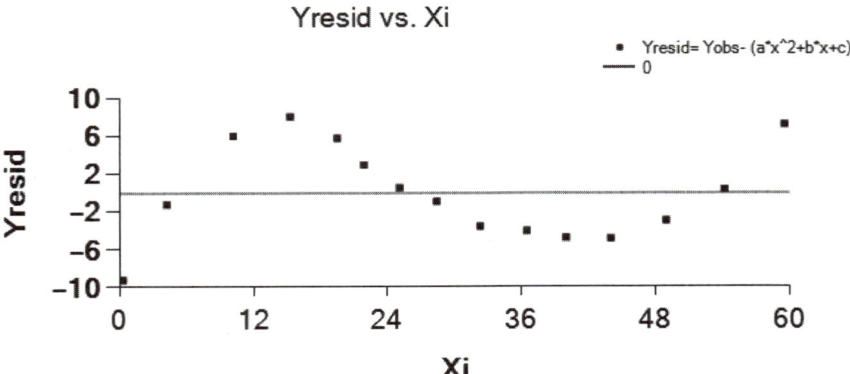

Figure 5.14 Distribution of residuals around their mean. The points (i.e., errors) are not randomly scattered around their mean and there exists such a pattern, say, tilde or S shape pattern with x_i. The plot was generated by Curve_It software (https://sites.google.com/site/almalahweb/software).

5.6 MATLAB's Curve Fitting Toolbox (cftool)

As it turns out that the errors are not randomly scattered around their mean, then it is time to seek or search for another model. This is called model customization. *First*, let us introduce the MATLAB Curve Fitting Toolbox. Such a tool can be invoked via:

```
>>    cftool
```

Figure 5.15*a*,*b* show the old and new look main windows for the Curve Fitting Toolbox.

NOTE The set of x and y data that we need to examine must be loaded first into the MATLAB environment in the form of vectors or matrices and their names should appear in the Workspace Browser window.

Second, define the dependent (y) and independent (x) variables. Click the **Data** button (shown as a framed button in Fig. 5.15). Figure 5.16 shows the data window that defines the x and y variables. With the new look of MATLAB Curve Fitting

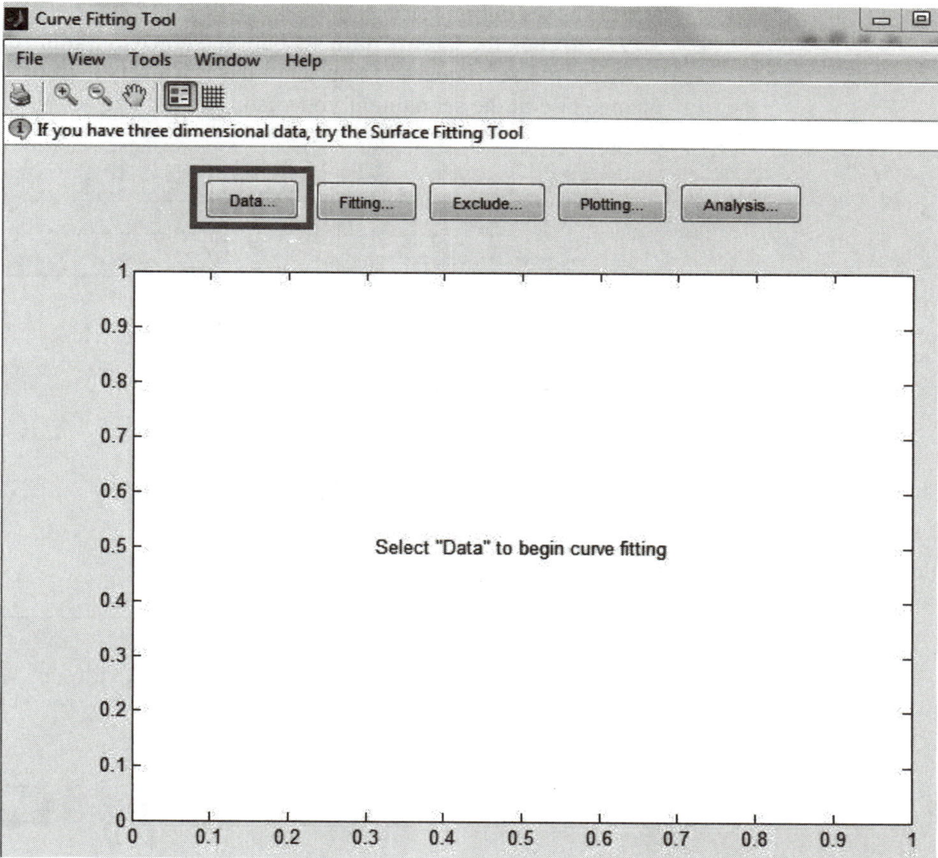

Figure 5.15(a) The main window for the MATLAB Curve Fitting Toolbox (old look).

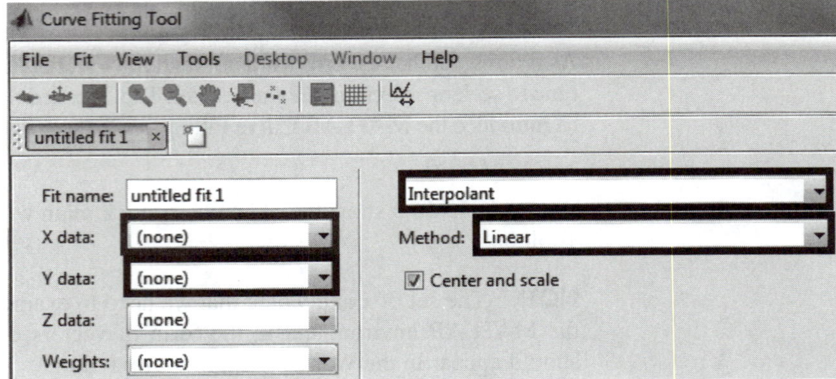

Figure 5.15(*b*) The main window for the MATLAB Curve Fitting Toolbox (new look).

Toolbox, you define the dependent variable (y), the independent variable (x), the model main category, and the method type from the drop-down lists (shown as framed boxes in Fig. 5.15*b*).

Third, click the **Create data set** button (shown as a framed button in Fig. 5.16) whereby the set name will appear in the **Data sets** textbox. Of course, you can change or edit the set name if you wish.

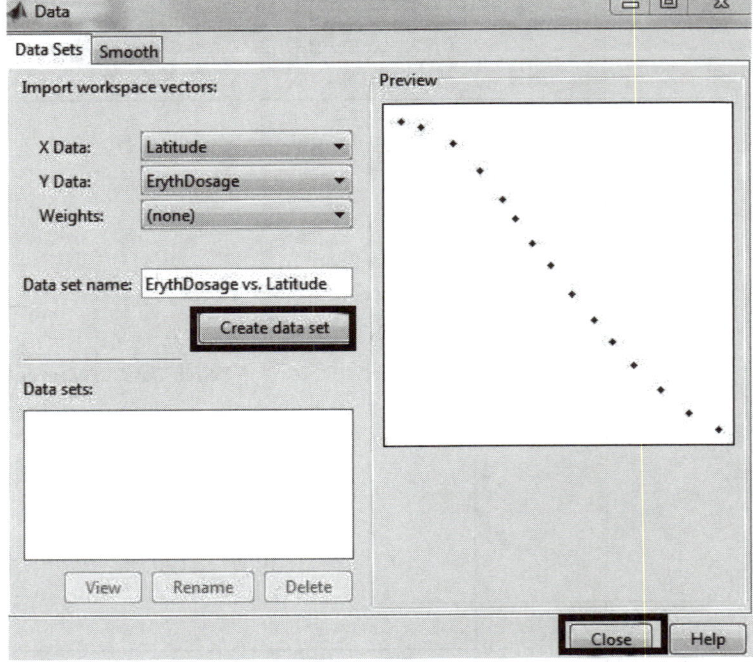

Figure 5.16 The **Data Sets** tab found on the **Data** window is the place to define the dependent (y) and independent (x) variables.

Fourth, click the **Close** button to finish the data input step. Figure 5.17 shows the main window of the Curve Fitting Toolbox, after finishing the data input step.

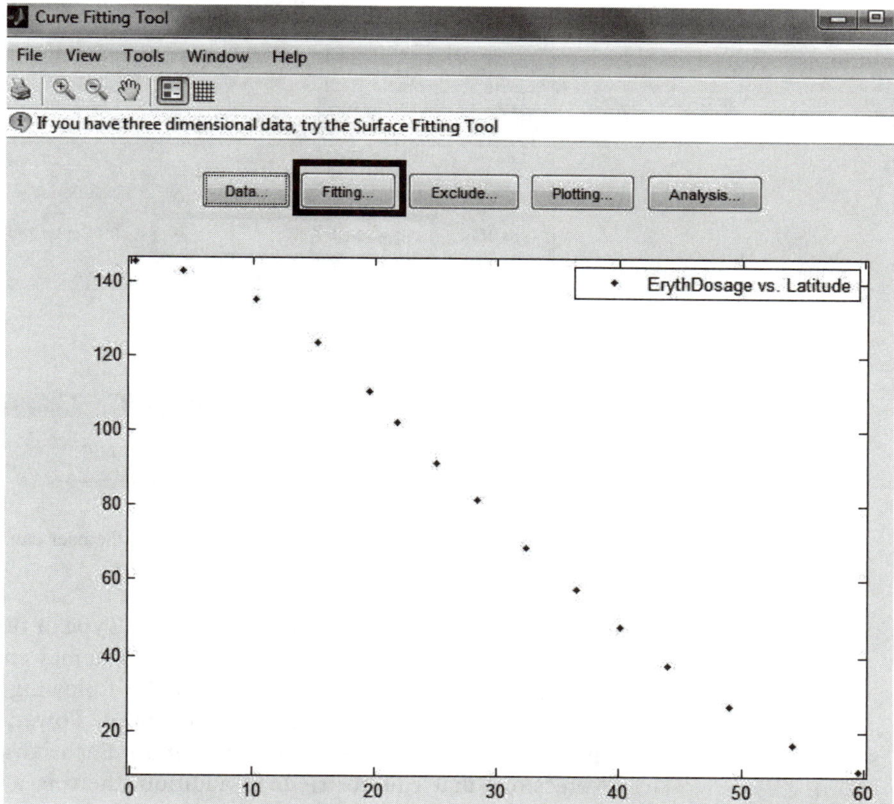

Figure 5.17 The MATLAB Curve Fitting Toolbox is ready for the next step, i.e., the fitting step, after properly defining the x and y variables.

Fifth, click the **Fitting** button (shown as a framed button in Fig. 5.17). This will pop up the Fitting window, as shown in Fig. 5.18.

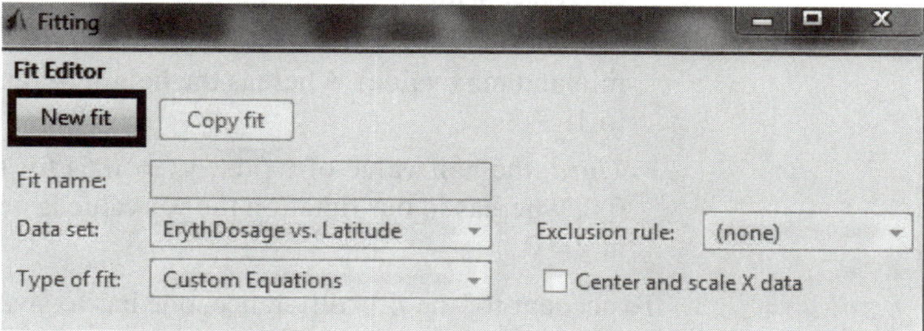

Figure 5.18 The Fitting window whereby the fitting model will be defined.

Sixth, click **New fit** button (shown as framed button in Fig. 5.18) to define the model type. Figure 5.19 shows the **Fitting** window where the user can define or input his or her model.

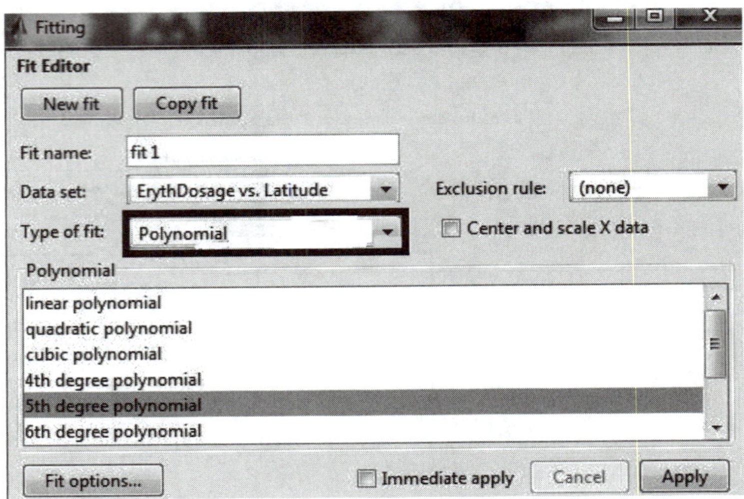

Figure 5.19 The Fitting window where the user can define his or her model.

Seventh, pick up the type of model via the **Type of fit** drop-down list (shown as the framed drop-down menu in Fig. 5.19). You may choose one of those built-in models. They are categorized under the following subsets: **Exponential, Fourier, Gaussian, Interpolant, Polynomial, Power, Rational, Smoothing Spline, Sum of Sin Functions,** and **Weibull**. Each subset has numerous model expressions that can be tried. In addition, there is a separate subset called **Custom Equations**. We are going to try a separate model called the Sigmoid (or, S-shaped) Function. The S-shape function has the following form as shown in Fig. 5.20.

Notice that there is a similarity between the sigmoid function and how erythemal UV dosage changes with latitude, except that:

- *First*, sigmoid function increases with x, whereas the erythemal UV dosage decreases with x.
- *Second*, the height of the sigmoid function is 1 (distance measured between min and max value), whereas the height of our function is finite but not equal to 1.
- *Third*, the half value of y (i.e., $y_{1/2} = 0.5$) for the sigmoid function lies at $x = 0.0$, whereas in our function the $y_{1/2}$ value is neither equal to 0.5 nor does it lie at $x = 0$.

To account for the first difference, one has to switch the sign of x in the exponential term; for the second difference, a curve-fitted parameter called $\boldsymbol{a}$ is thrown in the model; and for the third difference, because both the $y_{1/2}$ value and its location

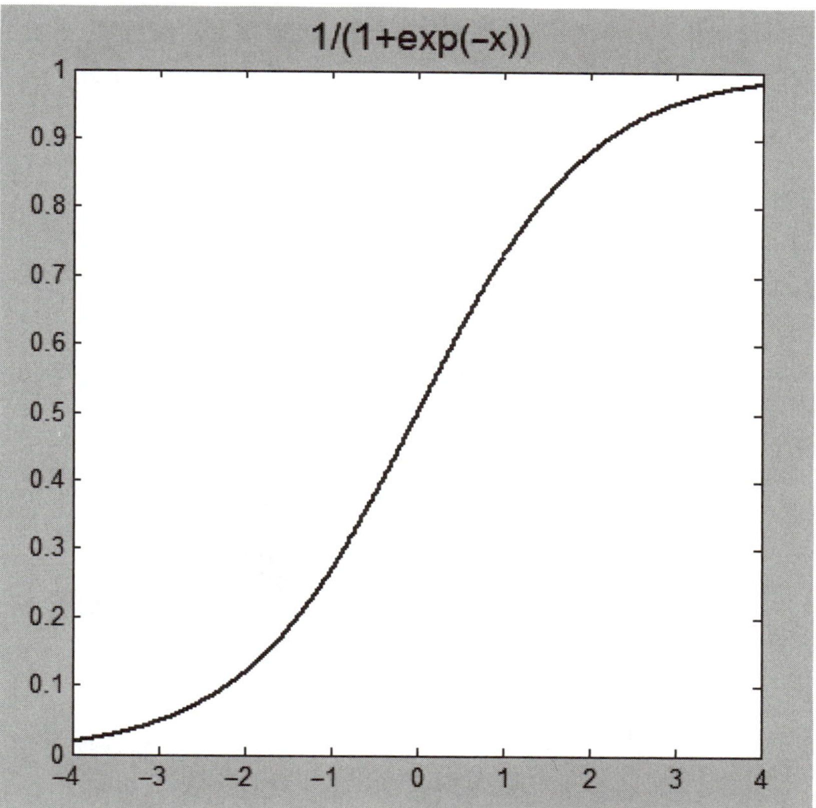

Figure 5.20 The sigmoid function, also called the sigmoidal curve.

are different, then two additional parameters are thrown in the model (i.e., parameters ***b*** and ***c***).

Therefore, the proposed model will look like:

$$y = a \times \left(\frac{1}{b + \exp(c * x)} \right) \tag{5.1}$$

This general sigmoid function can be used in any similar situation.

Eighth, pick up the type of model via the **Type of fit** drop-down list (shown as a framed drop-down list in Fig. 5.19) and choose the subset called **Custom Equations**. A new window will be opened (Fig. 5.21).

Ninth, click the **New** button (shown as a framed button in Fig. 5.21), the **New Custom Equation** window will show up as shown in Fig. 5.22.

Tenth, choose the **General Equations** tab (shown as a framed tab in Fig. 5.22). The default custom equation is shown in Fig. 5.23.

Eleventh, enter the general sigmoid function [Eq. (5.1)] in the dedicated textbox (shown as a framed textbox in Fig. 5.23). The **General Equations** tab found

Fitting

Fit Editor

New fit | Copy fit

Fit name: fit 1

Data set: ErythDosage vs. Latitude

Exclusion rule: (none)

Type of fit: Custom Equations

Center and scale X data

Custom Equations

Click "New" to create a custom equation.

New

Figure 5.21 The Fitting window that allows the user to pick the Customs Equations type of fit.

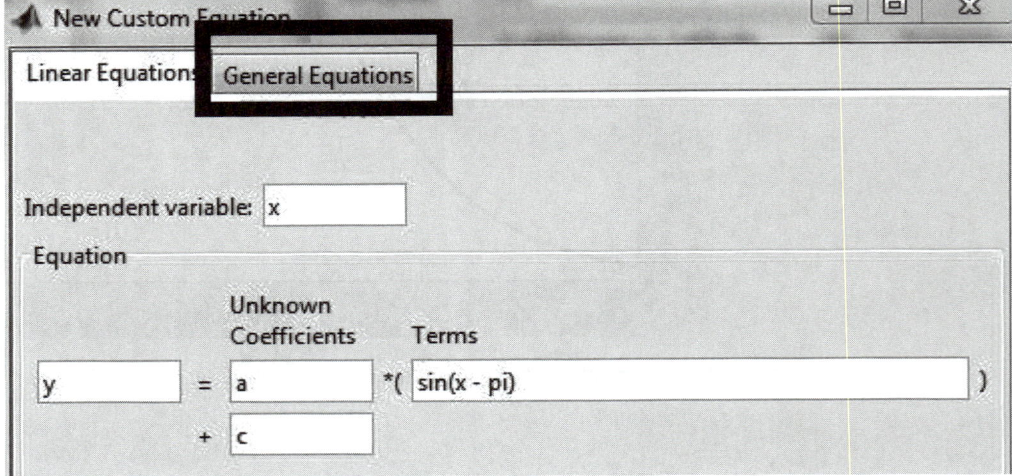

Figure 5.22 The **New Custom Equation** window where the user may input the model expression.

New Custom Equation

Linear Equations | General Equations

Independent variable: x

Equation: y = a*exp(-b*x)+c

Unknowns	StartPoint	Lower	Upper
a	0.129	-Inf	Inf
b	0.648	-Inf	Inf
c	0.473	-Inf	Inf

Figure 5.23 The default custom equation supplied by MATLAB.

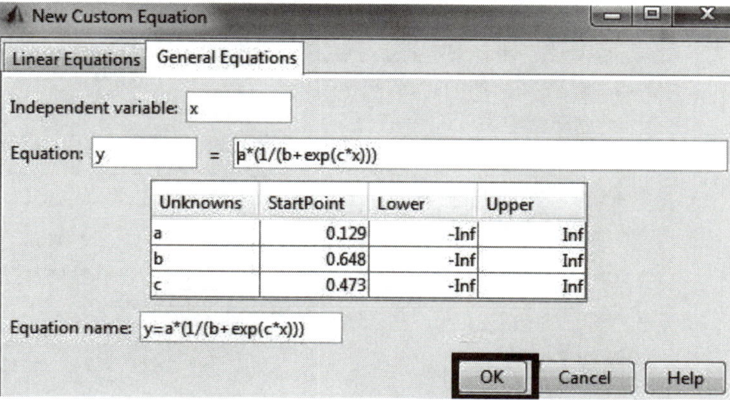

Figure 5.24 Defining our lovely general sigmoid function.

in the **New Custom Equation** window shows the user-defined sigmoid function (Fig. 5.24).

Twelfth, click the **OK** button (shown as a framed button in Fig. 5.24). MATLAB will return to the main Fit Editor window as shown in Fig. 5.25*a*. Figure 5.25*b* shows the new look of the Curve Fitting Toolbox window where the user selects Custom Equation from the main category drop-down list and then enters the expression for $y = f(x; a, b, c)$.

Thirteenth, click the **Apply** button as shown in Fig. 5.25*a*. MATLAB will start the computational process for finding the best estimated parameters. The results are shown in the following, as the *first trial* carried out by MATLAB.

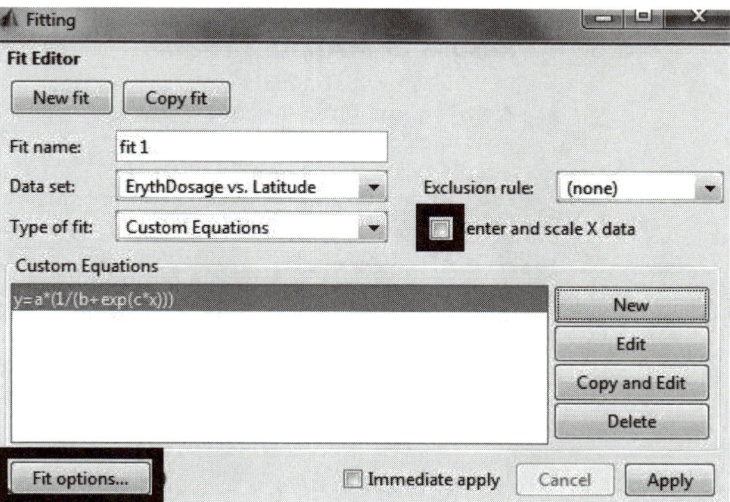

Figure 5.25(*a*) The old look of the MATLAB Curve Fitting Toolbox, where it is ready to start the process of curve fitting or nonlinear regression.

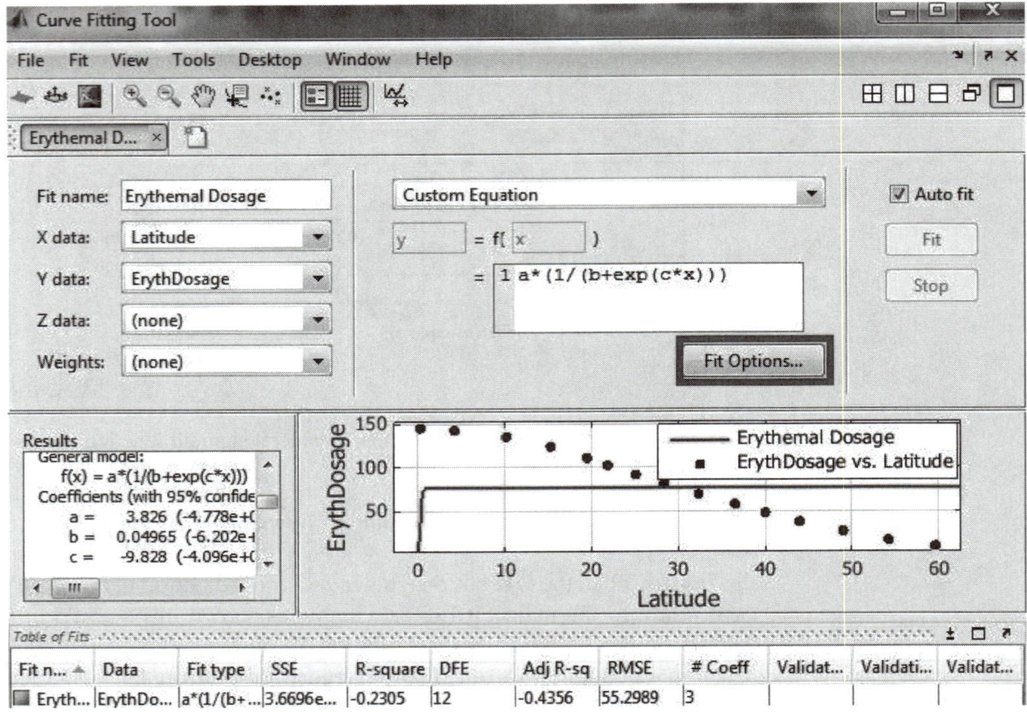

Figure 5.25(*b*) The new look of the MATLAB Curve Fitting Toolbox. After successfully entering the syntax error-free model (i.e., $y = f(x)$), MATLAB will either start the process of curve fitting or wait for the user's command if the **Auto fit** option is de-selected.

5.6.1 *Results of MATLAB's cftool*

```
Fit computation did not converge:
Fitting stopped because the number of iterations or function
evaluations exceeded the specified maximum.

Fit found when optimization terminated:

General model:
     f(x) = a*(1/(b+exp(c*x)))
Coefficients (with 95% confidence bounds):
       a =      22.58  (-9.716e+009, 9.716e+009)
       b =     0.2861  (-1.231e+008, 1.231e+008)
       c =     -8.362  (-1.411e+009, 1.411e+009)

Goodness of fit:
  SSE: 3.236e+004
  R-square: -0.08524
  Adjusted R-square: -0.2661
  RMSE: 51.93
```

```
Warning: A negative R-square is possible if the model
         does not contain a constant term and the fit
         is poor (worse than just fitting the mean).
        Try changing the model or using a different StartPoint.
```

Based on the aforementioned results, the model with the estimated parameters is definitely a misfit, as also shown in Fig. 5.26. The same terrible situation will occur with the new look MATLAB Curve Fitting Toolbox (see Fig. 5.25*b*).

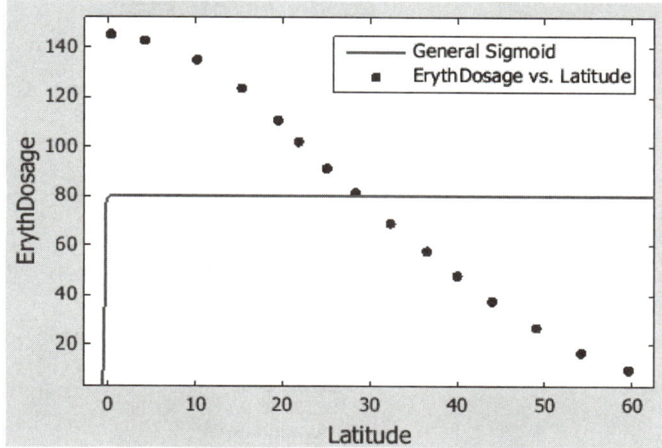

Figure 5.26 The plot of the General Sigmoid function with the current estimated parameters.

Before, we decide that such a model is terribly bad, let us make some cosmetic changes.

Fourteenth, click the **Center and scale X data** check (shown as a framed box in Fig. 5.25*a*). For the new look MATLAB Curve Fitting Toolbox (see Fig. 5.25*b*), click the **Fit Options** radio button. MATLAB will pop up a new small-size **Fit Options** window, similar to Fig. 5.32 for the old-look case, in which the user can tune up the algorithm (Trust Region or Levenberg-Marquardt) being exploited in the search for the optimum point and can also refine the bracketing interval for each parameter. Limiting the search to positive values of ***a***, ***b***, and ***c*** while using the **Trust Region** algorithm will result in curve-fitted parameters such that the "**goodness of fit**" parameters (namely, SSE, R^2, Adjusted-R^2, and RMSE) are exactly the same as those shown in Fig. 5.27 for both the old and new look of the MATLAB cftool.

NOTE The fact that the ***a***, ***b***, and ***c*** parameters are pairwise different in magnitude for each case means that the ***a***, ***b***, and ***c*** parameters are not truly independent of one another. From an optimization standpoint there exists more than one local minimum in ***a***, ***b***, ***c*** optimization coordinates that will satisfy the condition of minimizing SSE with the same magnitude. Any of those local minima will be a valid or representative point.

```
General model:
     f(x) = a*(1/(b+exp(c*x)))
        where x is normalized by mean
29.37 and std 17.97
Coefficients (with 95% confidence
bounds):
       a =         156.6   (147.5, 165.6)
       b =         0.9733  (0.8889, 1.058)
       c =         1.479   (1.392, 1.566)

Goodness of fit:
  SSE: 32.98
  R-square: 0.9989
  Adjusted R-square: 0.9987
  RMSE: 1.658
```

```
Results

General model:
  f(x) = a/(b+exp(c*x))
Coefficients (with 95% confidence bounds):
  a =      1755  (1429, 2082)
  b =      10.91  (8.565, 13.26)
  c =    0.0823  (0.07747, 0.08714)

Goodness of fit:
 SSE: 32.98
 R-square: 0.9989
 Adjusted R-square: 0.9987
 RMSE: 1.658
```

Figure 5.27 MATLAB's second-round estimated model parameters with the **Center and scale X data** checkbox being selected, in the old look (*top*) and after changing the type of optimization algorithm and search interval for each model parameter, in the new look (*bottom*).

Fifteenth, click the **Apply** button for the second time to give MATLAB a second chance to calculate the general sigmoid function parameters. Figure 5.27 shows the second-round estimated parameters.

Figure 5.28 shows the curve-fitted data of erythemal UV dosage vs. latitude.

Obviously, the fit now is almost perfect with an adjusted R^2 equal to 0.9987. Let us have a look at further analysis regarding the general sigmoid model goodness. Click the **Analysis** button as shown in Fig. 5.15. Figure 5.29 shows the **Analysis** window showing the 95% confidence interval for the estimated y-value at the given x_i.

Figure 5.30*a* shows the 95% confidence interval which brackets (or sandwiches) the curve. From a statistics point of view, the 95% confidence interval means that out of 100 samples being measured for y_i at the given x_i, 95 of them will have a value of y that lies within the range of y_i (i.e., between lower $f(x_i)$ and upper $f(x_i)$) at the given x_i. For the new look of the MATLAB Curve Fitting

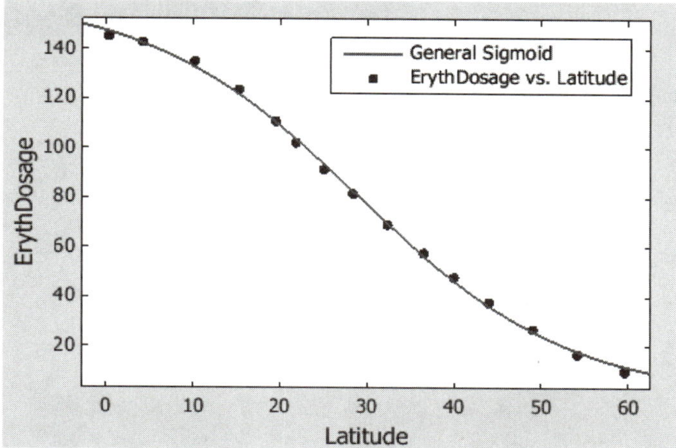

Figure 5.28 The curve-fitted data of erythemal UV dosage vs. latitude using the general sigmoid function.

Analysis

Fit to analyze: General Sigmoid...

Analyze at Xi = 0.304944:5.9307:59.6119

☑ Evaluate fit at Xi

Prediction or confidence bounds:

◯ None

◉ For function

◯ For new observation

Level 95 %

Xi	lower f(Xi)	f(Xi)	upper f(Xi)
0.304944	144.528	147.043	149.558
6.23564	137.663	139.503	141.343
12.1663	127.283	128.748	130.212
18.097	112.887	114.38	115.874
24.0277	95.2445	96.784	98.3234
29.9584	75.9629	77.3874	78.8118
35.8891	56.9549	58.3388	59.7227
41.8198	40.1208	41.64	43.1591
47.7505	26.7936	28.397	30.0005
53.6812	17.1877	18.705	20.2223

Figure 5.29 The lower and upper $f(x_i)$ values being calculated at the 95% confidence border.

Toolbox (see Fig. 5.25*b*), click on the **Tools** menu, followed by the **Prediction Bounds** submenu, and pick up, for example, a 95% confidence interval so that the 95% confidence envelope will sandwich the curve, as shown in Fig. 5.30*b*.

This being the case, such a fit is almost perfect and the departure or deviation from measured values is quite negligible. Back to the distribution of the residuals (i.e., errors) around their zero mean, the residual plot can be obtained via the **View** menu, as shown in the main window of the Curve Fitting Toolbox (see Fig. 5.15). Click **Residuals** from the drop-down list followed by the **Scatter Plot** option. Figure 5.31*a* shows the residuals plot. For the new look of the MATLAB Curve Fitting Toolbox (see Fig. 5.25*b*), click on the **View** menu followed by the **Residuals Plot** drop-down list to get the residuals plot (Fig. 5.31*b*).

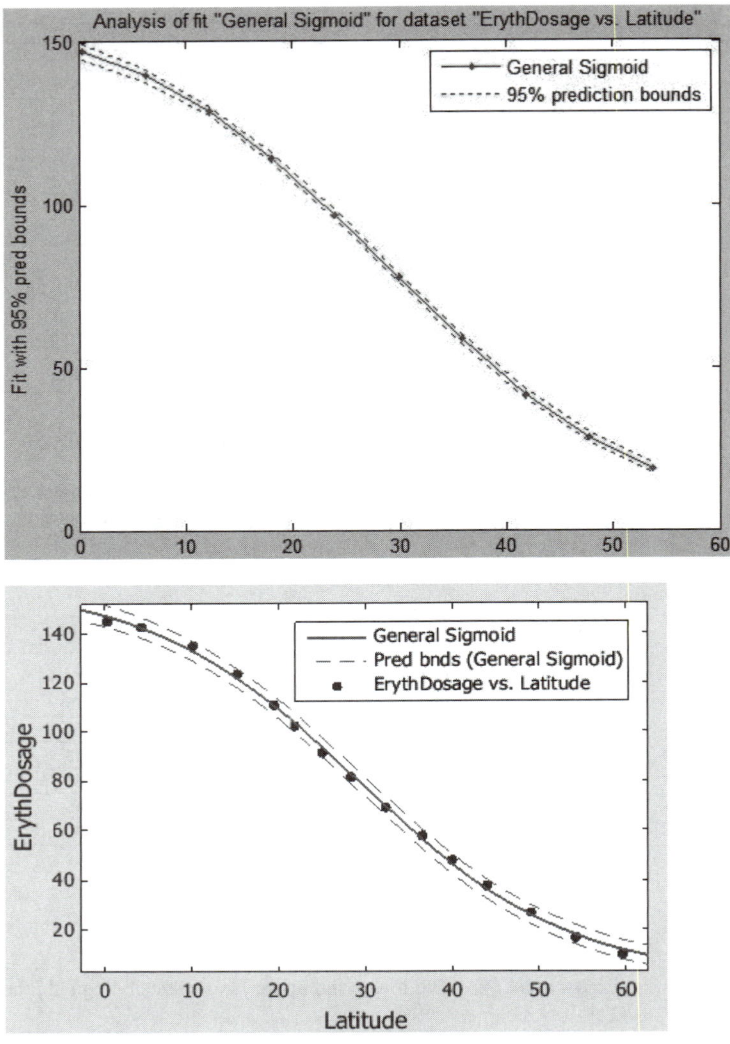

Figure 5.30 (*a*) The plot of 95% confidence interval that brackets the curve (old look). (*b*) The plot of the 95% confidence interval that brackets the curve (new look).

Notice here that the degree of scatter is more pronounced (more randomly distributed) than that of Fig. 5.14 and the magnitude (absolute value) of residual itself is smaller at each x_i value with a maximum value of 2.25 compared with 6 in the former case (the quadratic polynomial). The last thing to mention regarding the Curve Fitting Toolbox is that the user may have the chance and choice to change the way MATLAB finds the best estimated values for the model parameters. If the user clicks the **Fit options** (framed) button, as shown in Fig. 5.25*a*, the following window will pop up, as shown in Fig. 5.32. Here the user may decide on the algorithm to be used for minimizing the objective function. In addition, the

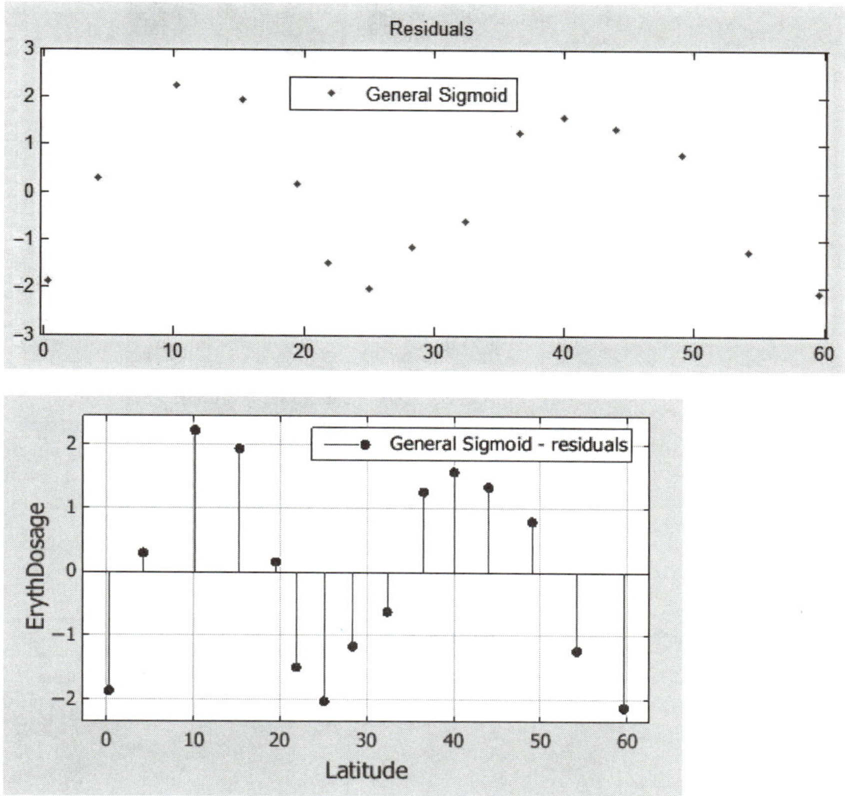

Figure 5.31 (*a*) Plot of residual $(y_i - \hat{y}_i)$ vs. x_i for the given data; where y_i represents the experimental value and $\hat{y}_i$ the predicted value for a given x_i (old look). (*b*) Plot of residual $(y_i - \hat{y}_i)$ vs. x_i for the given data; where y_i represents the experimental value and $\hat{y}_i$ the predicted value for a given x_i (new look).

number of iterations and function evaluations, convergence criteria, and the initial guess for each parameter appearing in the model can be entered in the dedicated textbox.

NOTE Do not forget to save your work. You can save your session with a file extension: **.cfit** using the **File** menu in the main window of the Curve Fitting Toolbox. You can also save the results of your curve fitting, as Fig. 5.33 shows. If you click on the *Save to workspace* button, you will be able to save the specific model, its goodness, and the fit output. Do not also forget to save the workspace itself. For the new look, you can save your session with a file extension **.sfit** using the "**File**" menu in the main window of the MATLAB Curve Fitting Toolbox (see Fig. 5.25*b*).

Finally, from the **File** menu in the main window of the Curve Fitting Toolbox, you may choose **Generate Code** from the drop-down list to create the M-file, which when executed will create a plot similar to the plot in the main Curve

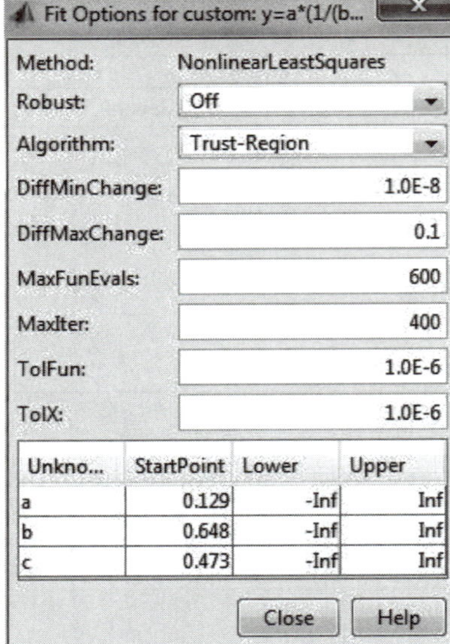

Figure 5.32 Fit options for the algorithm of optimization (i.e., nonlinear least squares) and other optimization settings.

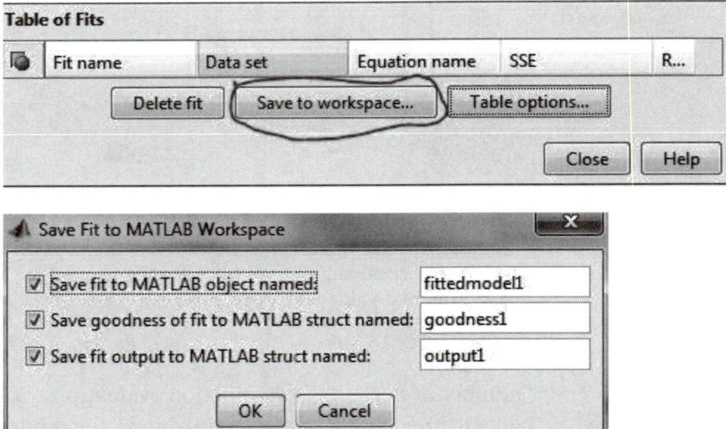

Figure 5.33 Save to the Workspace the specific model with a **.cfit** extension (*top*); its goodness and the fit output (*bottom*).

Fitting Toolbox, using the data that you provide as input. You can use this function with the same data you used with the MATLAB Curve Fitting Toolbox or different data sets. You may want to edit the function to customize the code.

5.7 MATLAB's Surface Fitting Toolbox (sftool)

The MATLAB Surface Fitting Toolbox (introduced in R2009a) is similar to the new look of the Curve Fitting Toolbox (see Fig. 5.25*b*). With the release of MATLAB R2011b and above, **sftool** will be abandoned; instead, **cftool** will be used.

MATLAB Surface Fitting Toolbox requires, however, that the model to be tried is a function of two linearly independent variables, not only a single variable, as is the case with the old MATLAB Curve Fitting Toolbox. The MATLAB Surface Fitting Tool allows the fitting of nonsymmetrical experimental data that may contain NAN (not number) values. The general formula for curve fitting is:

$$Z = f(X,Y)$$

where Z is the dependent variable and X and Y are the independent variables.

For MATLAB versions between R2009a and R2011a, such a tool can be invoked by typing at the command prompt:

```
>>    sftool
```

For MATLAB version R2011b or above, type at the command prompt:

```
>>   cftool
```

Figure 5.34 shows the default main window of the MATLAB Surface Fitting Toolbox, which is similar to that of Fig. 5.25*b* (a new look for the MATLAB Curve Fitting Toolbox).

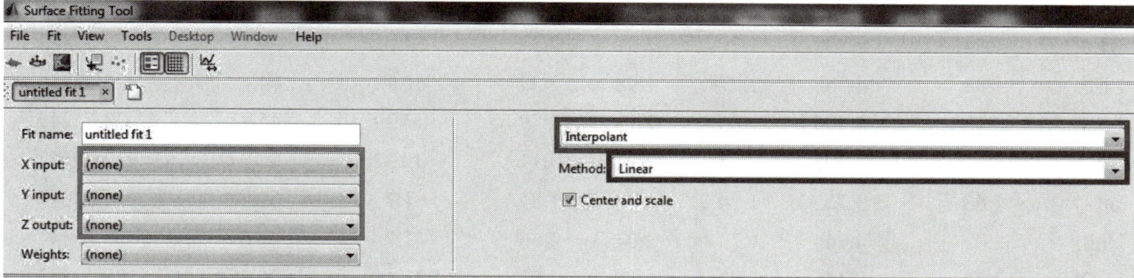

Figure 5.34 The default main window of the MATLAB Surface/Curve Fitting Toolbox.

The user needs to input or define X, Y, and Z vectors (see the framed drop-down list for each variable as shown in Fig. 5.34). The surface fit model to be used can be defined via either or both of the framed drop-down lists.

Let us demonstrate how to use the MATLAB Surface/Curve Fitting Toolbox via running the case study done by the author [cited in Al-Malah, K., Aqueous solubility of a diatomic molecule as a function of its size and electronegativity difference, *J. Mol Mod*, 17(2):325–331, (2011) DOI: 10.1007/s00894-010-0729-1]. The solubility (Z) was fitted as a function of both the molecular volume (X) and the electronegativity difference (Y). Different electronegativity difference models were tested for diatomic, univalent, and alkali halide molecules. We will take one subset and use it here to demonstrate the usage of the MATLAB Surface/Curve Fitting Toolbox.

Let us fit $Z = f(X,Y)$ using the following function: $Z = a + b * X + c * Y + d * Y^2 + e * (X/Y)$.

The columns of Table 5.2, except the first column, were imported to the MATLAB environment and converted into vectors (i.e., Workspace vectors). Figure 5.35 shows the MATLAB Workspace being populated by the vectors corresponding to the data columns of Table 5.2.

Table 5.2 Electronegativity difference (ΔX) for atoms making up the alkali halides, based on five different electronegativity scales, the solubility data, and the molecular volume, expressed in cubic angstroms (Å³).

Species	Δ Xpauling	Δ Xparr	Δ Xalrd-roch	Δ Xmll-jaf	Δ Xsndrsn	Solubility @20°C	Solubility @25°C	Vol(Å³)
CsBr	2.17	5.405	1.88	2.33	3		55.2	61.58291
CsCl	2.37	6.11	1.97	2.48	3.26	64.96	65.64	54.22093
CsF	3.19	8.23	3.24	3.29	3.78		85.1	34.64076
CsI	1.87	4.575	1.35	2.12	2.56	43.2	45.9	74.70004
KBr	2.14	5.165	1.83	2.22	2.77	39.4	40.4	46.61507
KCl	2.34	5.87	1.92	2.37	3.03	25.39	26.22	40.52542
KF	3.16	7.99	3.19	3.18	3.55	47.3	50.41	24.63676
KI	1.84	4.335	1.3	2.01	2.33	59	59.7	57.57882
LiBr	1.98	4.58	1.77	1.98	2.33	62.7	64.4	26.92715
LiCl	2.18	5.285	1.86	2.13	2.59	45.29	45.81	22.74249
LiF	3	7.405	3.13	2.94	3.11	0.131	0.134	12.29663
LiI	1.68	3.75	1.24	1.77	1.89	61.7	62.3	34.64076
NaBr	2.03	4.74	1.73	2.04	2.66	47.7	48.6	33.26447
NaCl	2.23	5.445	1.82	2.19	2.92	26.41	26.45	28.42832
NaF	3.05	7.565	3.09	3	3.44	3.89	3.97	16.13521
NaI	1.73	3.91	1.2	1.83	2.22	63.9	64.8	42.09398
RbBr	2.14	5.25	1.85	2.26	2.91	52.6	53.8	53.75216
RbCl	2.34	5.955	1.94	2.41	3.17	47.53	48.42	47.04142
RbF	3.16	8.075	3.21	3.22	3.69	75		29.35526
RbI	1.84	4.42	1.32	2.05	2.47	61.1	62.3	65.76847

After selecting what X, Y, and Z represent out of the list of vectors, choose *Custom Equation* from the drop-down list (see the framed drop-down list in Fig. 5.34). You will then be prompted to enter the function formula in the dedicated textbox (see Fig. 5.36).

Figure 5.37 shows a 3D plot for $Z = f(X,Y)$, where Z is surface-fitted to both X and Y. In addition, the figure shows the nonlinear regression results: estimated

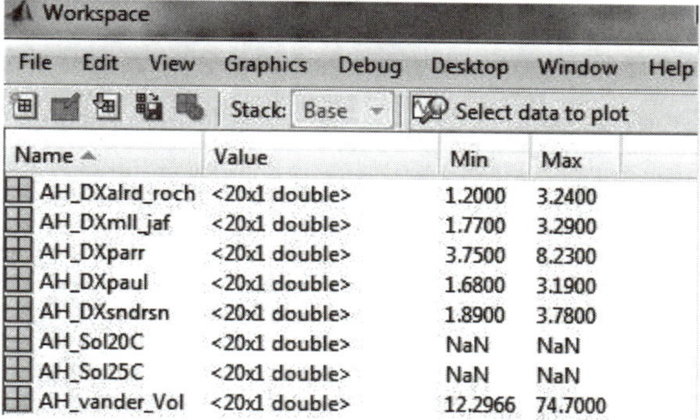

Figure 5.35 MATLAB Workspace containing a vector for each data column in Table 5.2.

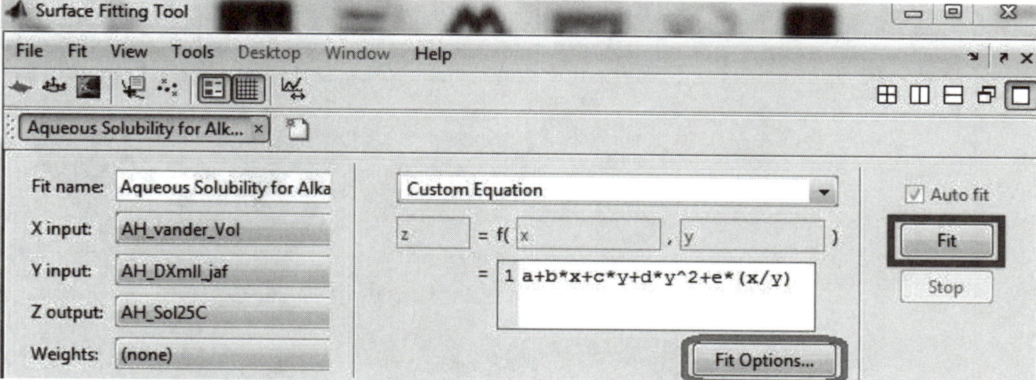

Figure 5.36 After defining what X, Y, and Z represent and entering the function statement, MATLAB is ready for surface fitting. You need to click the **Fit** (framed) button.

parameters, their 95% confidence interval limits, and parameters indicating model goodness; R^2, *Adj* R^2, *SSE,* and *RMSE.*

Figure 5.38 shows the contour (color-graded) plot for $Z = f(X,Y)$. For example, the color brown means a region with maximum solubility and dark blue means minimum. The contour plot can be shown via the **View** menu at the top of the main window of the surface-fitting tool (see Fig. 5.34). The contour plot can be presented with or without the surface plot, depending on your choice for the selected items to be viewed from the **View** menu.

The following points are worth mentioning:

1. You can change the view angle (the oblique view) for the 3D plot simply by hovering your mouse over the plot area. Notice that the cursor changes into a round or circled arrow. This means if your mouse is in the drag (left button pressed) mode, you will be able to flip the 3D plot in any direction you wish (right, left, up, and down).

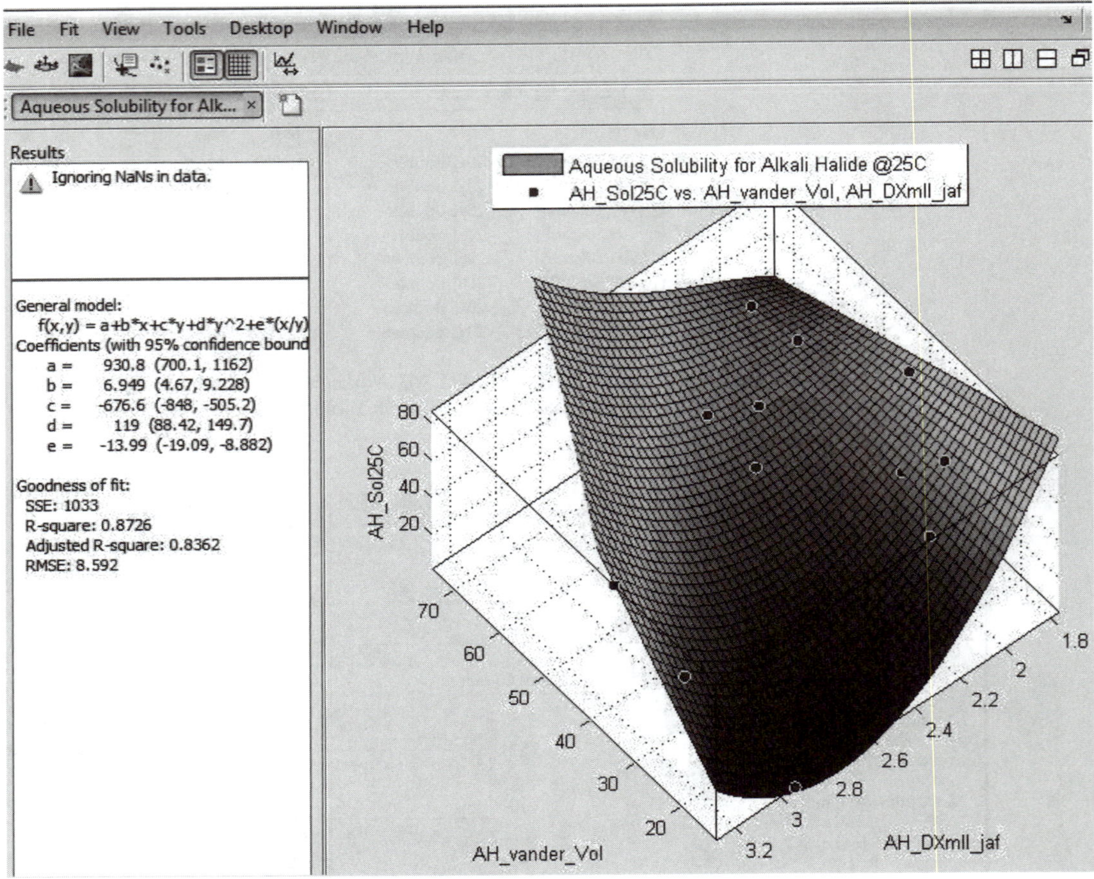

Figure 5.37 A 3D plot $Z = f(X,Y)$, where Z is surface-fitted both to X and Y, augmented by the nonlinear regression results (estimated parameters and model goodness).

2. Clicking the **Fit Options** button (shown in Fig. 5.36) will give the user more choices about how the MATLAB platform should tackle the optimization algorithm. For example, the user can choose between the Trust-Region or Levenberg-Marquardt algorithm. The user can also supply the initial guess, lower, and upper limit for each parameter plugged into the model.
3. NAN values are not a serious issue for the Surface/Curve Fitting Tool. For example, some solubility data at either 20 or 25°C are missing in Table 5.1. Nevertheless, MATLAB will take care of such a shortage in data and consider only fully populated rows. A fully populated row means all data entries are supplied for each column.
4. Save your work.

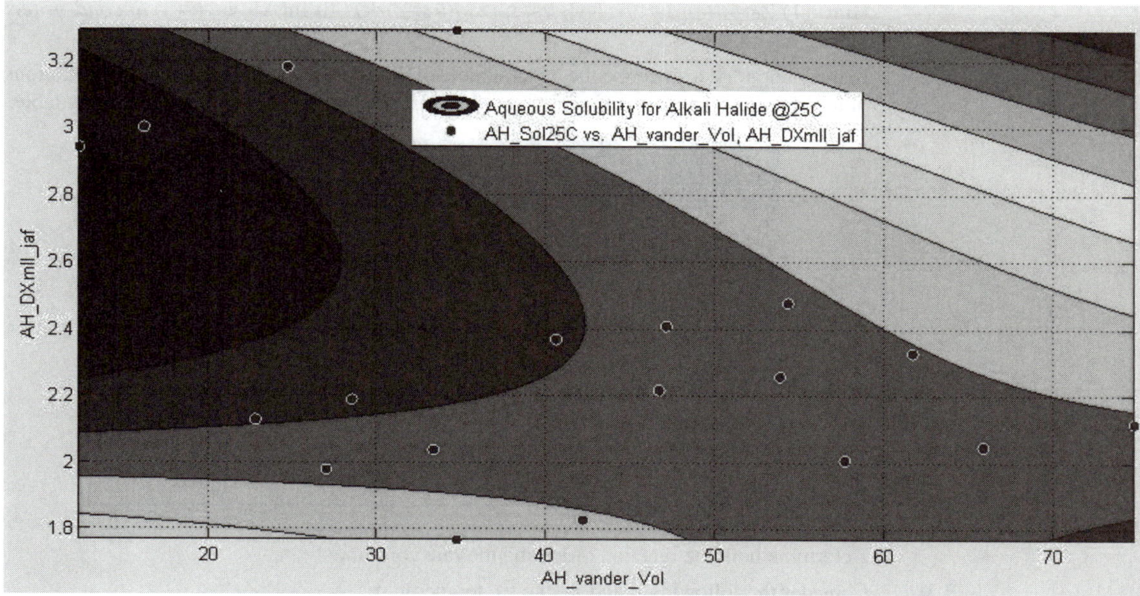

Figure 5.38 A contour (color-graded) plot for $Z = f(X,Y)$. In this plot, the color brown (*top right*) means a region with maximum solubility and dark blue (*top left*) means minimum.

5.8 Problems

5.1 Ten data points were taken in an experiment in which the independent variable x is the mole percentage of a reactant and the dependent variable Y is the yield (in percent):

x	Y	x	Y
20	73	50	87
20	78	50	86
30	85	50	91
40	90	60	75
40	91	70	65

Find a *quadratic* model with these data and determine the values of x that maximize the yield.

5.2 The following data were obtained on the reaction $2\ A \rightarrow B$:

Time, s	0	5	10	15	20
[A], mol L^{-1}	0.100	0.0141	0.0078	0.0053	0.004

Determine the order of the reaction and its rate constant.

Hint: (1) *Zeroth-order*: fit $[A]_t$ vs. t; (2) *First-order*: fit $\ln[A]_t$ vs. t; (3) *Second-order*: plot $1/[A]_t$ vs. t.

5.3 In the laboratory, a student collected a sample of data consisting of measured concentrations of a certain reactant A at different times. The sample data are shown in the following table:

Time (min)	Concentration (M)	Time (min)	Concentration (M)
0	0.906	35.2118	0.1495
0.9184	0.8739	42.973	0.1029
9.0875	0.5622	46.6555	0.086
11.2485	0.5156	50.3922	0.0697
17.5255	0.3718	55.4747	0.0546
23.9993	0.2702	61.827	0.0393
27.7949	0.2238	65.6603	0.0324
31.9783	0.1761	70.0939	0.026

Determine both the reaction order and the rate constant.

5.4 Consider the following reaction rate vs. temperature:

#	Temperature (K)	Rate Constant, k (s^{-1})
1	430	0.0026
2	440	0.0057
3	450	0.0118
4	460	0.0236
5	470	0.0460
6	480	0.0873
7	490	0.1800

Use the Arrhenius kinetic $k = k_o e^{\left(-\frac{E_A}{RT}\right)}$ to determine the pre-exponential factor, k_o, and the activation energy, E_A.

5.5 Consider this actual calibration data for the vortex flow meter from the frictional losses experiment in junior lab:

#	Voltage (x_i)	Flow rate, L/s (y_i)	#	Voltage (x_i)	Flow rate, L/s (y_i)
1	1.01	0.00	5	2.83	1.26
2	1.27	0.19	6	3.13	1.47
3	1.85	0.58	7	3.96	2.07
4	2.38	0.96	8	4.91	2.75

Use the simplest (straight-line) model to fit $y = f(x)$. Find the regressed parameters, their 95% confidence interval, R^2, Adj-R^2, SSE, and RMSE. Plot the residuals and see if the errors are randomly distributed around their zero-value mean.

5.6 A laboratory study of convective heat transfer to a cylinder in cross-flow was conducted in a small wind tunnel. The sensor was located at the forward stagnation line and tests were performed over a range of air speeds. For each test were the Reynolds (N_{Re}) and Nusselt (N_{Re}) numbers were calculated based on fluid, flow, and geometrical properties.

#	N_{Re}	N_{Nu}	#	N_{Re}	N_{Nu}
1	15025	117.1	9	35026	174.2
2	17527	127.1	10	37215	178.9
3	20007	133.6	11	40003	185.9
4	23331	143.3	12	42500	192
5	24880	148.8	13	44998	197.4
6	27196	153.8	14	47463	202.7
7	30008	161.1	15	49992	208.8
8	32583	168.1			

Your job is to express N_{Nu} as a function of N_{Re} using $N_{Nu} = a(N_{Re})^b \times (N_{Pr})^c$. Because the Prandtl number (N_{Pr}) in this study is close to unity, then the previous model can be reduced to:

$$N_{Nu} = a(N_{Re})^b$$

Based on the given experimental data, find the regressed point estimates for parameters a and b. Report all typical statistical parameters that pertain to model goodness. Plot the residuals and see if the errors are randomly distributed around their zero-value mean.

5.7 In an experiment to determine the friction factor for turbulent flow of water in a smooth wall tube, the pressure drop ΔP over a length L was measured over a range of flow rates. The following table shows the calculated results as friction factor $f = \left(\frac{L}{D}\right) \times \frac{\Delta P}{\left(\frac{1}{2}\rho v^2\right)}$ vs. Reynolds number $N_{Re} = \frac{\rho v D}{\mu}$.

#	N_{Re}	f	#	N_{Re}	f
1	35050	0.02295	10	63320	0.01992
2	38090	0.02215	11	67350	0.01971
3	39580	0.02221	12	70710	0.01958
4	42170	0.02213	13	74430	0.01933
5	44860	0.02131	14	78120	0.01926
6	47690	0.02118	15	81760	0.01928
7	49950	0.02124	16	87340	0.01895
8	52430	0.02100	17	89510	0.01882
9	59220	0.02032			

Your job is to express f as a function of N_{Re} using $f = a(N_{Re})^b$.

Based on the given experimental data, find the regressed point estimates for parameters a and b. Report all typical statistical parameters that pertain to model goodness. Plot the residuals and see if the errors are randomly distributed around their zero-value mean.

5.8 The following table represents temperature distribution $T(x, y)$ in a metal plate:

#	x	y	T(°C)	#	x	y	T(°C)
1	0.5	0.5	7.51	9	1.5	0.5	12.51
2	0.5	1	10.05	10	1.5	1	9.95
3	0.5	1.5	12.7	11	1.5	1.5	7.32
4	0.5	2	15.67	12	1.5	2	4.33
5	1	0.5	10	13	2	0.5	15
6	1	1	10	14	2	1	10
7	1	1.5	10	15	2	1.5	5
8	1	2	10	16	2	2	0

Your job is to express $T(K)$ as a function of x *and* y using different models.

a) First, try the case in which T is linear in both directions. Does the linear model predict the temperature profile?

b) Try the following permutations, reflecting different possible boundary conditions for x and y:

1) $T(x,y) = sin(a*x)*cosh(b*y) + c$
2) $T(x,y) = sin(a*x)*sinh(b*y) + c$
3) $T(x,y) = cos(a*x)*cosh(b*y) + c$
4) $T(x,y) = cos(a*x)*sinh(b*y) + c$
5) $T(x,y) = sin(a*y)*cosh(b*x) + c$
6) $T(x,y) = sin(a*y)*sinh(b*x) + c$
7) $T(x,y) = cos(a*y)*cosh(b*x) + c$
8) $T(x,y) = cos(a*y)*sinh(b*x) + c$

NOTE In the **Fit options**, with the **Trust-Region** algorithm selected, try the three robust cases (**Off**, **LAR**, and **Bisquare**) and see which case gives the best fit.

If any one of the afore-mentioned expressions fits well (i.e., highest R^2) the given data, then a *steady-state* temperature profile may exist. For the best fit, find the regressed point estimates for parameters: $\boldsymbol{a}$, $\boldsymbol{b}$, and $\boldsymbol{c}$. Report all typical statistical parameters that pertain to model goodness. Plot the residuals and see if the errors are randomly distributed around their zero-mean.

c) What do you conclude about the *steadiness* of temperature?

Numerical Integration

Chapter Outline

6.1 Trapezoid Rule

Both the trapezoid and Simpson's rules are based on the notion that the total area under the curve is split into slices (strips), with each slide having a width h and an average height of $\left(\frac{f_{i+1}+f_i}{2}\right)$. This is an approximation method in which a straight line is used to approximate $f(x)$ on each interval. This straight line is the *chord* joining the points $(xi, f(xi)$ and $(xi+1, f(xi+1)$ on the curve. The equation for this straight line can then be written as:

$$\frac{y-f_i}{x-x_i}=\frac{f_{i+1}-f_i}{x_{i+1}-x_i} \tag{6.1}$$

Using the notation that $h=x_{i+1}-x_i$ is the step in x, we can then write the straight-line approximation to $f(x)$ as:

$$y=f_i+\frac{1}{h}(f_{i+1}-f_i)(x-x_i) \tag{6.2}$$

The area of the *ith* slice (strip) is thus the area of the trapezoid:

$$\alpha_i = width \times average\ height = h \times \frac{(f_{i+1}+f_i)}{2} \tag{6.3}$$

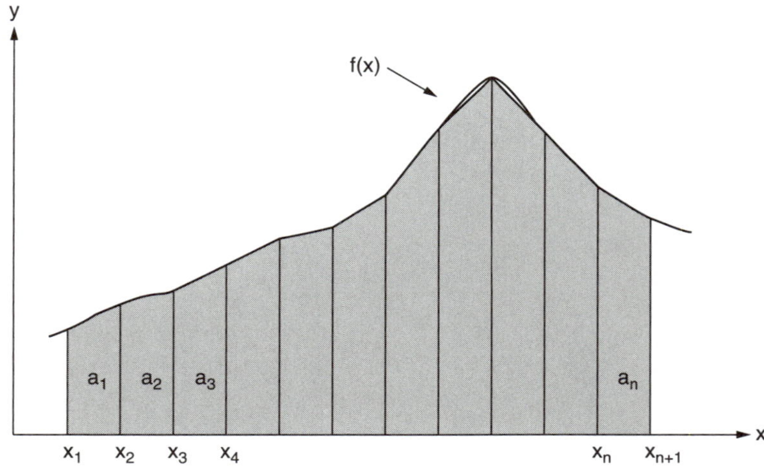

Figure 6.1 The total area under the curve is split into slices with each slide having a width h and an average height of $\left(\frac{f_{i+1}+f_i}{2}\right)$. Each slice has an area (a_i) equal to *width*height* $= h^* \left(\frac{f_{i+1}+f_i}{2}\right)=\alpha_i$.

Thus, the total area under the curve is equal to:

$$A=\sum_{i=1}^{n} a_i = \frac{h}{2}\sum_{i=1}^{n}(f_i+f_{i+1}) \tag{6.4}$$

We note that except for the two end values of i, i.e., $i = 1$ and $i = n$, each evaluation of $f(x)$ at a point x_i occurs twice. Thus, we can write the approximation to the integral, using the trapezoid rule in the form:

$$\int_{\alpha}^{\beta} f(x)dx \approx \frac{h}{2}[f_1 + 2f_2 + \cdots + 2f_n + f_{n+1}] = \frac{h}{2}\{f(\alpha) + f(\beta)\} + h\sum_{i=1}^{n-1} f(\alpha + ih) \tag{6.5}$$

In developing this approximation, we still have left some area under the $f(x)$ out of the sum (near the maximum of $f(x)$, for example), or included some area of trapezoids that lies above $f(x)$. Although we will not derive limits on this error in the trapezoidal approximation here, analysis shows that the error can be expressed in the following form:

$$E_{Trpzd} = -\frac{(\beta - \alpha)}{12} h^2 f^{II}(\omega) \tag{6.6}$$

where $f^{II}(\omega)$ is the second derivative of $f(x)$ evaluated at some point ω with $\alpha \le \omega \le \beta$. The error estimate is particularly useful in suggesting how to improve the accuracy of our result by decreasing the step-size, h. If we halve the step-size, the number of calculations required to compute the sum is doubled, but the error estimate E_{Trpzd} decreases by a factor of four. Hence, we can predict the trade-off between improved accuracy and the amount of work, as measured by the number of evaluations of the function $f(x)$.

It should be pointed out here that calculation of E_{Trpzd} was done based on defining numerically the second derivative of a given function. For further information, refer to the Numerical Differentiation PDF file attached with U_Integrate software (https://sites.google.com/site/almalahweb/software).

The trapezoidal numerical integration can be invoked by:

```
>>Z = trapz(Y)
>>Z = trapz(X,Y)
```

6.1.1 Description (MATLAB's Trapz Help)

Z = trapz(Y) computes an approximation of the integral of Y via the trapezoidal method (with unit spacing). To compute the integral for spacing other than one, multiply Z by the spacing increment. Input Y can be complex.

If Y is a vector, trapz(Y) is the integral of Y.

If Y is a matrix, trapz(Y) is a row vector with the integral over each column.

If Y is a multidimensional array, trapz(Y) works across the first nonsingleton dimension.

Z = trapz(X,Y) computes the integral of Y with respect to X using trapezoidal integration. Inputs X and Y can be complex.

If X is a column vector and Y an array whose first nonsingleton dimension is length(X), trapz(X,Y) operates across this dimension.

Example 6.1 The exact value of $\int_0^{\pi} \sin x\, dx$ is 2.

To approximate this numerically on a uniformly spaced grid, use:

```
>> X = 0:pi/100:pi;
>>Y = sin(X);
>> Z = trapz(X,Y)
Z =
    1.9998
```

Alternatively,

```
>> Z = pi/100*trapz(Y)
Z =
    1.9998
```

Example 6.2 The exact value of

$$\int_0^4 x^3 dx \text{ is } \left.\frac{x^4}{4}\right|_0^4 = \frac{256}{4} = 64.$$

```
>> X = 0:4/100:4;
>> Y = X.^3;
>> Z = trapz(X,Y)
Z =
   64.0064
```

Example 6.3 The exact value of

$$\int_0^4 e^x dx \text{ is } e^x\Big|_0^4 = e^4 - 1 = (2.718281825)^4 - 1 = 53.5982.$$

```
>> X = 0:4/100:4;
>> Y = exp(X);
>> Z = trapz(X,Y)
Z =
   53.6053
```

Obviously, you can increase the accuracy by decreasing the step-size of integration.

Example 6.4 Find the value of $\int_0^4 e^x\, dx$ using $h = 4/200$ instead of 4/100 and see how it affects the result.

```
>> X = 0:4/200:4;
>> Y = exp(X);
```

```
>> Z = trapz(X,Y)
Z =
    53.5999
```

Notice how the answer gets closer to the exact value.

The following M-file (**Trapezoid.m**) utilizes the trapezoid rule derived earlier [Eq. (6.5)].

```
function [ NumIntegRes ] = Trapezoid( lower,upper,numstep)
%The evaluation of the integrand between the lower and upper limit
%using Trapezoid Rule.
ytemp2=input('Enter y as a function of x: ', 's');
%In the following statement strrep means to replace '*' by '.*'to account
%for element-wise multiplication when the argument is a vector not scalar.
ytempi=strrep(ytemp2,'*','.*');
ytempj=strrep(ytempi,'/','./');
ytemp3=strrep(ytempj,'^','.^');
ytemp4=inline(ytemp3,'x');
mysum=0.0;
hval=(upper-lower)/numstep;
fterm=0.5*hval*(ytemp4(lower)+ytemp4(upper));
for kind=1:1:numstep-1
serterm=hval*ytemp4(lower+kind*hval);
mysum=mysum+serterm;
end
NumIntegRes=mysum+fterm;
display(['The value of integration: ',num2str(NumIntegRes)]);
end
```

Figure 6.2 M-file (**Trapezoid.m**) for trapezoid rule.

Example 6.5 The exact value of

$$\int_0^4 e^x dx \text{ is } e^x\Big|_0^4 = e^4 - 1 = (2.718281825)^4 - 1 = 53.5982.$$

```
>> Trapezoid(0,4,100);
Enter y as a function of x: exp(x)
```

The value of integration:

53.6053

which is similar to the answer of Example 6.3.

Example 6.6 Find the integral value for:

$$\int_0^{10} (x^3 + 2x^2 + 5)dx$$

```
>> Trapezoid(0,10,200);
```

Enter y as a function of x:

(x^3) + (2*x^2) + 5

The value of integration:

3216.7375

Using the MATLAB built-in **trapz** function:

```
>> x = 0:10/200:10;
>> y = (x.^3+2*x.^2+5);
>> trapz(x,y)
```

ans = 3.2167e + 003

Example 6.7 Find the integral value for:

$$\int_0^{10} (x^3 + 2x^2 + 5)e^x dx$$

```
>> Trapezoid(0,10,200);
```

Enter y as a function of x:

(x^3+2*x^2+5)*exp(x)

The value of integration:

20337514.233

Using the MATLAB built-in **trapz** function:

```
>> x = 0:10/200:10;
>> y = @(x) (x.^3+2*x.^2+5).*exp(x);
>> trapz(x,y(x))
```

ans = 2.0338e + 007

Notice here that a function handle (using @ followed by the variable in two brackets and finally the function expression or statement) for y was used. Of course, you can also use the previous way as shown here:

```
>> x = 0:10/200:10;
>> y = (x.^3+2*x.^2+5).*exp(x);
>> trapz(x,y)
```

ans = 2.0338e+007

6.2 Simpson's Rule

In the trapezoid rule, a linear approximation is used for the integrand on each step-sized interval. When this approximation is integrated, we obtain a quadratic expression for the piece of the integral over that step-sized interval. In deriving the first of Simpson's rules, we will approximate $f(x)$ over each pair of intervals by a quadratic function in the form:

$$\tilde{f}(x) = ax^2 + bx + c$$

Although it takes only two points to establish a straight line, it takes three points to establish the quadratic function, i.e., to establish the values of a, b, and c. Requiring that the quadratic function $\tilde{f}(x)$ equals the function $f(x)$ at each of x_i, x_{i+1}, and x_{i+2}, we can then form the set of three simultaneous linear equations for a, b, and c.

$$f_i = ax_i^2 + bx_i + c \tag{6.7}$$

$$f_{i+1} = ax_{i+1}^2 + bx_{i+1} + c \tag{6.8}$$

$$f_{i+2} = ax_{i+2}^2 + bx_{i+2} + c \tag{6.9}$$

Here, we use the notation that f_i means $f(x_i)$. Thus, we have three equations, with three unknown terms, which can be solved to obtain:

$$a = \frac{1}{2h^2}\{f_i - 2f_{i+1} + f_{i+2}\} \tag{6.10}$$

$$b = -\frac{1}{2h^2}\{2x_i(f_i - 2f_{i+1} + f_{i+2}) + h(3f_i - 4f_{i+1} + f_{i+2})\} \tag{6.11}$$

$$c = \frac{1}{2h^2}\{2h^2 f_i + hx_i(3f_i - 4f_{i+1} + f_{i+2}) + x_i^2(f_i - 2f_{i+1} + f_{i+2})\} \tag{6.12}$$

Having established the approximation $\tilde{f}(x)$ through these coefficients, we can then evaluate directly the contribution to the integral of $\tilde{f}(x)$ over the interval x_i to x_{i+2} of length $2h$:

$$\int_{x_i}^{x_{i+2}} \tilde{f}(x)dx = \int_{x_i}^{x_{i+2}} (ax^2 + bx + c)dx = \frac{h}{3}[f_i + 4f_{i+1} + f_{i+2}] \tag{6.13}$$

To evaluate the integral of $f(x)$ over the whole interval α to β, we must divide that interval into n equal panels, where n must be an even integer to accommodate the pairwise use of panels to develop the quadratic approximations. We then obtain the Simpson $\frac{1}{3}$ rule:

$$\int_{\alpha}^{\beta} f(x)\,dx \approx \frac{h}{3}[f_o + 4f_1 + 2f_2 + 4f_3 + \cdots + 2f_{n-2} + 4f_{n-1} + f_n]$$

$$= \frac{h}{3}\left[f(\alpha) + f(\beta) + 4\sum_{i=1,odd}^{n-1} f(\alpha + ih) + 2\sum_{i=2,even}^{n-2} f(\alpha + ih)\right] \quad (6.14)$$

The error in this approximation can be shown to be:

$$E_{1/3} = -\frac{(\beta - \alpha)}{180} h^4 f^{iv}(\omega) \quad (6.15)$$

where $f^{iv}(\omega)$ is the fourth derivative of *f(x)* evaluated at some point ω with $\alpha \le \omega \le \beta$. The Simpson 1/3 rule has the property of variation of its error term through h^4; that is, although halving the interval still requires twice as many evaluations of *f(x)*, it reduces the error term by a factor of 16, so that we expect to diminish the impact of error with this integration rule much faster than with the trapezoidal rule.

It should be pointed out here that calculation of $E_{1/3}$ was done based on numerically defining the fourth derivative of a given function. For further information, refer to the Numerical Differentiation PDF file attached to **U_Integrate** (http://almalah.0fees.net/Software.html) software.

One drawback of Simpson's 1/3 rule is the requirement that the number of panels be even. If the function, to be integrated, is given in a table, such as that obtained from experimental data, it may not be the case that the number of panels is an even number. Simpson's 3/8 rule solves this problem, being based on the use of a cubic polynomial approximation function, rather than a quadratic function. This approximating function takes the form:

$$\tilde{f}(x) = ax^3 + bx^2 + cx + d$$

where *a, b, c,* and *d* are constants to be determined by restricting that $f(x)$ and $\tilde{f}(x)$ are equal at four different points. Thus, it takes three panels to determine the cubic function, and therefore, the number of panels over the whole interval must be a multiple of three. The technique of deriving Simpson's 3/8 rule is similar to that for Simpson's 1/3 rule; the result is shown here:

$$\int_{\alpha}^{\beta} f(x)\,dx \approx \frac{3h}{8}[f_o + f_n + 3(f_1 + f_4 + \cdots + f_{n-2}) + 3(f_2 + f_5 + \cdots + f_{n-1})$$

$$+ 2(f_3 + f_6 + \cdots + f_{n-3})] = \frac{3h}{8}\left[f(\alpha) + f(\beta) + 3\sum_{i=1,4,7,\ldots}^{n-2} f(\alpha + ih)\right.$$

$$\left. + 3\sum_{i=2,5,8,\ldots}^{n-1} f(\alpha + ih) + 2\sum_{i=3,6,9,\ldots}^{n-3} f(\alpha + ih)\right] \quad (6.16)$$

The error term for Simpson's 3/8 rule differs from the error term for Simpson's 1/3 rule only in the magnitude of the numerical factor. In particular, the dependence on h remains the same.

NOTE Because the numerical factor is larger for Simpson's 3/8 rule, *it is best to use Simpson's* 1/3 *rule when possible.*

The problem of requiring an even number of panels with Simpson's 1/3 rule can now be handled when the number of panels is odd by using the 3/8 rule for the first three panels, and the 1/3 rule for the remaining panels, whose number is definitely even.

The MATLAB built-in Simpson's 1/3 rule is given by the **quad**ratic approximation: **quad** that numerically evaluates the integral using Simpson quadrature.

```
>>q = quad(fun,a,b)
>>q = quad(fun,a,b,tol)
>>q = quad(fun,a,b,tol,trace)
>>[q,fcnt] = quad(...)
```

Example 6.8 Compute the integral:

$$\int_0^2 \frac{1}{x^3-2x-5}dx$$

Write an M-file function `myfunc` that computes the integrand:

```
function y = myfunc(x)
y = 1./(x.^3-2*x-5);
```

Then pass @myfunc, a function handle to myfunc, to quad, along with the limits of integration, 0 to 2:

```
>> Q = quad(@myfunc,0,2)
```

$$Q = -0.4605$$

Alternatively, you can pass the integrand to quad as an anonymous function handle y:

```
>> y = @(x)1./(x.^3-2*x-5);
>> Q = quad(y,0,2)
```

$$Q = -0.4605$$

Example 6.9 Compute the integral:

$$\int_0^2 \frac{x}{x^3-2x-5}dx$$

```
>> y2 = @(x)x./(x.^3-2*x-5);
>>Q = quad(y2,0,2)
```

$$Q = -0.5515$$

Example 6.10 Numerically compute the following integral that has an analytical value of:

$$\int_0^\infty e^{-x}dx = -1\left(e^{-x}\Big|_0^\infty\right) = \left(e^{-x}\Big|_\infty^0\right) = 1 - 0 = 1$$

If you attempt to use ∞ as the upper limit for integration, you will get:

```
>> y3 = @(x)exp(-x);
>> Q = quad(y3,0,inf)
```

Warning: Infinite or Not-a-Number function value encountered.

> In quad at 109

Q = NaN

One way to avoid this mathematical hurdle is to replace ∞ by a sufficiently large number. For example, choose the upper limit to be 1000 and see what MATLAB says. Keep increasing the value until you end up with an error or warning. Stop there. The maximum value is:

```
>> Q = quad(y3,0,1e18)
```

Q = 1.0000

Example 6.11 (The Error Function)

erf(u) reads as the error function of the argument *u*, which is defined as:

$$erf(u) = \frac{2}{\sqrt{\pi}}\int_0^u e^{-x^2}dx$$

Notice that

$$erf(0) = 0.$$

Numerically compute the integral that has an analytical value of:

$$\frac{2}{\sqrt{\pi}}\int_0^\infty e^{-x^2}dx = erf(\infty) = 1.$$

Try the same approach as in Example 6.10.

```
>>  y4 = @(x)exp(-x.^2);
>> Q=(2/sqrt(pi))*quad(y4,0,1000)
```

Q = 1.0000

Also,

```
>>Q=(2/sqrt(pi))*quad(y4,0,1e18)
```

$$Q = 1.0000$$

The maximum number for replacement of ∞ is 1×10^{18}.

Example 6.12 Numerically compute the integral that has an analytical value of:

$$\frac{2}{\sqrt{\pi}}\int_{-\infty}^{0} e^{-x^2}dx = -\frac{2}{\sqrt{\pi}}\int_{\infty}^{0} e^{-u^2}du = +\frac{2}{\sqrt{\pi}}\int_{0}^{\infty} e^{-u^2}du = erf(\infty) = +1.$$

What we did in the previous step was to replace x by $-u$ and dx by $-du$.

Try the same approach as in Example 6.10.

```
>>  y4 = @(x)exp(-x.^2);
>> Q=(2/sqrt(pi))*quad(y4,-1000,0)
```

$$Q = 1.0000$$

Also,

```
>> Q=(2/sqrt(pi))*quad(y4,-1e18,0)
```

$$Q = 1.0000$$

The minimum number for replacement of $-\infty$ is -1×10^{18}.

Example 6.13 Numerically, compute the following integral that has an analytical value of:

$$\int_{3}^{0} x\,dx = \frac{x^2}{2}\bigg|_{3}^{0} = \frac{1}{2}(0-3^2) = -4.5$$

Let's try it using the MATLAB approach.

```
>> y5 = @(x)x;
>> Q=quad(y5,3,0)
```

$$Q = -4.5000$$

The following M-file (**Simpson_OneThird.m**) utilizes Simpson's 1/3 rule derived earlier [Eq. (6.14)].

```
function [ NumIntegRes ] = Simpson_OneThird( lower,upper,numstep)
%The evaluation of the integrand between the lower and upper limit
%using Simpson's 1/3 rule.

if mod(numstep+1,2) == 0
disp('Routine needs an even number of steps')
numstep=input('Enter even number of steps: ');
end
ytemp2=input('Enter y as a function of x: ', 's');
ytempi=strrep(ytemp2,'*','.*');
ytempj=strrep(ytempi,'/','./');
ytemp3=strrep(ytempj,'^','.^');
ytemp4=inline(ytemp3,'x');
hval=(upper-lower)/numstep;
x = lower:hval:upper;
N=numstep+1;
step_odd = 1:2:N;
step_even = 2:2:(N-1);
weights(step_odd) = 2; weights(1) = 1; weights(N) = 1;
weights(step_even) = 4;
NumIntegRes = hval/3*sum(weights.*ytemp4(x));
format long e
disp(['The numerical integration of ',ytemp3, ' from ', num2str(lower),...
    ' to ', num2str(upper),' is: ', num2str(NumIntegRes)])
end
```

Figure 6.3 M-file (**Simpson_OneThird.m**) for Simpson's 1/3 rule.

Example 6.14 Numerically, compute the following integral that has an analytical value of:

$$\int_0^{\pi} \sin(x)\,dx = -\cos(x)\big|_0^{\pi} = -(\cos(\pi) - \cos(0)) = -1(-1-1) = 2$$

>> Simpson_OneThird(0,pi,20)

Enter y as a function of x: sin(x)

The numerical integration of sin(x) from 0 to 3.1416 is: 2

ans = 2.000006784441801e + 000

Exercise 6.1 Create an M-file (or, function) that utilizes Simpson's 3/8 rule [Eq. (6.16)].

6.3 Symbolic Integration

MATLAB has the ability to perform symbolic integration. Using the **int** command, symbolic integration of a function can be done.

```
>>int(expr,var)
```

int(expr,var) computes the indefinite integral of **expr** with respect to the symbolic scalar variable **var**. Specifying the variable **var** is optional. If you do not specify it, **int** uses the default variable determined by **symvar**.

```
>>int(expr,var,a,b)
```

int(expr,var,a,b) computes the definite integral of **expr** with respect to **var** from **a** to **b**. If you do not specify it, **int** uses the default variable determined by **symvar**.

Example 6.15 Using MATLAB indefinite symbolic integration, carry out the integration step for the following *f(x)*:

$$I = \int f(x)\,dx = \int (ax^3 + bx^2 + cx + f)\,dx$$

```
>> syms a b c f x;
>> fx=a*x^3+b*x^2+c*x+f
fx =
a*x^3 + b*x^2 + c*x + f
```

The **syms** command is used to define a, b, c, f, and x as variables for symbolic purposes. At the command prompt, apply the **int** command:

```
>> Anal_Integ=int(fx)
Anal_Integ =
(a*x^4)/4 + (b*x^3)/3 + (c*x^2)/2 + f*x
```

MATLAB tells us that the analytical solution has the form:

$$I = \int (ax^3 + bx^2 + cx + f)\,dx = \frac{a}{4}x^4 + \frac{b}{3}x^3 + \frac{c}{2}x^2 + fx + C$$

Example 6.16 Integrate the following function:

$$I_2 = \int Q(x)\,dx = \int \frac{-2x}{(1+x^2)^2}\,dx$$

```
>> Integ_Exp=int(-2*x/(1 + x^2)^2)
Integ_Exp =
1/(x^2 + 1)
```

NOTE It is assumed that **x** is defined as symbolic variable, using the **syms** command. Alternatively, **x** can be defined as:

```
>> x=sym('x');
```

$$I_2 = \int Q(x)dx = \int \frac{-2x\,dx}{(1+x^2)^2} = \int \frac{-du}{(u)^2} = -\left(-\frac{1}{u}\right) = \frac{1}{u} = \frac{1}{(1+x^2)} + C$$

Here $(1+x^2)$ was replaced by u and $2x\,dx$ was replaced by du, integrated over du, and the solution $1/u$ was back-substituted to the x domain.

Example 6.17 Carry out the definite integration for the following function:

$$I_3 = \int_0^1 g(x)\,dx = \int_0^1 (5x^3 + 4x^2 + 3x + 2)\,dx$$

```
>> gx=5*x^3+4*x^2+3*x+2
gx =
5*x^3 + 4*x^2 + 3*x + 2
>> I3=int(gx,0,1)
I3 = 73/12
```

$$I_3 = \frac{5}{4}x^4 + \frac{4}{3}x^3 + \frac{3}{2}x^2 + 2x = \frac{5}{4}1^4 + \frac{4}{3}1^3 + \frac{3}{2}1^2 + 2\times 1 = \frac{15}{12} + \frac{16}{12} + \frac{18}{12} + \frac{24}{12} = \frac{73}{12}$$

6.4 Problems

NOTE Unless otherwise stated, choose the step size $\boldsymbol{n}$ such that $\boldsymbol{h} = \frac{\boldsymbol{b}-\boldsymbol{a}}{\boldsymbol{n}} \le \mathbf{0.1}$, where $\boldsymbol{b} > \boldsymbol{a}$.

6.1 Using the trapezoid rule calculate the integral of the quadratic equation $x^2 - 5x + 7$ between $x = 1$ and $x = 3$ (check your answer against the exact answer, given by symbolic or analytical solution).

6.2 Using Simpson's 1/3 rule integrate the cubic equation $x^3 - 2x + 5$ between $x = 0$ and $x = 1$ (check your answer against the exact answer, given by symbolic or analytical solution).

6.3 Using Simpson's 1/3 rule integrate the function $f(x) = sin(x)$ between $x = 0$ and $x = \pi$ (check your answer against the exact answer, given by symbolic or analytical solution). You may want to change the number of points you use and see what happens to the error.

6.4 Using Simpson's 1/3 rule integrate the function $f(x) = cos(x)$ between $x = 0$ and $x = \pi/2$ (check your answer against the exact answer, given by symbolic or analytical solution). You may want to change the number of points you use and see what happens to the error.

6.5 The expression for the length of the curve shown in Fig. 6.4 is given by:

$$L = \int_{\phi=\beta}^{\pi-\beta} \sqrt{1+cos^2(x)}dx \tag{6.17}$$

Determine the value numerically using the trapezoid rule for a variety of values of β.

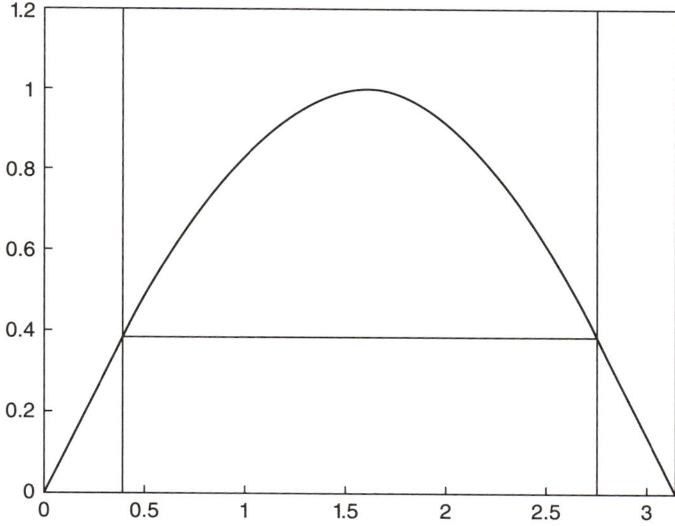

Figure 6.4 The curve given by the integral function [Eq. (6.17)].

6.6 Integrate the function ***x* ln *x*** (MATLAB: *x.*log(x)*) between 1 and 2 using the MATLAB built-in function **quad**.

6.7 Integrate the function ***x* ln *x*** (MATLAB: *x.*log(x)*) between 0 and 2 using the MATLAB built-in function **quad**. Compare this answer with that of Problem 6.6; Why is it smaller although the range of x is wider than that of Problem 6.6?

6.8 Integrate the function ***x* ln *x*** (MATLAB: *x.*log(x)*) between 0 and 1 using the MATLAB built-in function **quad**. The answer is negative. Did you realize why the answer in Problem 6.7 is smaller than that of Problem 6.6?

6.9 A homogeneous gas reaction A →3R has a reported rate at 215°C

$$-r_A = kCA^{0.5} = 10^{-2}C_A^{0.5} \quad \left[\frac{mol}{L.s}\right]$$

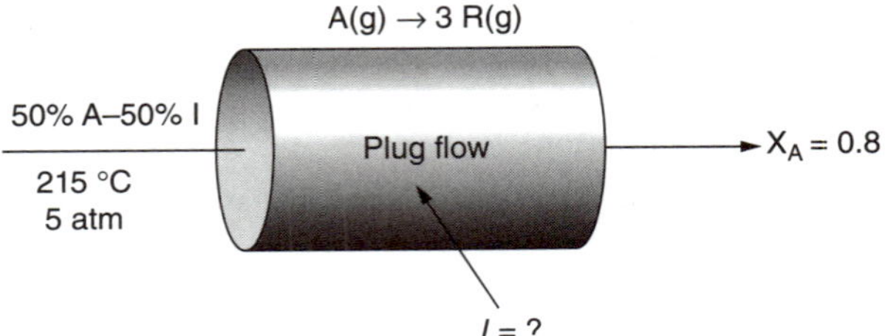

Figure 6.5

Find the space-time needed for 80% conversion of a 50% A-50% inert feed to a plug flow reactor (Fig. 6.5) operating at 215°C and 5 atm ($C_{Ao} = 0.0625$ mol/L).

Given that the volume expansion is $\epsilon_A = 1$, the space-time for the plug flow reactor (PFR) is given by:

$$\tau = C_{Ao}\int_0^{XA_f}\frac{dX_A}{-r_A} = C_{Ao}\int_0^{XA_f}\frac{dX_A}{kC_{Ao}^{1/2}\left(\dfrac{1-X_A}{1+\epsilon_A X_A}\right)^{1/2}} = \frac{C_{Ao}^{1/2}}{k}\int_0^{0.8}\left(\frac{1+X_A}{1-X_A}\right)^{1/2} dX_A$$

Use any of the numerical integration techniques to find the numerical value of the integral between 0 and 0.8. Then calculate the space time, τ, of the PFR. Compare your results with the analytical solution:

$$\int_0^{0.8}\left(\frac{1+X_A}{1-X_A}\right)^{1/2} dX_A = \left[\arcsin X_A - \sqrt{(1-X_A^2)}\right]\Big|_0^{0.8} = 1.328$$

6.10 One mole ($n = 1$) of an ideal gas undergoes an isothermal (constant-temperature) expansion at temperature $T = 300$ K, during which its volume changes from $V_1 = 10$ to $V_2 = 20$ L. How much work does the gas do?

$$W_{12} = \int_{V1}^{V2} P\,dV = \int_{V1}^{V2}\frac{nRT}{V}dV = RTln\left(\frac{V2}{V1}\right)$$

Use the trapezoid rule and symbolic integration to calculate the integral and contrast your answer vs. the analytical. $R = 8.314$ *J/(mol·K)*.

6.11 A sample of ideal gas is expanded to twice its original volume of 1.00 m^3 in a quasi-static process for which $P = \alpha V^2$, with $\alpha = 5\times10^5\ Pa/m^6$, as shown in Fig. 6.6.

a) How much work is done by the expanding gas?
b) Calculate the amount of work using Simpson's 1/3 rule and the symbolic integration.
c) Compare the answers vs. the analytical solution.

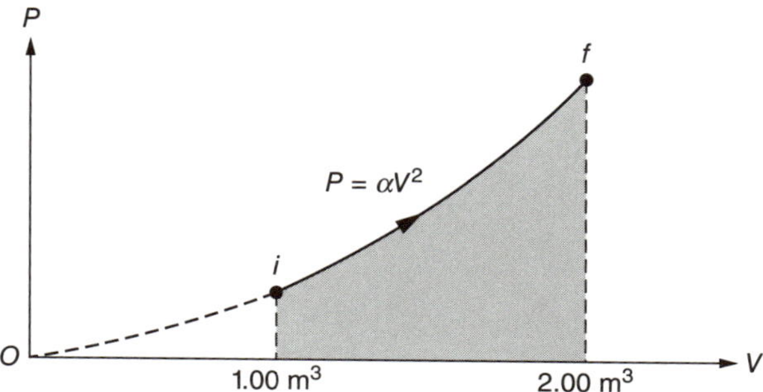

Figure 6.6

6.12 The Handbook (*Perry's Chemical Engineers' Handbook*) gives the dependence of the volume of mercury on temperature as:

$$\frac{d\bar{V}(T)}{dT} = f(T) = 7.35\times10^{-2}(0.18182\times10^{-3} + 0.0156\times10^{-6}T)$$

where $\bar{V}(T)$ is the specific volume (cm^3/g) of mercury at temperature $T(°C)$ and $7.35\times 10^{-2}\, cm^3/g$ is the specific volume of mercury at 0°C. Calculate the specific volume change of mercury if one gram is heated from 0 to 100°C. Calculate the total volume change of mercury if 10 grams are heated from 0 to 100°C. Use both numerical and symbolic integration.

6.13 Methane and oxygen react in the presence of a catalyst to form formaldehyde. Consider the heat capacity data for methane (CH_4) as $f(T)$ where T is in K. Calculate the amount of heat needed to heat up 100 moles of methane gas from 300 to 900 K at constant pressure.

$$\frac{C_P}{R} = 4.568 - 8.975\times10^{-3}T + 3.631\times10^{-5}T^2 - 3.407\times10^{-8}T^3 + 1.091\times10^{-11}T^4$$

Use both numerical and symbolic integration.

6.14 Evaluate the following integral:

$$\int_0^4 (2 - e^{-3x})\,dx$$

(a) Analytically/symbolic integration
(b) The trapezoidal rule with $n = 4$ [Eq. (6.5)]
(c) Simpson's 1/3 rule with $n = 4$ [Eq. (6.14)]
(d) Simpson's 3/8 rule with $n = 4$ [Eq. (6.16)]
(e) For each of the numerical estimates in (b) through (d), determine the percent relative error based on the exact solution (a).

7 Solving Differential Equations

Chapter Outline

Differential equations are mathematical descriptions of how the variables and their derivatives (rates of change) with respect to one or more independent variable affect each other in a dynamical way. Their solutions show us how the dependent variable(s) will change with the independent variable(s). Many problems in natural sciences and engineering fields are formulated into a scalar differential equation or a vector differential equation—that is, a system of differential equations.

In almost all cases, a differential equation will be chosen such that it has an analytical solution. This will give us the chance to draw a comparison between a true (analytical) and an approximate (numerical) answer. This chapter looks into several methods of obtaining the numerical solutions to ordinary differential equations (ODEs) in which all dependent variables (y) depend on a single independent variable (t). First, the initial value problems (IVPs) are handled with several methods including the Runge–Kutta method. The final section (Sec. 7.7) introduces the shooting method for solving the two-point boundary value problem (BVP). Ordinary differential equations are called an IVP if the values $y(t_0)$ of dependent variables are given at the initial point t_0 of the independent variable; they are called a BVP if the values $y(t_0)/y(t_f)$ are given at the initial/final points t_0 and t_f. *Regarding the solution of partial differential equations (PDEs), please refer to Secs. 10.7 and 10.8.*

7.1 Euler's Explicit Method

When it comes to the numerical solution of ODEs, everyone begins with Euler's method because it is easy to understand and simple to program. Even though its low accuracy keeps it from being widely used for solving ODEs, it gives us an idea of the basic concept of numerical solution for a differential equation simply and clearly. Let us consider a first-order differential equation:

$$\frac{dy}{dt} = f(y,t) \qquad y(0) = y_0 \tag{7.1}$$

Using the formal definition, we can write the derivative as:

$$\frac{dy}{dt} = \lim_{\Delta t \to 0} \frac{y(t+\Delta t) - y(t)}{\Delta t} \tag{7.2}$$

As Δt approaches zero, then Eq. (7.2) cools down (reduces to) to the meaning of the derivative. If Δt is chosen such that it is sufficiently small and if we denote $y(t+\Delta t)$ by y_{n+1} and $y(t)$ by y_n, then Eq. (7.2) can be written as:

$$\left.\frac{dy}{dt}\right|_{t=t_n} \cong \frac{y_{n+1} - y_n}{\Delta t} \tag{7.3}$$

From Eq. (7.1)

$$\left.\frac{dy}{dt}\right|_{t=t_n} = f(y,t)\big|_{t=t_n} = f(y_n, t_n) \tag{7.4}$$

Equating both equations, we get:

$$\frac{y_{n+1} - y_n}{\Delta t} \cong f(y_n, t_n) \tag{7.5}$$

Rearrangement of Eq. (7.5) yields:

$$y_{n+1} = y_n + f(y_n, t_n)\Delta t \tag{7.6}$$

If we use the Taylor series expansion for y_{n+1}:

$$y_{n+1} = y_n + \left.\frac{dy}{dt}\right|_{t=t_n} \Delta t + \left.\frac{d^2 y}{dt^2}\right|_{t=t_n} \Delta t^2 + \cdots + \left.\frac{d^n y}{dt^n}\right|_{t=t_n} \Delta t^n \tag{7.7}$$

Moreover, if we truncate all terms containing second- or higher-order derivatives and replace the first derivative by its equivalent from Eq. (7.4), then:

$$y_{n+1} = y_n + \left.\frac{dy}{dt}\right|_{t=t_n} \Delta t = y_n + f(y_n, t_n)\Delta t \tag{7.8}$$

which means that Euler's approach is essentially an approximate solution, like using a truncated Taylor's series expansion. Equation (7.6) or (7.8) is called an explicit Euler's method. "Explicit" here indicates that that the unknown term, that is, y_{n+1}, is explicitly (on the left-hand side) expressed in terms of known quantities (i.e., the old values being evaluated at both t_n and y_n) on the right-hand side.

Example 7.1 Consider the differential equation:

$$\frac{dy}{dt} = y \qquad \text{with } y(0) = 1 \tag{7.9}$$

to be integrated from 0 to 0.5.

This equation has an analytical solution of the form:

$$ln(y) = t + c \qquad y = Ce^t$$

Applying the initial condition yields: $y = e^t$.

Let us pick up a value of $\Delta t = 0.1$ and carry out five successive iterations which will demonstrate how to implement Euler's explicit method. Finally, a comparison will be conducted between the approximate and analytical solutions.

The *startup* step:

$$y_{t=0} = y(0) = 1$$

The *explicit propagation steps*:

$$y_{t=0.1} = y_{t=0} + f(y_0,t_0)\Delta t = 1 + y_0\Delta t = 1 + 1\times 0.1 = 1.1$$

$$y_{t=0.2} = y_{t=0.1} + f(y_{0.1},t_{0.1})\Delta t = 1.1 + y_{0.1}\Delta t = 1.1 + 1.1\times 0.1 = 1.21$$

$$y_2 = y_1 + y_1\Delta t = y_1(1+\Delta t)$$

$$y_{t=0.3} = y_{t=0.2} + f(y_{0.2},t_{0.2})\Delta t = 1.21 + y_{0.2}\Delta t = 1.21 + 1.21\times 0.1 = 1.331$$

$$y_3 = y_2 + y_2\Delta t = y_2(1+\Delta t) = y_1(1+\Delta t)(1+\Delta t) = y_1(1+\Delta t)^2$$

$$y_{t=0.4} = y_{t=0.3} + f(y_{0.3},t_{0.3})\Delta t = 1.331 + y_{0.3}\Delta t = 1.331 + 1.331\times 0.1 = 1.4641$$

$$y_4 = y_3 + y_3\Delta t = y_3(1+\Delta t) = y_1(1+\Delta t)^3$$

The *termination step*:

$$y_{t=0.5} = y_{t=0.4} + f(y_{0.4},t_{0.4})\Delta t = 1.4641 + y_{0.4}\Delta t = 1.4641 + 1.4641\times 0.1 = 1.6105$$

Numerical solution:
$$y_n = y_1(1+\Delta t)^{n-1} \tag{7.10}$$

The *exact* solution:

$$y_n = y_1 e^{(n-1)(\Delta t)} = y_1\left(1+(\Delta t)+\frac{(\Delta t)^2}{2!}+\cdots+\frac{(\Delta t)^n}{n!}\right)^{n-1} \tag{7.11a}$$

For $\Delta t \approx 0$:
$$y_n = y_1 e^{(n-1)(\Delta t)} \cong y_1(1+(\Delta t))^{n-1} \tag{7.11b}$$

In words, for a *sufficiently small value* of Δt, the term that contains Δt^2 or a higher power of Δt will be even smaller and smaller, and the numerical solution will be very close to the exact.

The percent relative error (PRE) is defined as:

$$PRE = \frac{|Analytic\ value - Numerical\ value|}{|Analytic\ value|}\times 100\% \tag{7.12}$$

$$PRE = \frac{\left(1+\Delta t+\frac{\Delta t^2}{2!}+\cdots+\frac{\Delta t^n}{n!}\right)^{n-1} - (1+\Delta t)^{n-1}}{\left(1+\Delta t+\frac{\Delta t^2}{2!}+\cdots+\frac{\Delta t^n}{n!}\right)^{n-1}} = \left[1-\frac{(1+\Delta t)^{n-1}}{\left(1+\Delta t+\frac{\Delta t^2}{2!}+\cdots+\frac{\Delta t^n}{n!}\right)^{n-1}}\right]\times 100\% \tag{7.13}$$

Obviously, the PRE diminishes when *a value of* Δt *is sufficiently small.* Hence, the smaller the Δt, the higher the accuracy of the approximate solution. On the other hand, the number of steps and function evaluations will increase as Δt is reduced, which means a longer computational (CPU) time. There is a trade-off between these two extremities. Table 7.1 compares Euler's explicit method with the analytical method at each step of integration, using the PRE as the indicator of goodness.

Table 7.1 The solution for $\frac{dy}{dt} = y$ using Euler's explicit and the analytical method.

Step	t	y	Euler's Explicit Method	Analytical Method $y = e^t$	PRE (%)
0	0	$y_{t=0} = y(0) = 1$	1	1.0000000	0.000
1	0.1	$y_{t=0.1}$	1.1	1.1051709	0.468
2	0.2	$y_{t=0.2}$	1.21	1.2214027	0.934
3	0.3	$y_{t=0.3}$	1.331	1.3498588	1.397
4	0.4	$y_{t=0.4}$	1.4641	1.4918247	1.858
5	0.5	$y_{t=0.5}$	1.6105	1.6487213	2.318

PRE is the percent relative error for the numerically calculated value.

From Table 7.1, one may conclude that the numerical solution is very close to the analytical, with the maximum PRE less than 2.4%.

7.2 Euler's Implicit Method

To demonstrate Euler's implicit method we will use the same ODE used in Example 7.1.

Example 7.2 Consider the differential equation:

$$\frac{dy}{dt} = y \qquad \text{with } y(0) = 1$$

to be integrated from 0 to 0.5.

After applying the initial condition, the solution is: $y = e^t$.

Let us pick up a value of $\Delta t = 0.1$ and carry out five successive iterations which will demonstrate how to implement *Euler's implicit method.* Finally, a comparison will be conducted between the approximate and analytical solutions.

The *startup step*:

$$y_{t=0} = y(0) = 1$$

The *implicit propagation steps*:

$$y_1 = y_{t=0.1} = y_{t=0} + f(y_1, t_{0.1})\Delta t = 1 + y_1 \Delta t$$

$$\Rightarrow y_1(1-\Delta t) = 1 \Rightarrow y_1 = \frac{1}{(1-\Delta t)} = \frac{1}{0.9} = 1.111$$

$$y_2 = y_{t=0.2} = y_{t=0.1} + f(y_2, t_{0.2})\Delta t = y_1 + y_2 \Delta t$$

$$\Rightarrow y_2(1-\Delta t) = y_1 \Rightarrow y_2 = \frac{y_1}{(1-\Delta t)} = \frac{1.111}{0.9} = 1.23457$$

$$y_3 = y_{t=0.3} = y_{t=0.2} + f(y_3, t_{0.3})\Delta t = y_2 + y_3 \Delta t$$

$$\Rightarrow y_3(1-\Delta t) = y_2 \Rightarrow y_3 = \frac{y_2}{(1-\Delta t)} = \frac{y_1}{(1-\Delta t)^2} = \frac{1.111}{0.9^2} = 1.3716$$

$$y_4 = \frac{y_3}{(1-\Delta t)} = \frac{y_1}{(1-\Delta t)^3} = \frac{1.111}{0.9^3} = 1.5240$$

$$y_5 = \frac{y_4}{(1-\Delta t)} = \frac{y_1}{(1-\Delta t)^4} = \frac{1.111}{0.9^4} = 1.6933$$

$$y_n = \frac{y_{n-1}}{(1-\Delta t)} = \frac{y_1}{(1-\Delta t)^{n-1}} \tag{7.14}$$

The percent relative error (PRE) is defined as:

$$PRE = \frac{|Analytic\ \ value - Numerical\ \ value|}{|Analytic\ \ value|} \times 100\%$$

$$PRE = \frac{\left(1+\Delta t+\frac{\Delta t^2}{2!}+\cdots+\frac{\Delta t^n}{n!}\right)^{n-1} - \frac{1}{(1-\Delta t)^{n-1}}}{\left(1+\Delta t+\frac{\Delta t^2}{2!}+\cdots+\frac{\Delta t^n}{n!}\right)^{n-1}} = \left[1 - \frac{(1-\Delta t)^{1-n}}{\left(1+\Delta t+\frac{\Delta t^2}{2!}+\cdots+\frac{\Delta t^n}{n!}\right)^{n-1}}\right] \times 100\% \tag{7.15}$$

Table 7.2 compares Euler's implicit with the analytical method at each step of integration, using the PRE as the indicator of goodness.

Table 7.2 The solution for $\frac{dy}{dt} = y$ using Euler's implicit and the analytical method.

Step	t	y	Euler's Implicit Method	Analytical Method $y = e^t$	PRE (%)
0	0	$y_{t=0} = y(0) = 1$	1	1.0000000	0.000
1	0.1	$y_{t=0.1}$	1.1111	1.1051709	0.536
2	0.2	$y_{t=0.2}$	1.2346	1.2214027	1.080
3	0.3	$y_{t=0.3}$	1.3716	1.3498588	1.611
4	0.4	$y_{t=0.4}$	1.5240	1.4918247	2.157
5	0.5	$y_{t=0.5}$	1.6933	1.6487213	2.704

PRE is the percent relative error for the numerically calculated value.

NOTE

1. For a given method, PRE depends on both Δt and t itself. Euler's explicit method can be superior to Euler's implicit method at low values of t and becomes inferior at high values of t. The same is true with Δt.
2. Using the *implicit* method may result in a nonlinear algebraic equation that requires an iterative (or trial and error) technique to solve it.
3. Generally speaking, for both of Euler's numerical methods, the larger Δt is, the larger will be the deviation from the exact solution. Depending on the right-hand side of ODE, certain Δt values may blow up the solution (i.e., the solution becomes numerically unstable). For example, taking $\Delta t = 1$ makes the solution nonrealistic, as given by the implicit method [Eq. (7.14)], whereas the solution given by the explicit method [Eq. (7.10)] in this case does not suffer from numerical instability.

Exercise 7.1 Consider the differential equation:

$$\frac{dy}{dt} = 1 - y \qquad \text{with } y(0) = 0$$

to be integrated from 0 to 4. Try different values of Δt (i.e., $\Delta t = 0.5, 1$, and 2) and see how both methods behave with respect to the step size Δt. The analytical solution is: $y = 1 - e^{-t}$.

7.3 Runge–Kutta (R–K) Method

We show here the idea behind Runge–Kutta (fourth-order) approximation for solving an ODE like Eq. (7.1):

$$\frac{dy}{dt} = f(y,t)$$

$$t_{n+1} = t_n + \Delta t$$

$$k_1 = f(y_n, t_n)\Delta t$$

$$k_2 = f\left(y_n + \frac{1}{2}k_1, t_n + \frac{1}{2}\Delta t\right)\Delta t$$

$$k_3 = f\left(y_n + \frac{1}{2}k_2, t_n + \frac{1}{2}\Delta t\right)\Delta t$$

$$k_4 = f\left(y_n + k_3, t_n + \Delta t\right)\Delta t$$

$$y_{n+1} = y_n + \frac{1}{6}(k_1 + 2k_2 + 2k_3 + k_4)$$

Example 7.3 Let us carry out the first step with $\Delta t = 0.1$ for ODE in Example 7.1:

$$\frac{dy}{dt} = y \qquad y(0) = 1$$

$$k_1 = f(y_n, t_n)\Delta t = 1 \times \Delta t = 1 \times 0.1 = 0.1$$

$$k_2 = f\left(y_n + \frac{1}{2}k_1, t_n + \frac{1}{2}\Delta t\right)\Delta t = \left(1 + \frac{1}{2}k_1\right)\Delta t = (1.05) \times 0.1 = 0.105$$

$$k_3 = f\left(y_n + \frac{1}{2}k_2, t_n + \frac{1}{2}\Delta t\right)\Delta t = \left(1 + \frac{1}{2}k_2\right)\Delta t = \left(1 + \frac{1}{2} \times 0.105\right)0.1 = 0.10525$$

$$k_4 = f(y_n + k_3, t_n + \Delta t)\Delta t = (1 + k_3)\Delta t = (1 + 0.10525)\Delta t = 0.110525$$

$$y_{t=0.1} = y_{t=0} + \frac{1}{6}(k_1 + 2k_2 + 2k_3 + k_4)$$

$$y_{t=0.1} = 1 + \frac{1}{6}(0.1 + 2 \times 0.105 + 2 \times 0.10525 + 0.110525) = 1.051708$$

This value is extremely close to that of the analytical, that is, 1.1051709 (from Table 7.1 or 7.2).

Obviously, the R–K method is very accurate. It is the choice to be plugged into computer packages like MATLAB. Nevertheless, it is not suitable for manual calculations because one jump costs us four function evaluations and one summation step. It should be pointed out that the CPU time required to run Euler with a step size Δt would be the same as the time required to run R–K with a step size of $4 \times \Delta t$, with the accuracy of R–K being a slightly higher under the aforementioned condition.

Let us show how to use the MATLAB ODE solver. MATLAB's intrinsic functions **ode23** and **ode45** exploit Runge–Kutta techniques. These routines are more accurate than the Euler schemes, explained in the preceding sections. In order to use the MATLAB ODE solver, it is necessary to write a code that evaluates the right-hand side of the equation. **ode23, ode45, ode113, ode15s, ode23s, ode23t,** and **ode23tb** can be used to solve initial value problems for ordinary differential equations.

```
Syntax
>>[T,Y] = solver(odefun,tspan,y0)
>>[T,Y] = solver(odefun,tspan,y0,options)
>>[T,Y,TE,YE,IE] = solver(odefun,tspan,y0,options)
>>sol = solver(odefun,[t0 tf],y0...)
```

where *solver*, one of **ode45, ode23, ode113, ode15s, ode23s, ode23t,** or **ode23tb**, is chosen based on the following matrix shown in Table 7.3.

Table 7.3 Selection of the ODE solver is made based on problem type and order of accuracy.

Solver	Problem Type	Order of Accuracy	Method
ode45	Nonstiff differential equations	Medium	Runge–Kutta
ode23	Nonstiff differential equations	Low	Runge–Kutta
ode113	Nonstiff differential equations	Low to high	Adams
ode15s	Stiff diff. eq. and Diff. Algebraic Equations (DAEs)	Low to medium	NDFs (BDFs)
ode23s	Stiff differential equations	Low	Rosenbrock
ode23t	Moderately stiff diff. eq. and DAEs	Low	Trapezoidal rule
ode23tb	Stiff differential equations	Low	TR-BDF2

In mathematics, a stiff equation is a differential equation for which certain numerical methods for solving the equation are numerically unstable, unless the step size is taken to be extremely small. The main idea is that the equation includes some terms that can blow up the solution (i.e., the values of the dependent variable become infeasible or physically non-sense).

When integrating a differential equation numerically, one would expect the suggested step size to be relatively small in a region in which the solution curve displays much variation and to be relatively large where the solution curve straightens out to approach a line with a slope of nearly zero. Unfortunately, this is not always the case. The ODEs that make up the mathematical models of most chemical engineering systems usually represent a collection of fast and slow dynamics. For instance, in a typical distillation tower, the liquid mechanics (e.g., flow, hold-up) is considered as fast dynamics (time constant ≈ seconds), compared with the tray composition slow dynamics (time constant ≈ minutes). Systems with such a collection of fast and slow ODEs are denoted *stiff systems*.

NOTE As a rule of thumb, try first the recommended MATLAB solvers for nonstiff problems. If they do not work properly (i.e., the computer freezes or erroneous results are produced) then move to solvers recommended for stiff cases.

Example 7.4 Solve the same differential equation, used in previous examples:

$$\frac{dy}{dt} = y \qquad \text{with} \qquad y(0) = 1$$

then to be integrated from 0 to 0.5 using the MATLAB ODE solver.

First, let us define the ODE via the following function M-file, called **myode.m** (Fig. 7.1):

```
function dy = myode(t,y)
% This defines the function myode which has two input arguments which are:
% t and y and the output is the derivative of y, dy.
% Since we have only a single ode, we initialize dy as 1 by 1 matrix (i.e.,
% a column vector)dy = zeros(1,1);      % a column vector
dy = zeros(1,1);
dy(1)  = y; % Defining dy/dt
```

Figure 7.1 M-file that defines the **myode** function (i.e., the derivative *dy/dt*).

Second, define the "options" attribute, which is a structure of optional parameters that change the default integration properties.

```
>> options = odeset('RelTol',1e-4,'AbsTol',1e-4);
```

Third, invoke the proper solver; the default is **ode45**. There is a function handle (i.e., @myode) or function reference to myode which defines the derivative of *y*, *dy/dt*. The initial t_0 and t_f are defined within the matrix [0 0.5]. The initial condition, $y(0) = 1$, is defined via the matrix [1].

```
>>[T,Y] = ode45(@myode,[0 0.5],[1],options);
```

The output will be two vectors: one for *t* and another one for *y* (i.e., the solution *y* at the given *t*).

Fourth, you can inspect the numerical value of *y* at a given *t*. See Table 7.4 for ode45 results. Notice, however, that the MATLAB ode45 solver used $\Delta t = 0.0125$ not $\Delta t = 0.1$.

Table 7.4 The solution for $\frac{dy}{dt} = y$ using the MATLAB ode45 solver and the analytical.

Step	t	y	MATLAB's ode45	Analytical Method $y = e^t$	PRE (%)
0	0	$y_{t=0} = y(0) = 1$	1	1.0000000	0.000
8	0.1	$y_{t=0.1}$	1.1051709	1.1051709	0.000
16	0.2	$y_{t=0.2}$	1.2214027	1.2214027	0.000
24	0.3	$y_{t=0.3}$	1.3498588	1.3498588	0.000
32	0.4	$y_{t=0.4}$	1.4918247	1.4918247	0.000
40	0.5	$y_{t=0.5}$	1.6487213	1.6487213	0.000

PRE is the percent relative error for the numerically calculated value. Notice that $\Delta t = 0.0125$; hence, not all y values are shown here. Only those that go with the given t are shown.

Fifth, you may plot Y vs. T via the following command:

```
>> plot(T,Y'k-','LineWidth',2)
>> ylabel('\bf y value'),xlabel('\bf time, t');
```

This will generate a plot as shown in Fig. 7.2.

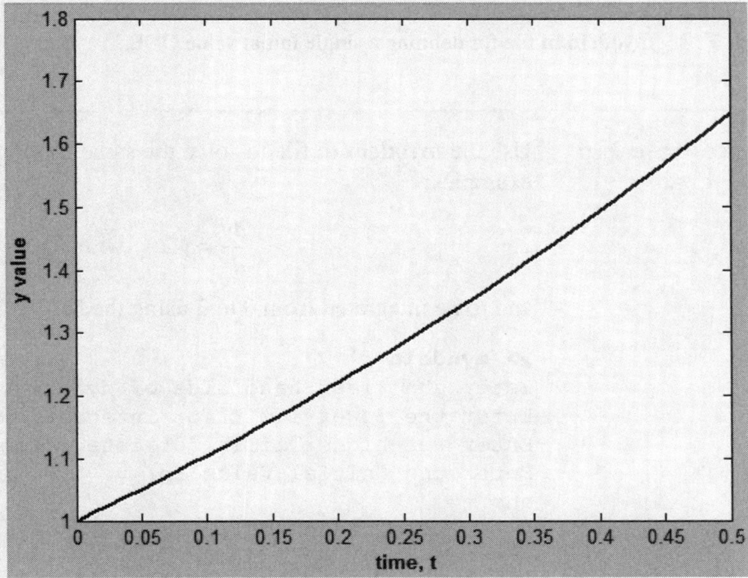

Figure 7.2 Plot of y vs. t as a solution for $dy/dt = y$ using MATLAB **ode45** solver with $\Delta t = 0.0125$.

Figure 7.3 shows the **myodein.m** file that can be used to solve a single initial value ODE between the lower and upper limits of integration.

```
function dy = myodein
% This defines the function myode which has two input arguments which are:
% t and y and the output is the derivative of y, dy.
% Since we have only a single ode, we initialize dy as 1 by 1 matrix (i.e.,
% a column vector)dy = zeros(1,1); % a column vector
clear all;;
ytemp2=input('Enter the right hand side of dy/dt: ', 's');
ytempi=strrep(ytemp2,'*','.*');
ytempj=strrep(ytempi,'/','./');
ytemp3=strrep(ytempj,'^','.^');
ytemp4=inline(ytemp3,'t','y');
dy(1) = ytemp4; % Defining dy/dt=y
options = odeset('RelTol',1e-4,'AbsTol',1e-4);
lower=input('Enter the lower limit for integration: ');
upper=input('Enter the higher limit for integration: ');
initialval=input('Enter the initial value for y: ');
[T,Y] = ode45(dy,[lower upper],[initialval],options);
plot(T,Y,'k-','LineWidth',2);
get(gcf,'CurrentAxes');
h=gca;
set(h,'YGrid','on');
ylabel('\fontsize{14} \bf y value'),xlabel('\fontsize{14} \bf time, t');
```

Figure 7.3 **myodein.m** file for defining a single initial value ODE.

Example 7.5 Use the **myodein.m** file to solve the same differential equation, used in previous examples:

$$\frac{dy}{dt} = y \qquad \text{with} \qquad y(0) = 1$$

and to be integrated from 0 to 3 using the MATLAB ode45 solver.

```
>> myodein
Enter the right hand side of dy/dt: y
Enter the lower limit for integration: 0.0
Enter the higher limit for integration: 3.0
Enter the initial value for y: 1.0
ans =
     Inline function:
     ans(t,y) = y
```

The plot command in the **myodein.m** file (Fig. 7.3) will generate y vs. t, as shown in Fig. 7.4.

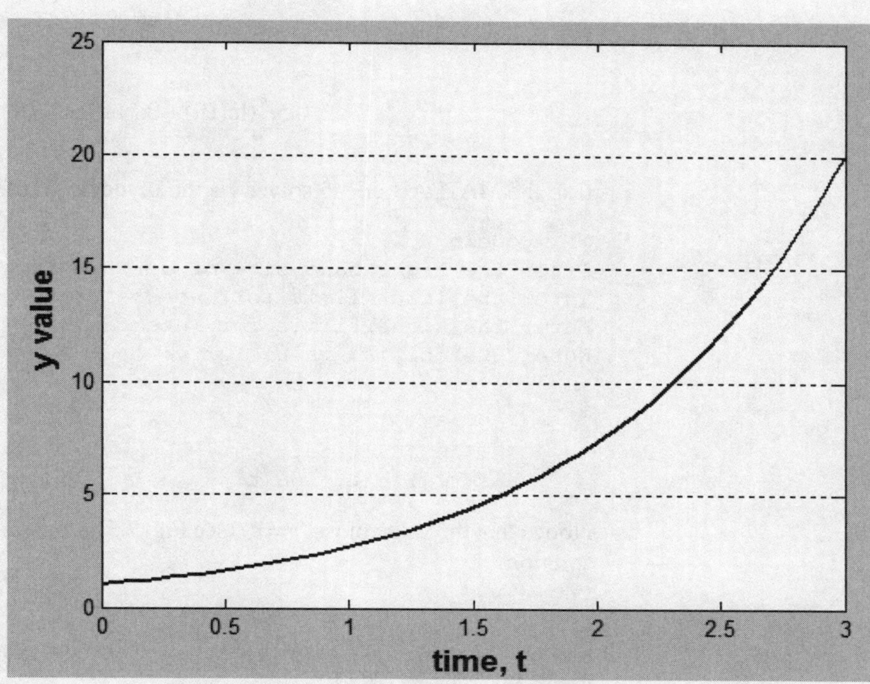

Figure 7.4 The plot of y vs. t as a solution for $dy/dt = y$ with $y(0) = 1$ using MATLAB's ode45 solver.

Example 7.6 Use the **myodein.m** file to solve the differential equation:

$$\frac{dy}{dt} = tln(t) \qquad \text{with} \qquad y(1) = 1 \tag{7.16}$$

and to be integrated from 1 to 10 using the MATLAB ode solver.

The analytical solution for this ODE:

$$y = \int tln(t)\,dt + C = \frac{t^2}{2} \times \ln(t) - \int 0.5 \times t^2 \times \left(\frac{1}{t}\right) dt + C = 0.5t^2 \ln(t) - 0.25t^2 + C$$

$$y = 0.5t^2(\ln(t) - 0.5) + C$$

Applying the initial condition yields:

$$y = 0.5t^2(\ln(t) - 0.5) + 1.25 \tag{7.17}$$

For example, at $t = 10$,

$$y = 0.5t^2(\ln(10) - 0.5) + 1.25 = 91.38$$

Call the MATLAB **ode45** solver via the **myodein.m** file:

```
>> myodein
Enter the right hand side of dy/dt: t*log(t)
Enter the lower limit for integration: 1.0
Enter the higher limit for integration: 10.0
Enter the initial value for y: 1.0

ans =

    Inline function:
    ans(t,y) = t.*log(t)
```

Notice that the value of y at $t = 10$ (see Fig. 7.5) is the same as that of the analytical solution.

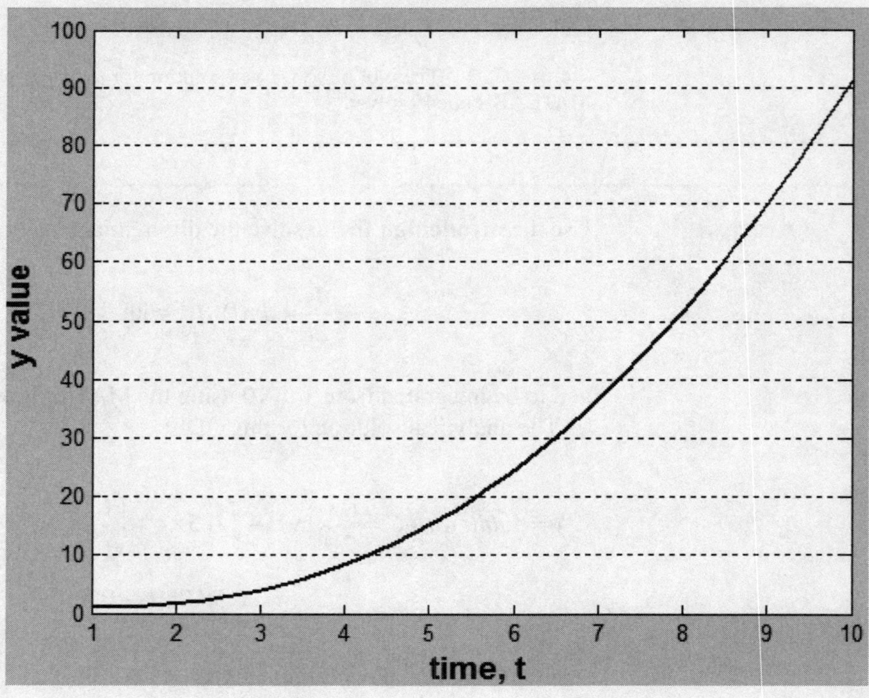

Figure 7.5 The plot of y vs. t as a solution for $\frac{dy}{dt} = t\ln(t)$ with $y(1) = 1$ using MATLAB **ode45** solver.

7.4 Solution of a Set of Ordinary Differential Equations

Example 7.7 Consider the following set of ordinary differential equations:

$$\frac{dy_1}{dt} = 2y_2y_3 \qquad y_1(0) = 0 \tag{7.18a}$$

$$\frac{dy_2}{dt} = -3y_1y_3 \qquad y_2(0) = 1 \tag{7.18b}$$

$$\frac{dy_3}{dt} = -2.5 \times y_2y_3 \qquad y_3(0) = 1 \tag{7.18c}$$

First, we need to define the derivatives of *y* via an M-file called **multiode.m**, as shown in Fig. 7.6.

```
function [dy] = multiode(t,y)
% This defines the function myode which has two input arguments which are:
% t and y and the output is the derivative of y, dy.
% Since we have three ode's, we initialize dy as 3 by 1 matrix (i.e.,
% a column vector)dy = zeros(3,1)
dy = zeros(3,1);
dy(1) = 2.0* y(2) * y(3);
dy(2) = -3.0* y(1) * y(3);
dy(3) = -2.50 * y(2) * y(3);
```

Figure 7.6 **multiode** function M-file that defines the derivative of each *y*.

Second, create **callmultiode.m** (a script M-file) that contains the initial value for each *y*; the lower and upper limit for integration; the "options" attributes, which is a structure of optional parameters that change the default integration properties; a call to the MATLAB ode45 solver; and a plot command for *y* as a function of *t*. Figure 7.7 embodies such a list of commands.

```
clear all;
for index=1:3
myprompt2=['Enter the initial value for y:(',num2str(index),'): '];
initialval(index)=input(myprompt2);
end
lower=input('Enter the lower limit for integration: ');
upper=input('Enter the higher limit for integration: ');
options = odeset('RelTol',1e-4,'AbsTol',1e-4);
```

Figure 7.7 **callmultiode.m** (script M-file) that defines integration-related parameters and calls the MATLAB ode45 solver using a function handle (i.e., @multiode) to multiode function.

```
[T,Y] = ode45(@multiode,[lower upper],initialval,options);
tf=T(end)
Y1_at_tf=Y(end,1)
Y2_at_tf=Y(end,2)
Y3_at_tf=Y(end,3)
plot(T,Y(:,1),'k-',T,Y(:,2),'k-.', ...
T,Y(:,3),'k.','LineWidth',2)
ylabel('\bf y value'),xlabel('\bf time, t');
hleg = legend('y1','y2','y3','Location','Best');
```

Figure 7.7 (*Continued*).

```
>> callmultiode
Enter the initial value for y:(1): 0
Enter the initial value for y:(2): 1
Enter the initial value for y:(3): 1
Enter the lower limit for integration: 0
Enter the higher limit for integration: 10
```

Figure 7.8 shows the output, that is, the plot of *y* vs. *t* for the three *y*s.

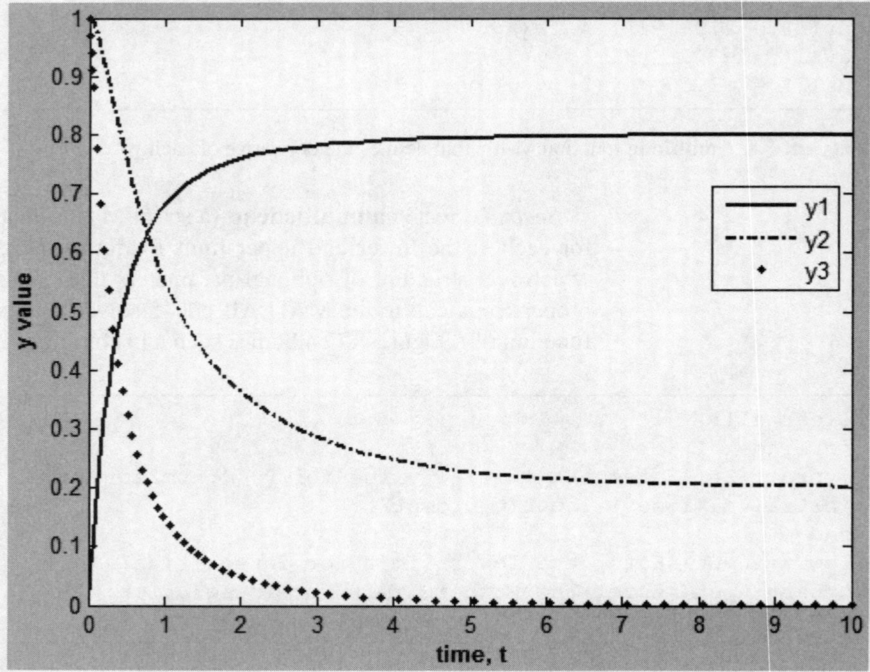

Figure 7.8 The solution for a set of three ODEs using the MATLAB **ode45** solver.

7.5 Solution of a Higher-Order Ordinary Differential Equation

Example 7.8 Consider the following differential equation:

$$\frac{d^2y}{dt^2} - 5\frac{dy}{dt} + 6y = 0.0 \tag{7.19a}$$

$$y(0) = 0 \tag{7.19b}$$

$$y'(0) = 0.078762 \tag{7.19c}$$

This has a characteristic equation:

$$D^2 - 5D + 6 = 0 \Rightarrow (D-3)(D-2) = 0$$

which has two real distinct roots. Thus the solution will be of the form:

$$y = C_1 e^{3t} + C_2 e^{2t}$$

Applying the first boundary condition [Eq. (7.19*b*)]:

$$0 = C_1(1) + C_2(1) \Rightarrow C_2 = -C_1$$

Thus the solution reduces to:

$$y = C_1 e^{3t} - C_1 e^{2t} = C_1(e^{3t} - e^{2t})$$

Applying the second boundary condition (7.19*c*):

$$0.078762 = y'\big|_{t=0} = C_1(3e^{3t} - 2e^{2t})\big|_{t=0} \Rightarrow C_1 = 0.078762$$

The final (analytical) solution will be:

$$y_{anal} = 0.078762(e^{3t} - e^{2t}) \tag{7.20}$$

Let us numerically solve the aforementioned ode using the **ode45** solver.

First, the *second-order* ode will be transformed into a set of two *first-order* ODEs via the following transformation:

$$y_1 = y$$

$$y_2 = \frac{dy_1}{dt} = \frac{dy}{dt} \Rightarrow \frac{dy_2}{dt} = \frac{d^2y}{dt^2}$$

Thus, Eq. (7.19*a*–*c*) becomes

$$\frac{dy_2}{dt} - 5y_2 + 6y_1 = 0 \Rightarrow \frac{dy_2}{dt} = 5y_2 - 6y_1 \tag{7.21a}$$

$$\frac{dy_1}{dt} = y_2 \tag{7.21b}$$

$$y_1(0) = 0 \tag{7.21c}$$

$$y'(0) = y_2(0) = 0.078762 \tag{7.21d}$$

Thus Eq. (7.19*a*) was transformed into a set of two first-order ODEs coupled with two initial values. To solve such a set of ODEs, see Example 7.7 in Sec. 7.4.

Figure 7.9 shows **example7_8.m** file that defines the set of two odes [Eqs. (7.21*a*–*b*)].

```
function [dy] = example7_8(t,y)
% This defines the function myode which has two input arguments which are:
% t and y and the output is the derivative of y, dy.
% Since we have three ode's, we initialize dy as 3 by 1 matrix (i.e.,
% a column vector)dy = zeros(3,1);
dy = zeros(2,1);
dy(1) = y(2);
dy(2) = 5.0* y(2) -6.0*y(1);
```

Figure 7.9 **example7_8.m** file defining the set of two odes [Eqs. (7.20, 7.21)].

Figure 7.10 shows the **callmultiode7_8.m** file that calls upon MATLAB's **ode45** solver and makes a plot of the variables *y*'s vs. *t*.

```
clear all;
for index=1:2
myprompt2=['Enter the initial value for y:(',num2str(index),'): '];
initialval(index)=input(myprompt2);
end
lower=input('Enter the lower limit for integration: ');
upper=input('Enter the higher limit for integration: ');
options = odeset('RelTol',1e-4,'AbsTol',1e-4);
[T,Y] = ode45(@example7_8,[lower upper],initialval,options);
Yanal=0.078762*(exp(3.0.*T)-exp(2.0.*T));
display(['tfinal: ',num2str(T(end)),' yfinal: ',num2str(Y(end,1))]);
plot(T,Y(:,1),'k*',T,Y(:,2),'k.',T,Yanal,'k-','LineWidth',2)
hleg = legend('y1','y2','Yanal','Location','Best');
ylabel('\bf y value'),xlabel('\bf time, t');
```

Figure 7.10 **callmultiode7_8.m** file that defines integration-related parameters and calls the MATLAB **ode45** solver using a function handle (i.e., @example7_8) to **example7_8** function.

```
>> callmultiode7_8
Enter the initial value for y:(1): 0
Enter the initial value for y:(2): 0.078762
Enter the lower limit for integration: 0
Enter the higher limit for integration: 1

tfinal: 1 yfinal: 1
```

Figure 7.11 shows the plot of y_1, y_2, and y_{anal} as a function of t. Notice how y_I exactly matches with y_{anal} [Eq. (7.20)].

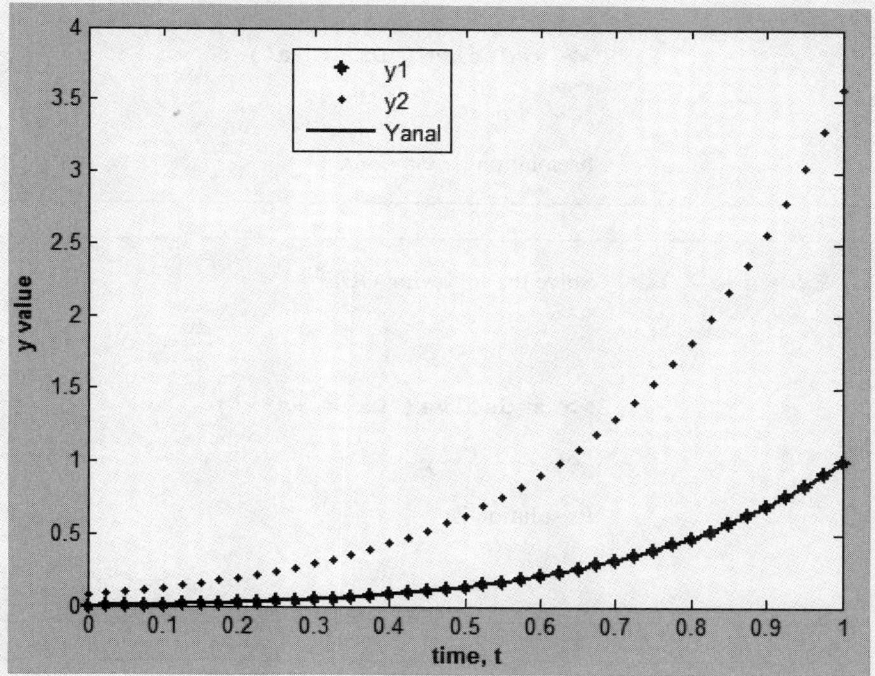

Figure 7.11 The plot of y_1 , y_2, and y_{anal} as a function of t. Notice how y_1 exactly matches with y_{anal} [Eq. (7.20)]. y_1 and y_2 were obtained via ode45 solver.

7.6 ODE and System Solver Using MATLAB dsolve

MATLAB can solve ordinary differential equations symbolically with or without boundary conditions or initial value parameters. The '**dsolve**' command is used for this purpose. Within **dsolve**, the letters '***Dij***' are used to indicate a derivative, where ***i*** is the order of differentiation and ***j*** is the dependent variable. '***D***' implicitly specifies a first-order derivative, '***D2***' signifies a second-order derivative, and so on. The letter *t* is the default independent variable for **dsolve**. So, *D2y* is analogous to $\frac{d^2y}{dt^2}$.

```
>>S = dsolve(eq,cond,var)
```

solves the ordinary differential equation **eq** with the initial or boundary condition **cond** with respect to the variable **var**. If you do not specify **var**, the solver uses the default variable **t**.

Example 7.9 Solve the following ODE:

$$\frac{dx}{dt} = -a$$

```
>> x=dsolve('Dx = -a')
x =
 C4 - a*t
```

Its solution is: $x = C_1 - at$.

Example 7.10 Solve the following ODE:

$$\frac{dx}{dt} = -ax$$

```
>> x=dsolve('Dx = -a*x')
x =
 C2/exp(a*t)
```

Its solution is:

$$x = Ce^{-at} = C/e^{at}$$

Example 7.11 Solve the following ODE:

$$\frac{dy}{dt} = ay \qquad y(0) = b$$

```
>> y=dsolve('Dy = a*y','y(0)=b')
y =
b*exp(a*t)
```

Its solution is:

$$y = be^{at}.$$

Example 7.12 Solve the following ODE:

$$\frac{d^2y}{dt^2}=6y-\frac{dy}{dt} \quad \leftrightarrow \quad \frac{d^2y}{dt^2}+\frac{dy}{dt}-6y=0$$

```
>> y=dsolve('D2y=6*y-Dy','x')
y =
C4*exp(2*x) + C5/exp(3*x)
```

To solve it analytically, we use the characteristic equation: $(D^2+D-6)y=0 \rightarrow (D+3)(D-2)=0 \rightarrow D=-3, D=2$. Hence, the solution takes the form: $y=C_Ie^{2x}+C_{II}e^{-3x}$.

Notice that MATLAB shows the last term as $\frac{C_{II}}{e^{3x}}$, which is essentially the same.

Example 7.13 Consider the following set of three ordinary differential equations:

$$\frac{dy_1}{dt}=2y_2+y_3 \qquad y_1(0)=0 \tag{7.22a}$$

$$\frac{dy_2}{dt}=-3y_1+y_3 \qquad y_2(0)=1 \tag{7.22b}$$

$$\frac{dy_3}{dt}=-2.5\times y_1+y_2 \qquad y_3(0)=1 \tag{7.22c}$$

```
>> Z=dsolve('Dy1=2*y2+y3','Dy2=-3*y1+y3','Dy3=-2.5*y1+y2',
'y1(0)=0','y2(0)=1','y3(0)=1')
Z =
    y2: [1x1 sym]
    y1: [1x1 sym]
    y3: [1x1 sym]
```

For example, to see the expression for y1, enter:

```
>>Z.y1

ans =
(4941200*exp(t/((8^(1/2)*253^(1/2))/8 + 4)^(1/3))^(5/2))/(93
*exp(t*((8^(1/2)*253^(1/2))/8 + 4)^(1/3))* ....
```

The y_1 expression is very long and only a portion of it is shown above.

To see the result on the screen in the form of a plot, execute the following commands:

```
>> t=0:0.05:3;
>> plot(t,eval(Z.y1),'k-',t,eval(Z.y2), ...
'k-.',t,eval(Z.y3),'k.','LineWidth',2)
>> hleg = legend('y1','y2','y3','Location','Best');
>> xlabel('\bf t'); ylabel(['\bf y1,','\bf y2,','\bf y3']);
```

NOTE The solution Z is 1×1 struct that has three symbolic variables: $y1$, $y2$, and $y3$. After defining t as a row vector, the MATLAB built in **eval** function is used to evaluate the symbolic expression of y into numbers as a function t.

Figure 7.12 shows the plot of y_1, y_2, and y_3 [solution for ODEs, given by Eqs. (7.22*a*–*c*)].

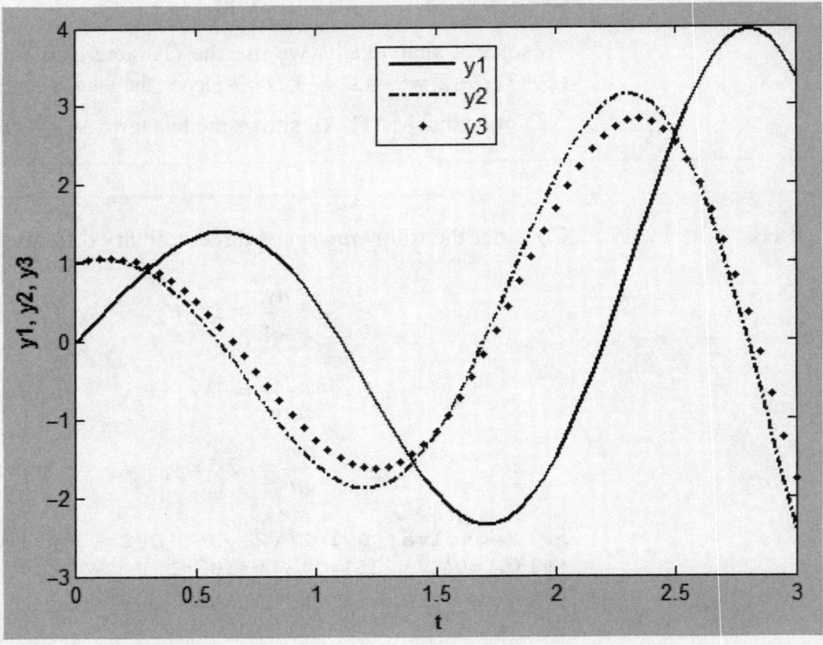

Figure 7.12 The plot of y_1, y_2, and y_3 as a function of t, which represents the solution of ODEs given by Eqs. (7.22*a*–*c*).

7.7 The Shooting Method for a Higher-Order Ordinary Differential Equation

Example 7.14 Consider the same differential equation in Example 7.8 but this time with different boundary conditions:

$$\frac{d^2y}{dt^2} - 5\frac{dy}{dt} + 6y = 0.0 \tag{7.23a}$$

$$y(0) = 0 \tag{7.23b}$$

$$y(1) = 1 \tag{7.23c}$$

This has a characteristic equation:

$$D^2 - 5D + 6 = 0 \Rightarrow (D-3)(D-2) = 0$$

which has two real distinct roots. Thus the solution will be of the form:

$$y = C_1 e^{3t} + C_2 e^{2t}$$

Applying the first boundary condition [Eq. (7.23*b*)]:

$$0 = C_1(1) + C_2(1) \Rightarrow C_2 = -C_1$$

Thus the solution reduces to:

$$y = C_1 e^{3t} - C_1 e^{2t} = C_1(e^{3t} - e^{2t})$$

Applying the second boundary condition (7.23*c*):

$$1 = C_1(e^3 - e^2) \Rightarrow C_1 = \frac{1}{(e^3 - e^2)}$$

The final (analytical) solution will be:

$$y = \frac{(e^{3t} - e^{2t})}{(e^3 - e^2)} = 0.078762(e^{3t} - e^{2t}) \tag{7.24}$$

Let us numerically solve the aforementioned ode using the MATLAB **ode45** solver. First, the *second-order* ODE will be transformed into a set of two *first-order* ODEs via the following transformation:

$$y_1 = y$$

$$y_2 = \frac{dy_1}{dt} = \frac{dy}{dt} \Rightarrow \frac{dy_2}{dt} = \frac{d^2y}{dt^2}$$

Thus, transformation results are:

$$\frac{dy_2}{dt} - 5y_2 + 6y_1 = 0 \Rightarrow \frac{dy_2}{dt} = 5y_2 - 6y_1 \tag{7.25a}$$

$$\frac{dy_1}{dt} = y_2 \tag{7.25b}$$

$$y_1(0) = 0 \tag{7.25c}$$

$$y_1(1) = 1 \tag{7.25d}$$

To solve such a set of ODEs, see Example 7.8 in Sec. 7.5. However, one should notice that $y'^{(0)} = y_2(0)$ is not given; thus, we cannot proceed with the solution until we get the value of $y_2(0) = y'^{(0)}$. One way to tackle this problem is as follows:

We will assume that $y_2(0) = y'^{(0)} = U$. U is an unknown quantity; however, it will be iteratively computed (i.e., its value is adjusted) such that the boundary condition given by Eq. (7.25*d*) is met.

1. Use the previous set of equations given by (7.25*a-d*), except for the last equation where it is replaced by:

$$y_2(0) = y'^{(0)} = U \qquad (7.25e)$$

MATLAB numerical ODE solvers solve only initial value problems; thus, assume that $U = 10$:

```
>> callmultiode7_8
Enter the initial value for y:(1): 0
Enter the initial value for y:(2): 10
Enter the lower limit for integration: 0
Enter the higher limit for integration: 1
```

NOTE @$t = 1.0$ $y_1 = 126.965$, which is obviously larger than 1.0, the correct value.

2. Assume that $U = 0$

```
>> callmultiode7_8
Enter the initial value for y:(1): 0
Enter the initial value for y:(2): 0.0
Enter the lower limit for integration: 0
Enter the higher limit for integration: 1
```

NOTE @$t = 1.0$ $y_1 = 0.00$, which is smaller than 1.0, the correct value.

Obviously, the correct value should lie between 0 and 10. The new guess should be made in light of how close the calculated answer is to the correct value.

3. Let us take $U = 1$

```
> callmultiode7_8
Enter the initial value for y:(1): 0
Enter the initial value for y:(2): 1
Enter the lower limit for integration: 0
Enter the higher limit for integration: 1
```

NOTE @$t = 1.0$ $y_1 = 12.696$, which is still *larger* than 1.0, the correct value.

The correct value must lie (if MATLAB does not lie) between 0 and 1. The new guess should be made in light of how close the calculated answer is to the correct value.

4. Let us take $U = 0.2$

```
>> callmultiode7_8
Enter the initial value for y:(1): 0
Enter the initial value for y:(2): 0.2
Enter the lower limit for integration: 0
Enter the higher limit for integration: 1
```

NOTE @$t = 1.0$ $y_1 = 2.5393$, which is still *larger* than 1.0, the correct value.

5. Let us take $U = 0.1$

```
>> callmultiode7_8
Enter the initial value for y:(1): 0
Enter the initial value for y:(2): 0.1
Enter the lower limit for integration: 0
Enter the higher limit for integration: 1
```

NOTE @$t = 1.0$ $y_1 = 1.269$, which is *larger* than 1.0, the correct value. You have to be patient.

So, we keep narrowing and squeezing the interval for the right answer. If you keep doing so, you will reach the right answer, that is, $U = 0.078762$. This is basically the second boundary condition in Example 7.8 [Eq. (7.19*c*)].

```
>> callmultiode7_8
Enter the initial value for y:(1): 0
Enter the initial value for y:(2): 0.078762
Enter the lower limit for integration: 0
Enter the higher limit for integration: 1
```

NOTE Notice now, with $U = 0.078762$, that @$t = 1.0$, $y_l = 1.000$, which is *exactly* 1.0, the correct value. The approach that we have adopted in this problem is called the "shooting method." It is as if you were to shoot at a target, with a given height equal to one meter (the height is dictated by the value of *y* at the final condition/boundary). This target lies far away from you (in this case, one unit of time or space, depending on the final value of *t* or *x*). You throw a rock at the target and it falls either before or after the target. You keep adjusting (tuning up) your initial power (here, initial value of U) until you exactly hit the tip of the target at the right spot (i.e., one meter above the ground, no more, no less).

7.8 Problems

NOTE It should be noted that even if the system does not explicitly vary with *t* (that is a dependence of the right-hand side on *t*) the first line of the corresponding function should include *t* in its arguments. For example, in solving the system:

$$\frac{dy}{dt} = y$$

The corresponding function would be:

```
function [value] = myfunc(t,y)
value = y;
```

7.1 The following code describes the solution for the first-order ODE using Euler's explicit method:

$$\frac{dy}{dt} = (t - t^2)y$$

Subject to $y(0) = 1$ and to be integrated between the limits 0 and 5.

The analytical solution is given by: $lny=\frac{t^2}{2}-\frac{t^3}{3}+C_1 \Rightarrow y=C_2e^{(0.5t^2-0.333t^3)}$

Applying the initial condition yields: $y=e^{(0.5t^2-0.333t^3)}$

```
% Euler's explicit method for Problem 7.1
dt = 0.1;
t = 0.0:dt:5.0;
y = zeros(size(t));
% MATLAB array indices are one-based; this will match with
t=0; i.e., y(0);
y(1) = 1;
for j=1:(length(t)-1)
y(j+1) = y(j) + dt * (t(j)-t(j)^2)*y(j);
end
exact = exp(0.5.*t.^2-0.333.*t.^3);
figure(1)
plot(t,y,'*',t,exact,'-')
```

Modify the above code to solve the equation:

$$\frac{dy}{dt}=-yt$$

Subject to the boundary condition $y(0) = 1$ from $t = 0$ to $t = 5$

You should be able to get the exact solution for this simple ODE and compare your answers. Calculate the percent relative error (PRE) for the estimated value of y at $t=5$.

7.2 Using Euler's explicit method, calculate the PRE involved in the solution of ODE:

$$\frac{dy}{dt}=yt$$

Subject to the boundary condition $y(0) = 1$ between $t = 0$ and $t = 2$, using $\Delta t = 0.50$. (You should manually carry out such calculations using either a calculator or MATLAB command prompt.) Repeat the calculations for the following ODE, using the same settings for t_0, t_f, and Δt. Compare PRE values in both cases based on the corresponding analytical solution.

$$\frac{dy}{dt}=-yt$$

7.3 Modify the code in Problem 7.1 to solve the following ODE:

$$\frac{dy}{dt}=sint-y$$

Subject to the condition that $y(0) = 2$ between $t = 0$ and $t = 12\pi$. Use $\Delta t = 0.1$, 0.5, 1.0, 1.5, 2.0, 2.5 and see how the numerical solution starts diverging from the analytical with increasing Δt. You may get the solution based on your knowledge in a differential equations course, or use MATLAB built-in **dsolve** to get the analytical solution to plug it in replacement of the *exact* definition appearing in the Problem 7.1 code.

7.4 Use the MATLAB ode45 solver to solve Problem 7.3. First, define the derivative using the function M-file named Pr7_4.m. Second, write the calling code (main code) that defines the initial value of y, the time interval, calls the MATLAB **ode45** solver, defines the exact solution from Pr. 7.3, and finally plots both the numerical and analytical solutions on one plot that shows the title of the x- and y-axes and the legend of the figure.

7.5 Use both MATLAB built-in **ode45** and **dsolve** solvers for the following ODE:

$$\frac{dy}{dt} = t^2 + y$$

from $t = 0$ to $t = 5$ subject to the boundary condition that $y(0) = 0$.

Write the calling code (main code) that defines the initial value of y, the time interval, calls the MATLAB **ode45** solver, defines the exact solution from **dsolve**, and finally plots both the numerical and analytical solutions on one plot that shows the title of the x- and y-axes and the legend of the figure.

7.6 Given the following ODE:

$$\frac{dy}{dt} = -y + t$$

Subject to the condition that $y(0) = 0$ and integrated from $t = 0$ to $t = 3$.

Write the calling code (main code) that defines the initial value of y, the time interval, calls the MATLAB **ode45** solver, defines the exact solution from MATLAB built-in **dsolve**, and finally plots both the numerical and analytical solutions on one plot that shows the title of x- and y-axes and the legend of the figure.

7.7 Solve numerically (using MATLAB **ode45** solver) the following second-order ordinary differential equation, making, first, the proper transformation into a set of two first-order ODEs $y'' + 5y = t$.

Subject to the initial conditions:

$$y(0) = 0$$

$$y'(0) = 0$$

where t lies between 0 and 1.

Find the analytical solution using either the technique(s) for solving second-order, nonhomogeneous ODE, or MATLAB built-in **dsolve**. Calculate PRE at $t = 1$.

7.8 Using the shooting method, solve the following two-point boundary value problem (BVP):

a) $y'' = t \cos t$ $\quad y(0) = 1$ $\quad y(1) = 0$

b) $y'' = t \cos t$ $\quad y'(0) = 0$ $\quad y(1) = 0$

c) $y'' - y' = 0$ $\quad y(0) = 0$ $\quad y(1) = 1$

Compare your numerical solution in each case with the analytical solution, obtained via MATLAB built-in **dsolve**.

7.9 Solve numerically (using MATLAB **ode45** solver) the first-order differential equation:

$$(t^3 + 1)y' = 3t^2 y$$

Subject to the condition that $y(0) = 1$ integrated from $t = 0$ to $t = 3$.

Find the analytical solution using either the technique(s) for solving first-order, separable ODE, or MATLAB built-in **dsolve**. Calculate PRE at $t = 3$.

7.10 Solve the third-order differential equation:

$$y''' + 3y'' - y' + 4y = 0$$

Subject to the following conditions:

$$y(0) = 0$$
$$y'(0) = 0$$
$$y''(0) = 1$$

by transforming it into a system of three first-order ODEs; where t lies between 0 and 1.

7.11 Consider the *barometric equation*, in a fluid mechanics course, which describes the change in barometric pressure with elevation (altitude):

$$\frac{dP}{dz} = -\rho g \tag{7.26}$$

where positive z is measured upward from sea level ($z = 0$).

If air forms an ideal gas mixture where $\rho = \dfrac{PMW}{RT}$, then the barometric equation becomes:

$$\frac{dP}{dz} = -\rho g = -\frac{PMW}{RT} g \rightarrow\rightarrow\rightarrow \qquad \frac{dP}{dz} = -\frac{gMW}{RT} P \tag{7.27}$$

NOTE Use the following numerical values: g = 9.806 m/s^2; MW = 0.029 kg/mol; R = 8.314 J/(mol·K); and z in m.

a) The *constant-density* model: $\dfrac{dP}{dz} = -\rho_o g = -\dfrac{1.22 \times g}{101325} \rightarrow dP = -1.1807 \times 10^{-4}\, dz.$ (7.28)

b) Find the final value, P_f by *analytically integrating* Eq. (7.28) from $z = 0$ where $P_o = 1$ atm to z_f = 1000, 3000, and 10,000 m.

c) The *isothermal* model: $\dfrac{dP}{dz} = -\dfrac{gMW}{RT_o} P.$ (7.29)

Find the final value, P_f, by *numerically integrating* Eq. (7.29) from $z = 0$ where $P_0 = 1$ atm to z_f = 1000, 3000, and 10,000 m. $T_0 = 288.15$ K.

d) Compare the result at each z_f versus that of the *analytical solution* that can be obtained by either solving a separable first-order ODE or using MATLAB built-in **dsolve**.

e) The *real* model: Within the lower layer of the atmosphere, called the *troposphere*, T decreases with z, according to the following equation: $T = 288\text{K} - \left(\dfrac{71.6\text{K}}{11018.5m}\right) z = 288 - 0.0065z$. Eq. (7.27) becomes: $\dfrac{dP}{dz} = -\dfrac{gMW}{R(288 - 0.0065z)} P$ (7.30)

Find the final value, P_f, by numerically integrating Eq. (7.30) from $z = 0$ where $P_0 = 1$ atm to z_f = 1000, 3000, and 10,000 m.

f) Compare the result at each z_f versus that of the *analytical solution* that can be obtained by either solving a separable first-order ODE or using MATLAB built-in **dsolve**.

g) The *constant-density* model is a good approximate for the *real* model up to what z_f?

h) The *isothermal* model is a good approximate for the *real* model up to what z_f?

7.12 Consider the continuous stirred tank reactor (CSTR) with entering fluid having concentration C_{A0} and flow rate F, as shown in the figure below. A liquid-phase first-order reaction takes place: $A \xrightarrow{k} B$.

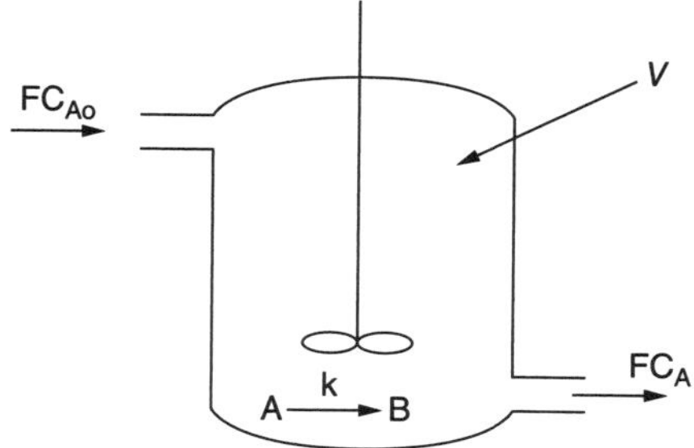

The flow rate out is also F and the volume of the tank is V. If the tank is completely mixed, the concentration in the tank is C_A and the concentration of the fluid leaving the tank is also C_A. The differential equation that describes such a system is given by carrying out A-component material balance around CSTR.

$$FC_{Ao} - FC_A - kC_AV = V\frac{dC_A}{dt}$$

Divide by V, introduce the residence time $\tau = V/F$, and combine C_A–containing terms, we reach at:

$$\frac{1}{\tau}C_{Ao} - \left(\frac{1}{\tau} + k\right)C_A = \frac{dC_A}{dt} \tag{7.31}$$

Use the following values:

$$\tau = 3s;\ C_{Ao} = 10\frac{\text{mol}}{\text{L}};\text{and}\ k = 0.1s^{-1}$$

Find the final value, CA_f, by *numerically integrating* Eq. (7.31) from $t = 0$ where $CA_0 = 10$ mol/L to $t_f = 12$ s.

Compare the result at t_f versus that of the analytical solution that can be obtained by either solving a first-order ODE, or using MATLAB built-in **dsolve**. What is the steady-state value of C_A (i.e., C_A^{ss})?

8 A System of Algebraic Equations

Chapter Outline

In this chapter, we look at the solution of n linear, algebraic equations in n unknowns. There is a good reason for this: it is almost impossible to carry out numerical analysis of any sort without encountering simultaneous equations. Moreover, equation sets arising from physical problems are often very large, consuming many computational resources. It is usually possible to reduce the storage requirements and the run time by exploiting special properties of the coefficient matrix, such as sparseness (most elements of a sparse matrix are zero). Hence, there are many algorithms dedicated to the solution of large sets of equations, each one being tailored to a particular form of the coefficient matrix (symmetric, banded, sparse, etc.). We cannot possibly discuss all the special algorithms in the limited space available.

The best we can do is to present the basic methods of solution and compare the manually obtained results with those obtained via MATLAB.

8.1 The Determinant of a Matrix and Its Inverse

The determinant of a matrix up to 3×3 will be analytically shown here; matrices larger than 3×3 will require exhaustive work; hence, computational techniques (e.g., MATLAB) should be used to find the inverse of a matrix that is larger than 3×3.

8.1.1 (2 × 2) Square Matrix

If $A=\begin{bmatrix} a & b \\ c & d \end{bmatrix}$

$$[A]^{-1}=inv(A)=\frac{1}{\det(A)}\times Adj(A) \tag{8.1}$$

$$[A]^{-1}=\frac{1}{\det(A)}\times Adj(A)=\frac{1}{ad-bc}\times\begin{bmatrix} d & -b \\ -c & a \end{bmatrix} \tag{8.2}$$

Example 8.1 Consider $A=\begin{bmatrix} 1 & 2 \\ 3 & 4 \end{bmatrix}$

$$[A]^{-1}=\frac{1}{1\times4-2\times3}\times\begin{bmatrix} 4 & -2 \\ -3 & 1 \end{bmatrix}=-\frac{1}{2}\begin{bmatrix} 4 & -2 \\ -3 & 1 \end{bmatrix}=\begin{bmatrix} -2.0 & 1.00 \\ 1.50 & -0.50 \end{bmatrix}$$

Using MATLAB

```
>> A=[1 2; 3 4]

A =

     1     2
     3     4
```

```
>> inv(A)

ans =

   -2.0000    1.0000
    1.5000   -0.5000
```

8.1.2 (3 × 3) Square Matrix

$$A=\begin{bmatrix} a & b & c \\ d & e & f \\ g & h & i \end{bmatrix}$$

$$\begin{aligned}|A| &= \det(A) \\ &= (-1)^{1+1}a\{(e\times i)-(f\times h)\}+(-1)^{1+2}b\{(d\times i)-(f\times g)\} \\ &\quad +(-1)^{1+3}c\{(d\times h)-(e\times g)\}=a\{(e\times i)-(f\times h)\} \\ &\quad -b\{(d\times i)-(f\times g)\}+c\{(d\times h)-(e\times g)\}\end{aligned} \tag{8.3}$$

$$Cofactor(A)=\begin{bmatrix} +\begin{vmatrix} e & f \\ h & i \end{vmatrix} & -\begin{vmatrix} d & f \\ g & i \end{vmatrix} & +\begin{vmatrix} d & e \\ g & h \end{vmatrix} \\ -\begin{vmatrix} b & c \\ h & i \end{vmatrix} & +\begin{vmatrix} a & c \\ g & i \end{vmatrix} & -\begin{vmatrix} a & b \\ g & h \end{vmatrix} \\ +\begin{vmatrix} b & c \\ e & f \end{vmatrix} & -\begin{vmatrix} a & c \\ d & f \end{vmatrix} & +\begin{vmatrix} a & b \\ d & e \end{vmatrix} \end{bmatrix} \tag{8.4}$$

$$[A]^{-1}=\frac{1}{\det(A)}\times Adj(A)=\frac{1}{\det(A)}\times Transpose\ of\ Cofactor(A) \tag{8.5}$$

Example 8.2 Consider $A=\begin{bmatrix} a & b & c \\ d & e & f \\ g & h & i \end{bmatrix}=\begin{bmatrix} 1 & 2 & 3 \\ 8 & 6 & 2 \\ 5 & 1 & 3 \end{bmatrix}$

$$Cofactor(A)=\begin{bmatrix} +\begin{vmatrix} e & f \\ h & i \end{vmatrix} & -\begin{vmatrix} d & f \\ g & i \end{vmatrix} & +\begin{vmatrix} d & e \\ g & h \end{vmatrix} \\ -\begin{vmatrix} b & c \\ h & i \end{vmatrix} & +\begin{vmatrix} a & c \\ g & i \end{vmatrix} & -\begin{vmatrix} a & b \\ g & h \end{vmatrix} \\ +\begin{vmatrix} b & c \\ e & f \end{vmatrix} & -\begin{vmatrix} a & c \\ d & f \end{vmatrix} & +\begin{vmatrix} a & b \\ d & e \end{vmatrix} \end{bmatrix}=\begin{bmatrix} +\begin{vmatrix} 6 & 2 \\ 1 & 3 \end{vmatrix} & -\begin{vmatrix} 8 & 2 \\ 5 & 3 \end{vmatrix} & +\begin{vmatrix} 8 & 6 \\ 5 & 1 \end{vmatrix} \\ -\begin{vmatrix} 2 & 3 \\ 1 & 3 \end{vmatrix} & +\begin{vmatrix} 1 & 3 \\ 5 & 3 \end{vmatrix} & -\begin{vmatrix} 1 & 2 \\ 5 & 1 \end{vmatrix} \\ +\begin{vmatrix} 2 & 3 \\ 6 & 2 \end{vmatrix} & -\begin{vmatrix} 1 & 3 \\ 8 & 2 \end{vmatrix} & +\begin{vmatrix} 1 & 2 \\ 8 & 6 \end{vmatrix} \end{bmatrix}$$

$$Cofactor(A) = \begin{bmatrix} 16 & -14 & -22 \\ -3 & -12 & 9 \\ -14 & 22 & -10 \end{bmatrix}$$

$$Cofactor(A)^T = \begin{bmatrix} 16 & -3 & -14 \\ -14 & -12 & 22 \\ -22 & 9 & -10 \end{bmatrix}$$

$$|A| = \det(A) = a\{(e \times i) - (f \times h)\} - b\{(d \times i) - (f \times g)\} + c\{(d \times h) - (e \times g)\}$$

$$|A| = \det(A) = 1\{(6 \times 3) - (2 \times 1)\} - 2\{(8 \times 3) - (2 \times 5)\} + 3\{(8 \times 1) - (6 \times 5)\}$$

$$|A| = \det(A) = 16 - 28 - 66 = -78$$

$$[A]^{-1} = \frac{1}{-78} \times Cofactor(A)^T = \frac{1}{-78} \times \begin{bmatrix} 16 & -3 & -14 \\ -14 & -12 & 22 \\ -22 & 9 & -10 \end{bmatrix}$$

$$[A]^{-1} = \begin{bmatrix} -0.2051 & 0.0385 & 0.1795 \\ 0.1795 & 0.1538 & -0.2821 \\ 0.2821 & -0.1154 & 0.1282 \end{bmatrix}$$

Let's verify the answer using MATLAB:

```
>> A=[1 2 3;8 6 2; 5 1 3]
A =
     1     2     3
     8     6     2
     5     1     3

>> det(A)
ans =
   -78

>> inv(A)
ans =

   -0.2051    0.0385    0.1795
    0.1795    0.1538   -0.2821
    0.2821   -0.1154    0.1282
```

The answers are essentially the same.

8.2 A System of ($n \times n$) Linear Algebraic Equations

A system of $(n \times n)$ algebraic equations has the form:

$$a_{11}x_1 + a_{12}x_2 + \cdots + a_{1n}x_n = b_1 \tag{8.6a}$$

$$a_{21}x_1 + a_{22}x_2 + \cdots + a_{2n}x_n = b_2 \tag{8.6b}$$

$$a_{31}x_1 + a_{32}x_2 + \cdots + a_{3n}x_n = b_3 \tag{8.6c}$$

$$\downarrow \qquad\qquad\qquad\qquad \downarrow$$

$$a_{n1}x_1 + a_{n2}x_2 + \cdots + a_{nn}x_n = b_n \tag{8.6z}$$

where the coefficients a_{ij} and the constants b_j are known, and x_i represents the unknowns. In matrix notation the equations are written as:

$$\begin{bmatrix} a_{11} & \cdots & a_{1n} \\ \vdots & \ddots & \vdots \\ a_{n1} & \cdots & a_{nn} \end{bmatrix} \begin{bmatrix} x_1 \\ \vdots \\ x_n \end{bmatrix} = \begin{bmatrix} b_1 \\ \vdots \\ b_n \end{bmatrix} \tag{8.7}$$

or, simply $\boldsymbol{Ax} = \boldsymbol{b}$.

8.2.1 A System of ($n \times n$) Linearly Independent Equations ($\Rightarrow |A| \neq 0$)

A system of n linear equations in n unknowns will have a unique solution, if the *determinant* of the coefficient matrix is *nonsingular*, i.e., if $|A| \neq 0$. The rows and columns of a nonsingular matrix are *linearly independent* in the sense that no row (or column) is a linear combination of other rows (or columns). If the coefficient matrix is *singular*, the equations may have an infinite number of solutions, or no solutions at all, depending on the constant vector. As an illustration, consider the equations:

$$\begin{bmatrix} 3x + y = 4 \\ 6x + 2y = 8 \end{bmatrix} \tag{8.8a-b}$$

Because the second equation can be obtained by multiplying the first equation by two, any combination of x and y that satisfies the first equation will be a solution for the second equation. The number of such combinations is infinite. Equations given by (8.8*a*–*b*) are called linearly dependent.

On the other hand, the equations:

$$\begin{bmatrix} 3x + y = 4 \\ 6x + 2y = 6 \end{bmatrix} \tag{8.9a-b}$$

have no solution because the second equation, being equivalent to $3x + y = 3$, contradicts the first one. Therefore, any solution that satisfies one equation cannot satisfy the other one.

For a linearly independent system of algebraic equations, the solution will take the form:

$$Ax = b$$

Multiplying both sides by A^{-1}, we get:

$$A^{-1}Ax = A^{-1}b \tag{8.10}$$

Given that:

$$A^{-1}A = I \tag{8.11}$$

We reach the solution, given that the coefficient matrix, A, is *nonsingular*; i.e., $|A| \neq 0$:

$$Ix = x = A^{-1}b \tag{8.12}$$

Example 8.3 Consider the following system of linear algebraic equations:

$$\begin{bmatrix} 3x + y = 9 \\ 2x + 3y = 13 \end{bmatrix}$$

Find the values of x and y that will satisfy the above system.

Let us check the determinant of A:

$$\begin{bmatrix} 3 & 1 \\ 2 & 3 \end{bmatrix} \begin{bmatrix} x \\ y \end{bmatrix} = \begin{bmatrix} 9 \\ 13 \end{bmatrix}$$

where A is $\begin{bmatrix} 3 & 1 \\ 2 & 3 \end{bmatrix}$.

$$|A| = \det(A) = 3 \times 3 - 2 \times 1 = 7$$

Let us use MATLAB to find some useful properties and eventually solve the system.

```
>> A=[3 1; 2 3]
A =
     3     1
     2     3
>> det(A)
ans =       7
```

Because $|A| \neq 0$, there exists a solution given by Eq. (8.12):

```
>> inv(A)
ans =
    0.4286   -0.1429
   -0.2857    0.4286
```

NOTE You can also calculate the inverse of A manually using Eq. (8.2).

```
>> [x]=inv(A)*[9;13]
x =
     2.0000
     3.0000
>> A=[3 1;2 3]
A =
      3     1
      2     3
```

Also, using the *backslash* operation in MATLAB:

```
>> b=[9;13]
b =
      9
     13
>> x=A\b
x =
      2
      3
```

This is equivalent to:

$$\begin{bmatrix} x \\ y \end{bmatrix} = \begin{bmatrix} 0.4286 & -0.1429 \\ -0.2857 & 0.4286 \end{bmatrix} \begin{bmatrix} 9 \\ 13 \end{bmatrix} = \begin{bmatrix} 0.4286\times 9 - 0.1429\times 13 \\ -0.2857\times 9 + 0.4286\times 13 \end{bmatrix} = \begin{bmatrix} 1.9997 \\ 3.0005 \end{bmatrix}$$

Example 8.4 Consider the following system of linear algebraic equations:

$$\begin{bmatrix} x+y+z=15 \\ 2x+y+3z=32 \\ 3x+2y-z=12 \end{bmatrix}$$

Find the values of x, y, and z that will satisfy the above system.

Let us check the determinant of A:

$$\begin{bmatrix} 1 & 1 & 1 \\ 2 & 1 & 3 \\ 3 & 2 & -1 \end{bmatrix} \begin{bmatrix} x \\ y \\ z \end{bmatrix} = \begin{bmatrix} 15 \\ 32 \\ 12 \end{bmatrix}$$

where $A = \begin{bmatrix} 1 & 1 & 1 \\ 2 & 1 & 3 \\ 3 & 2 & -1 \end{bmatrix}$.

$$|A| = \det(A) = (-1)^{1+1}\{(1\times -1) - (3\times 2)\} + (-1)^{1+2}\{(2\times -1) - (3\times 3)\}$$
$$+ (-1)^{1+3}\{(2\times 2) - (1\times 3)\} = 1\{-7\} - 1\{-11\} + 1\{1\} = 5$$

Let us use MATLAB to find some useful properties and eventually solve the system.

```
>> A=[1,1,1;2,1,3;3,2,-1]
A =
     1     1     1
     2     1     3
     3     2    -1
>> det(A)
ans =
    5.0000
```

Since $|A| \neq 0$, there exists a solution given by Eq. (8.7):

```
>> inv(A)
ans =
   -1.4000    0.6000    0.4000
    2.2000   -0.8000   -0.2000
    0.2000    0.2000   -0.2000
```

NOTE You can also calculate the inverse of A manually using Eq. (8.5).

```
>> [x]=inv(A)*[15;32;12]
x =
    3.0000
    5.0000
    7.0000
```

Also, using the *backslash* operation in MATLAB:

```
>> b=[15;32;12]
b =
    15
    32
    12
>> x=A\b
x =
    3.0000
    5.0000
    7.0000
```

This is equivalent to:

$$\begin{bmatrix} x \\ y \\ z \end{bmatrix} = \begin{bmatrix} -1.4 & 0.6 & 0.4 \\ 2.2 & -0.8 & -0.2 \\ 0.2 & 0.2 & -0.2 \end{bmatrix} \begin{bmatrix} 15 \\ 32 \\ 12 \end{bmatrix} = \begin{bmatrix} -1.4\times 15+0.6\times 32+0.4\times 12 \\ 2.2\times 15-0.8\times 32-0.2\times 12 \\ 0.2\times 15+0.2\times 32-0.2\times 12 \end{bmatrix} = \begin{bmatrix} 3 \\ 5 \\ 7 \end{bmatrix}$$

Exercise 8.1 Consider the following system of linear algebraic equations:

$$\begin{bmatrix} x+y+z=12 \\ 2x+y+3z=26 \\ 3x+2y-z=8 \end{bmatrix}$$

Find the values of x, y, and z that will satisfy the above system.

Try to solve it by hand (using information given in Sec. 8.1) and verify the result using MATLAB.

8.2.2 Gauss Elimination Method with Partial Pivoting

Gaussian elimination is considered the workhorse of computational science for the solution of a system of linear equations. Karl Friedrich Gauss, a great nineteenth-century mathematician, suggested this elimination method as a part of his proof of a particular theorem. Gaussian elimination is a systematic application of elementary row operations to a system of linear equations in order to convert the system to upper triangular form. Once the coefficient matrix is in upper triangular form, we use back substitution to find a solution. The general procedure for Gaussian elimination can be summarized in the following steps:

1. Write the augmented matrix for the system of linear equations.
2. Use elementary row operations on the augmented matrix $[\boldsymbol{A}|\boldsymbol{b}]$ to transform $\boldsymbol{A}$ into upper triangular form. If a zero is located on the diagonal, switch the rows until a nonzero is in that place (This process is called partial pivoting). The *kth* row is switched by a row below it having the element of the largest absolute value in the *kth* column. This procedure is recommended for reducing the round-off error even in the case where the *kth* pivot a_{kk}^{k-1} is not zero (at the *kth* stage, where a_{kk}^{k-1} is the diagonal element in the *kth* row). If you are unable to do so, stop; the system has either infinite or no solutions.
3. Use back substitution to find the solution of the problem.

Use the Gaussian elimination method to solve the system of equations:

$$\begin{bmatrix} a_{11} & \cdots & a_{1n} \\ \vdots & \ddots & \vdots \\ a_{n1} & \cdots & a_{nn} \end{bmatrix} \begin{bmatrix} x_1 \\ \vdots \\ x_n \end{bmatrix} = \begin{bmatrix} b_1 \\ \vdots \\ b_n \end{bmatrix} \tag{8.13}$$

The augmented matrix is written as

$$\left[\begin{array}{ccc|c} a_{11} & \cdots & a_{1n} & b_1 \\ \vdots & \ddots & \vdots & \vdots \\ a_{n1} & \cdots & a_{nn} & b_n \end{array}\right] \tag{8.14}$$

The elementary row operations are performed to put the augmented matrix into the upper triangular form (i.e., all elements below the diagonal are zero):

$$\left[\begin{array}{cccc|c} a_{11}^{*} & a_{12}^{*} & \cdots & a_{1n}^{*} & b_1^{*} \\ 0 & a_{22}^{*} & \cdots & a_{2n}^{*} & b_2^{*} \\ \vdots & \vdots & \ddots & \vdots & \vdots \\ 0 & 0 & \cdots & a_{nn}^{*} & b_n^{*} \end{array}\right] \tag{8.15}$$

Solve the equation of the *nth* row for x_n, then substitute back into the equation of the $(n-1)th$ row to obtain a solution for x_{n-1}, etc., according to the formula:

$$x_i = \frac{1}{a_{ii}^{*}}\left[b_i^{*} - \sum_{j=i+1}^{n} a_{ij}^{*} x_j\right] \tag{8.16}$$

Example 8.5 Solve the linear system using the Gaussian elimination method.

$$\begin{bmatrix} y+z=2 \\ 2x+3z=5 \\ x+y+z=3 \end{bmatrix}$$

Solution The augmented matrix is:

$$\begin{bmatrix} 0 & 1 & 1 & 2 \\ 2 & 0 & 3 & 5 \\ 1 & 1 & 1 & 3 \end{bmatrix}$$

1. Because $a_{11} = 0$, we have to switch the second row and the first row:

$$\begin{bmatrix} 2x+3z=5 \\ y+z=2 \\ x+y+z=3 \end{bmatrix} \quad \begin{bmatrix} 2 & 0 & 3 & 5 \\ 0 & 1 & 1 & 2 \\ 1 & 1 & 1 & 3 \end{bmatrix}$$

2. Divide the first equation by 2:

$$\begin{bmatrix} x+\frac{3}{2}z=\frac{5}{2} \\ y+z=2 \\ x+y+z=3 \end{bmatrix} \quad \begin{bmatrix} 1 & 0 & \frac{3}{2} & \frac{5}{2} \\ 0 & 1 & 1 & 2 \\ 1 & 1 & 1 & 3 \end{bmatrix}$$

3. Add $-1 \times$ the first equation to the third equation and insert the result in the third row (i.e., the third equation):

$$\begin{array}{r} -x-\frac{3}{2}z=-\frac{5}{2} \\ x+y+z=3 \\ \hline 0+y-\frac{1}{2}z=\frac{1}{2} \end{array}$$

So, the equations become:

$$\begin{bmatrix} x+\frac{3}{2}z=\frac{5}{2} \\ y+z=2 \\ y-\frac{1}{2}z=\frac{1}{2} \end{bmatrix} \quad \begin{bmatrix} 1 & 0 & \frac{3}{2} & \frac{5}{2} \\ 0 & 1 & 1 & 2 \\ 0 & 1 & -\frac{1}{2} & \frac{1}{2} \end{bmatrix}$$

4. Add $-1 \times$ the second equation to the third equation and insert the result in the third row (i.e., the third equation):

$$\begin{array}{r} -y-z=-2 \\ y-\frac{1}{2}z=\frac{1}{2} \\ \hline 0+0-\frac{3}{2}z=-\frac{3}{2} \end{array}$$

So, the final equations are:

$$\begin{bmatrix} x+\frac{3}{2}z=\frac{5}{2} \\ y+z=2 \\ -\frac{3}{2}z=-\frac{3}{2} \end{bmatrix} \quad \begin{bmatrix} 1 & 0 & \frac{3}{2} & \frac{5}{2} \\ 0 & 1 & 1 & 2 \\ 0 & 0 & -\frac{3}{2} & -\frac{3}{2} \end{bmatrix}$$

The last row says that:

$$-\frac{3}{2}z=-\frac{3}{2}$$

which implies that $z=1$.

From the second equation (or, second row), we have:

$$y+z=2$$

which implies that $y=1$.

Finally, from the first equation (or, first row), we have:

$$x+\frac{3}{2}z=\frac{5}{2}$$

which implies that $x=1$.

Exercise 8.2 Consider the following system of linear algebraic equations:

$$\begin{bmatrix} 9x+3y+4z=7 \\ 4x+3y+4z=8 \\ x+y+z=3 \end{bmatrix}$$

Find the values of x, y, and z that will satisfy the above system.

Try to solve it by hand (Sec. 8.1), using Gaussian elimination (Sec. 8.2.2), and verify the results using MATLAB.

Figure 8.1 shows an M-file (**gauss_elim.m**) for Gaussian elimination with partial pivoting.

```
function x = gauss_elim(A,B)
%The sizes of matrices A,B are supposed to be NA x NA and NA x NB.
%This function solves Ax = B by Gauss elimination algorithm.
NA = size(A,2); [NB1,NB] = size(B);
if NB1 ~= NA, error('A and B must have compatible dimensions'); end
N = NA + NB; AB = [A(1:NA,1:NA) B(1:NA,1:NB)]; % Augmented matrix
epss = eps*ones(NA,1);
for k = 1:NA
%Scaled Partial Pivoting at AB(k,k)
[akx,kx] = max(abs(AB(k:NA,k))./ ...
max(abs([AB(k:NA,k + 1:NA) epss(1:NA - k + 1)]'))');
if akx < eps, error('Singular matrix and No unique solution'); end
mx = k + kx - 1;
if kx > 1 % Row change if necessary
tmp_row = AB(k,k:N);
AB(k,k:N) = AB(mx,k:N);
AB(mx,k:N) = tmp_row;
end
% Gauss forward elimination
AB(k,k + 1:N) = AB(k,k+1:N)/AB(k,k);
AB(k,k) = 1; %make each diagonal element one
for m = k + 1: NA
AB(m,k+1:N) = AB(m,k+1:N) - AB(m,k)*AB(k,k+1:N);
AB(m,k) = 0;
end
end
%backward substitution for a upper-triangular matrix eqation
% having all the diagonal elements equal to one
x(NA,:) = AB(NA,NA+1:N);
for m = NA-1: -1:1
x(m,:) = AB(m,NA + 1:N)-AB(m,m + 1:NA)*x(m + 1:NA,:);
end
```

Figure 8.1 **gauss_elim** M-file for calculating the unknown vector **x** using the Gaussian elimination method with partial pivoting.

Example 8.6 Consider the following system of linear algebraic equations:

$$\begin{bmatrix} 2x+3y+4z=4 \\ x+3y-2z=9 \\ x+y+z=2 \end{bmatrix}$$

Find the values of x, y, and z that will satisfy the above system.

Here is the solution using three different methods, including Gaussian elimination:

```
>> A = [2 3 4;1 3 -2;1 1 1]; b = [4 9 2]';
>> x = gauss_elim(A,b)
x =
     1
     2
    -1

>> X1 = inv(A)*b

X1 =

    1.0000
    2.0000
   -1.0000

>> x1 = A\b

x1 =

     1
     2
    -1
```

8.2.3 *Ill-Conditioning*

The big question is: What happens when the coefficient matrix is almost singular; i.e., if |**A**| is very small? In order to determine whether the determinant of the coefficient matrix is "small," we need a reference against which the determinant can be measured. This reference is called the *norm* of the matrix, denoted by $\|A\|$. We can then say that the determinant is small if

$$|A| \ll \|A\| \tag{8.17}$$

Several norms of a matrix are defined in the existing literature.
The Frobenius norm is defined as:

$$\|A\|_F = \sqrt{\sum_{j=1}^{n} \sum_{j=1}^{n} A_{ij}^2} \tag{8.18}$$

The maximum absolute column sum norm is defined as:

$$\|A\|_1 = max|_{over\ j} \sum_{i=1}^{n} |a_{ij}| \tag{8.19}$$

The maximum absolute row sum norm is defined as:

$$\|A\|_\infty = max|_{over\ i} \sum_{j=1}^{n} |a_{ij}| \tag{8.20}$$

A formal measure of conditioning is the *matrix condition number*, defined as:

$$cond(A) = \|A\| \times \|A^{-1}\| \tag{8.21}$$

If this number is close to unity, the matrix will be well conditioned. The condition number increases with the degree of ill-conditioning, reaching infinity for a singular matrix. Note that the condition number is not unique, but depends on the choice of the matrix norm. Nevertheless, the condition number is time-consuming (CPU time) to compute for large matrices. In most cases, it is sufficient to inspect conditioning by comparing the determinant with the magnitudes of the elements in the matrix.

Example 8.7 Consider the system of linear algebraic equations used in Example 8.4:

$$\begin{bmatrix} x + y + z = 15 \\ 2x + y + 3z = 32 \\ 3x + 2y - z = 12 \end{bmatrix}$$

Find the values of different norms and verify that such a system of equations is well conditioned.

$$\begin{bmatrix} 1 & 1 & 1 \\ 2 & 1 & 3 \\ 3 & 2 & -1 \end{bmatrix} \begin{bmatrix} x \\ y \\ z \end{bmatrix} = \begin{bmatrix} 15 \\ 32 \\ 12 \end{bmatrix}$$

where $A = \begin{bmatrix} 1 & 1 & 1 \\ 2 & 1 & 3 \\ 3 & 2 & -1 \end{bmatrix}$.

Let us use MATLAB to calculate values of different norms. The `norm` function calculates several different types of matrix and vector norms. If the input is a vector or a matrix:

n = norm(X,1) returns the 1-norm of X.

n = norm(X,Inf) returns the infinity norm of X.

n = norm(X,'fro') returns the Frobenius norm of X.

$$\|A\|_F = Af$$

```
>>Af=norm(A,'fro')
Af =
     5.5678
```

$$\|A\|_1 = A1$$

```
>> A1=norm(A,1)
A1 =
     6
```

$$\|A\|_\infty = Ainf$$

```
>> Ainf=norm(A,'inf')
Ainf =
     6
>> Ainv=inv(A)
Ainv =
   -1.4000    0.6000    0.4000
    2.2000   -0.8000   -0.2000
    0.2000    0.2000   -0.2000

>> Ainvf=norm(Ainv,'fro')
Ainvf =

     2.8496
```

So, the condition number based on Frobenius will be:

$$cond(A) = \|A\| \times \|A^{-1}\| = \text{Af * Ainvf=5.5678*2.8496=15.8660}$$

This can be directly found using the MATLAB built-in function:

```
>> cond_num_fro=cond(A,'fro')
cond_num_fro =
   15.8657
```

From MATLAB online built-in product help:

cond(X,P) = norm(X,P) * norm(inv(X),P).

where P = 1, 2, inf, or 'fro'.

Exercise 8.3 Evaluate *cond(A)* using other norms.

NOTES

1. As the size of the matrix grows, round-off errors are inevitable and will propagate in matrix operations to the point that zero may appear to be an absolutely small number, or a nonzero number very close to zero may appear to be zero. Therefore, it is not a simple task to determine whether a zero or a number very close to zero is a real zero or not.
2. There are several criteria by which we judge the degree of ill-condition, such as how different $A * A^{-1}$ is from the identity matrix I, how far $det(A) \times det(A^{-1})$ is from one, in addition to the condition number given by Eq. (8.21).

If the equations are ill conditioned, small changes in the coefficient matrix will result in large changes in the solution. As an illustration, consider the following example:

Example 8.8 Consider the system of linear algebraic equations:

$$\begin{bmatrix} 3x + y = 4 \\ 3x + 1.001y = 0 \end{bmatrix}$$

$$\begin{bmatrix} 3 & 1 \\ 3 & 1.001 \end{bmatrix} \begin{bmatrix} x \\ y \end{bmatrix} = \begin{bmatrix} 4 \\ 0 \end{bmatrix}$$

$$A = \begin{bmatrix} 3 & 1 \\ 3 & 1.001 \end{bmatrix}$$

Let us use MATLAB to solve such a system and see how a slight change in the coefficient matrix will result in a drastic change in the solution.

```
>> A=[3,1;3,1.001]
A =
    3.0000    1.0000
    3.0000    1.0010

>> b=[4;0]

b =
     4
     0
>> [x]=inv(A)*b

x =

  1.0e+003 *

    1.3347
   -4.0000
```

Let us repeat the same set but change the last coefficient from 1.001 to 1.002, without changing the **b** (right-hand side) values.

```
>> A=[3,1;3,1.002]

A =

    3.0000    1.0000
    3.0000    1.0020

>> [x]=inv(A)*b

x =

   1.0e+003 *

    0.6680
   -2.0000
```

Notice how the solution dramatically changed; x_1 has changed from 1.3347×10^3 to 0.668×10^3, and x_2 from -4.0×10^3 to -2.0×10^3.

8.3 Problems

8.1 Consider the system of linear algebraic equations:

$$\begin{bmatrix} x+y=13 \\ x-y=3 \end{bmatrix}$$

Solve the aforementioned equations by hand (and calculator). Use $\boldsymbol{x} = \boldsymbol{inv}(\mathbf{A}) * \boldsymbol{b}$ to solve for x and y. See Sec. 8.1 on how to find the matrix inverse.

8.2 Consider the system of linear algebraic equations:

$$\begin{bmatrix} 2y+z=4 \\ x+y+2z=6 \\ 2x+y+z=7 \end{bmatrix} \quad \begin{matrix} (8.22) \\ (8.23) \\ (8.24) \end{matrix}$$

Solve the aforementioned equations by hand (and calculator). Eliminate x between Eqs. (8.23 and 8.24) and use the resultant equation, which contains y and z, and the first equation to solve for both y and z. Finally, use either Eq. (8.23) or (8.24) to find x.

8.3 Consider the same system of linear algebraic equations in Problem 8.2:

Solve the aforementioned equations using MATLAB. Use all of the methods that have been previously mentioned. Use Gaussian elimination, the inverse of the matrix of coefficients (**inv(A)∗b**), and the backslash operation (**A\b**) method.

8.4 Consider the system of linear algebraic equations:

$$\begin{bmatrix} x+2y+z=6 \\ x+y-2z=-5 \\ -2x+y+z=7 \end{bmatrix}$$

Solve the aforementioned equations using MATLAB. Use Gaussian elimination, the inverse of the matrix of coefficients (**inv(A)*b**), and the backslash operation (**A\b**) method.

8.5 Consider the system of linear algebraic equations:

$$\begin{bmatrix} x+2y-z=-8 \\ -y-z=-1 \\ x+y+2z=-1 \end{bmatrix}$$

Solve the aforementioned equations using MATLAB. Use Gaussian elimination, the inverse of the matrix of coefficients (**inv(A)*b**), and the backslash operation (**A\b**) method.

8.6 Consider the system of linear algebraic equations:

$$\begin{bmatrix} x+2y-z=-8 \\ -y+2z=-1 \\ x+y+2z=-1 \end{bmatrix}$$

Solve the aforementioned equations using MATLAB. Use Gaussian elimination, the inverse of the matrix of coefficients (**inv(A)*b**), and the backslash operation (**A\b**) method.

8.7 Consider the system of linear algebraic equations:

$$\begin{bmatrix} x+2y+3z=-8 \\ -y-z=-1 \\ x+y+2z=-1 \end{bmatrix}$$

Solve the aforementioned equations using MATLAB. Use Gaussian elimination, the inverse of the matrix of coefficients (**inv(A)*b**), and the backslash operation (**A\b**) to show that such a system *does not have a solution.*

8.8 The figure below shows a gas stream that contains 18.0 mole percent hexane; the remainder is nitrogen. The stream flows to a condenser, where its temperature is reduced and some of the hexane is liquefied. The hexane mole fraction in the gas stream leaving the condenser is 0.0500. Liquid hexane condensate is recovered at a rate of 10.0 mol/min. Find the flow rate of the gas stream entering and leaving the condenser in mol/min. Carry out both steady-state hexane and nitrogen balance around the condenser so that you will end up with two equations with two unknown terms ($\dot{n}_1$ and $\dot{n}_2$).

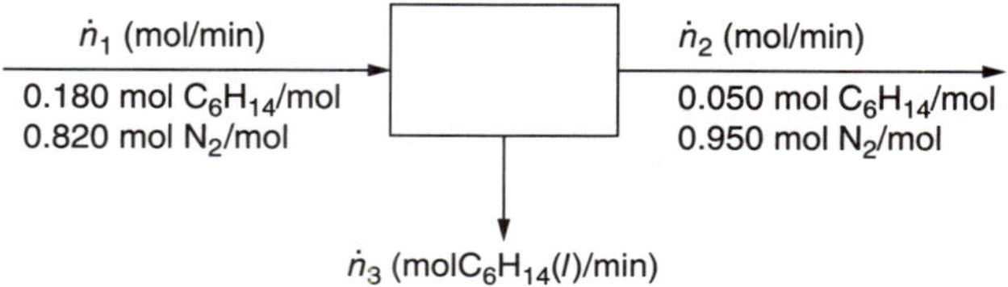

Define **A** and **b**, then use Gaussian elimination, the inverse of the matrix of coefficients (**inv(A)*b**), and the backslash operation (**A\b**) to find the unknown quantities: ($\dot{n}_1$ and $\dot{n}_2$).

8.9 An isothermal CSTR has the following series reaction:

$$A \xrightarrow{k_1} B \xrightarrow{k_2} C$$

Pure A enters the reactor with inlet concentration C_{Ao} and flow rate F. The outlet stream, F, contains all three components A, B, and C, but with different concentrations.

The dynamic equations that govern the system are given by carrying out the molar balance for each component: *Accumulation = In − Out + Generation − Consumption*

$$V\frac{dC_A}{dt} = FC_{Ao} - FC_A - k_1 C_A V \;\rightarrow\; \frac{dC_A}{dt} = \frac{F}{V}C_{Ao} - \frac{F}{V}C_A - k_1 C_A \rightarrow$$

$$\frac{dC_A}{dt} = \frac{1}{\tau}C_{Ao} - \left(\frac{1}{\tau} + k_1\right)C_A \tag{8.25}$$

$$V\frac{dC_B}{dt} = 0 - FC_B + k_1 C_A V - k_2 C_B V \;\rightarrow\; \frac{dC_B}{dt} = -\frac{F}{V}C_B + k_1 C_A - k_2 C_B \rightarrow$$

$$\frac{dC_B}{dt} = k_1 C_A - \left(\frac{1}{\tau} + k_2\right)C_B \tag{8.26}$$

$$V\frac{dC_C}{dt} = 0 - FC_C + k_2 C_B V \;\rightarrow \frac{dC_C}{dt} = -\frac{F}{V}C_C + k_2 C_B \rightarrow$$

$$\frac{dC_C}{dt} = k_2 C_B - \frac{1}{\tau}C_C \tag{8.27}$$

If we set up the left-hand side of each equation to zero, then we will end up with the counterpart version (i.e., a set of three linear algebraic equations) that describes the CSTR under steady-state condition.

Define **A** and **b**, then use Gaussian elimination, the inverse of the matrix of coefficients (**inv(A)*b**), and the backslash operation (**A\b**) to solve the three equations with three unknown terms (C_A, C_B, and C_C) for the following conditions: $\tau = 1s$; $C_{Ao} = 100\frac{\text{mol}}{\text{L}}$; $k_1 = 1s^{-1}$; and $k_2 = 2s^{-1}$.

8.10 A cascade of three isothermal CSTRs in series is used to carry out the following liquid-phase reaction, as shown in the figure below:

$$A \xrightarrow{k} B$$

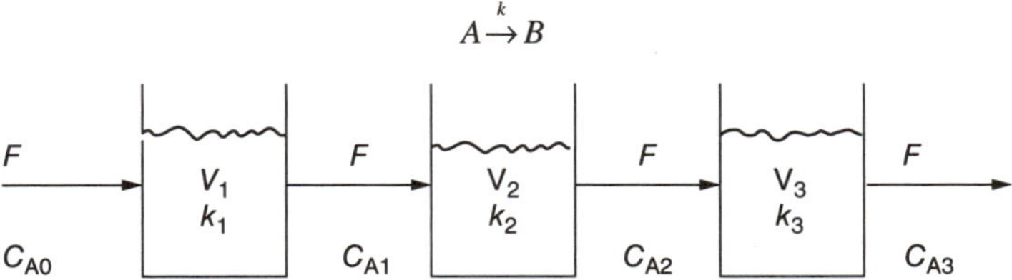

The dynamic equations that govern the system (under constant flow rates) are given by carrying out the molar balance for component A for each CSTR:
Accumulation = In − Out + Generation − Consumption

$$V_1\frac{dC_{A1}}{dt} = FC_{Ao} - FC_{A1} - k_1 C_{A1} V_1 \;\rightarrow\; \frac{dC_{A1}}{dt} = \frac{F}{V_1}C_{Ao} - \frac{F}{V_1}C_{A1} - k_1 C_{A1} \rightarrow$$

$$\frac{dC_{A1}}{dt} = \frac{1}{\tau_1}C_{Ao} - \left(\frac{1}{\tau_1} + k_1\right)C_{A1} \tag{8.28}$$

$$V_2 \frac{dC_{A2}}{dt} = FC_{A1} - FC_{A2} - k_2 C_{A2} V_2 \rightarrow \frac{dC_{A2}}{dt} = \frac{F}{V_2} C_{A1} - \frac{F}{V_2} C_{A2} - k_2 C_{A2} \rightarrow$$

$$\frac{dC_{A2}}{dt} = \frac{1}{\tau_2} C_{A1} - \left(\frac{1}{\tau_2} + k_2 \right) C_{A2} \tag{8.29}$$

$$V_3 \frac{dC_{A3}}{dt} = FC_{A2} - FC_{A3} - k_1 C_{A3} V_3 \rightarrow \frac{dC_{A3}}{dt} = \frac{F}{V_3} C_{A2} - \frac{F}{V_3} C_{A3} - k_3 C_{A3} \rightarrow$$

$$\frac{dC_{A3}}{dt} = \frac{1}{\tau_3} C_{A2} - \left(\frac{1}{\tau_3} + k_3 \right) C_{A3} \tag{8.30}$$

If we set up the left-hand side of each equation to zero, then we will end up with the counterpart version (i.e., a set of three linear algebraic equations) that describes the three CSTRs under steady-state condition. Notice that each CSTR has its own residence time (i.e., τ_i) and temperature (i.e., k_i).

Define **A** and **b**, then use Gaussian elimination, the inverse of the matrix of coefficients (**inv(A)*b**), and the backslash operation (**A\b**) to solve the three equations with three unknown terms (C_{A1}, C_{A2}, and C_{A3}) for the following conditions:

$$\tau_1 = 1s;\ \tau_2 = 2s;\ \tau_3 = 3s;\ C_{Ao} = 100 \frac{\text{mol}}{\text{L}};\ k_1 = 1s^{-1};\ k_2 = 2s^{-1};\ \text{and}\ k_3 = 3s^{-1}.$$

9 Statistics

Chapter Outline

In Secs. 1.7 and 5.4 some statistical definitions were introduced. In this chapter, we look at some common distribution functions and the usage of the MATLAB Statistics Toolbox.

9.1 Probability

A *random experiment* is defined as a process or action whose outcome cannot be predicted with certainty and would likely change when the experiment is repeated. The variability in the outcomes might arise from many sources: slight errors in measurements, choosing different objects for testing, etc. The ability to model and analyze the outcomes from experiments is at the heart of statistics. Some examples of random experiments that arise in different disciplines are given below:

- *Engineering:* Data are collected on the number of failures of piston rings in the legs of steam-driven compressors. Engineers are interested in determining the probability of piston failure in each leg and whether the failure varies among the compressors.
- *Medicine:* The oral glucose tolerance test is a diagnostic tool for early diabetes mellitus. The results of the test are subject to variation because of different rates at which people absorb the glucose, and the variation is particularly noticeable in pregnant women. Scientists are interested in analyzing and modeling the variation of glucose before and after pregnancy.
- *Manufacturing:* Manufacturers of cement are interested in the tensile strength of their product. The strength depends on many factors, one of which is the length of time the cement is dried. An experiment is conducted where different batches of cement are tested for tensile strength after different drying times. Engineers like to determine the relationship between drying time and tensile strength of the cement.
- *Software Engineering:* Engineers measure the central processing unit (CPU) time in MATLAB, for example, as an indicator for the effectiveness of different numerical, or computational methods/algorithms. Different cases are tested for the same method and CPU time is recorded for each case.

The sample space is the set of all possible outcomes from an experiment. It is possible sometimes to list all outcomes in the sample space. This is especially true in the case of some discrete random variables. Examples of these sample spaces are:

- When tossing a coin, the sample space is the face image or the rear image (the other face of the coin). If we represent the case that the face is up by one then if it is down it will be zero. Hence, the sample space will be $\{0, 1\}$.
- If rolling a six-sided dice and counting the number of dots on the face, then the sample space will be $\{1, 2, 3, 4, 5, 6\}$.

The outcomes from random experiments are often represented by an uppercase variable such as X. This is called a *random variable*, and its occurrence value is described by a probabilistic model, or probability density function. Formally, a random variable is a real-valued function defined on the sample space. Using our examples of experiments, a random variable X might represent the CPU time of a software method, or the glucose level of a patient. The observed value of a random variable X is denoted by a lowercase x. A random variable X might represent the number of failures of piston rings in a compressor, and would indicate that we observed, for example, 7 piston ring failures.

Random variables can be discrete or continuous. A *discrete random variable* can take on values from a limited or unlimited set of numbers. Examples of discrete random variables are the number of defective parts or the number of typographical errors on a page. On the other hand, a *continuous random variable* is one that can take on values from an interval of real numbers. Examples of continuous random variables are the times of planes staying in air, the weight of a newborn baby, the weight of a piece of stone picked up off the surface of earth, the weight in grams or milligrams of a product in a pharmaceutical or food production line, or the height of a human being expressed in centimeters or inches.

An *event* is a subset of outcomes in the sample space. An event might be that a product item is defective or that the weight of the "fresh" baby is between 2 and 5 kg. The probability of an event is assigned a value between 0 and 1.

Probability is a measure of the likelihood that some event will occur. It is also a way to quantify or to gauge the likelihood that an observed measurement or random variable will take on values within some set or range of values. Probabilities usually range between 0 and 1. A probability distribution of a random variable describes the probabilities associated with each possible value for the random variable. For example, letting 1 represent a defective food product and letting 0 represent a good food product, the probability of the event that a food item is defective will be written as: $P(X = 1)$. Let's consider another example. If X denotes the weight of a new baby. The probability that an observed baby weight is in the range 2 to 5 kg_f is expressed as: $P(2\ kg_f \leq X \leq 5\ kg_f)$.

In probability theory a probability density function (pdf), or density of a continuous random variable, is a function that describes the relative likelihood for this random variable to occur at a given point. The probability for the random variable to fall within a particular region is given by the integral of this variable's density over the region. The probability density function is non-negative everywhere, and its integral over the entire space must add to one.

To find the desired probability that an event occurs use a *probability density function* when we have continuous random variables or a *probability mass function* in the case of discrete random variables. In this text, $f(x)$ is used to represent the probability mass or density function for either discrete or continuous random variables, respectively. We now discuss how to find probabilities using these functions, first for the continuous case and then for discrete random variables. To find the probability that a continuous random variable falls in a particular interval of real numbers, we have to calculate the appropriate area under the curve of $f(x)$. Thus, we have to evaluate the integral of $f(x)$ over the interval of random variables corresponding to the event of interest. This is represented by:

$$P(a \leq X \leq b) = \int_a^b f(x)\,dx \tag{9.1}$$

The area under the curve of $f(x)$ between a and b represents the probability that an observed value of the random variable X will assume a value between a and b.

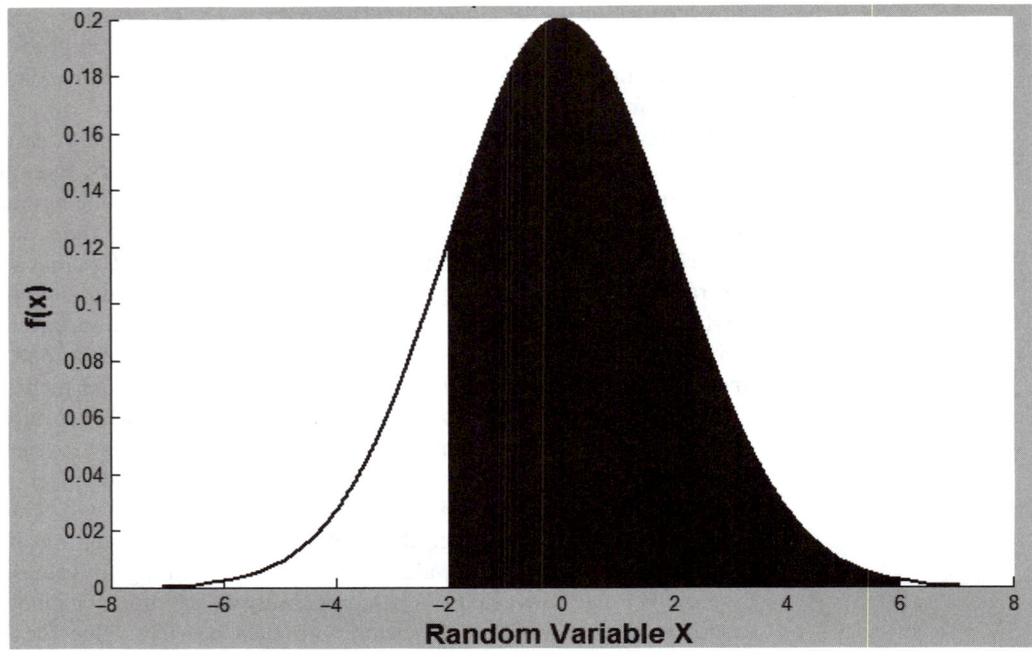

Figure 9.1 The area under the curve of $f(x)$ between –2 and 6 is the same as the probability that an observed value of the random variable will assume a value in the same interval; i.e., $P(-2 \le X \le 6)$.

This concept is illustrated in Fig. 9.1 where the shaded area represents the desired probability.

It should be noted that a valid probability density function should be nonnegative, and the total area under the curve must equal 1.

$$P(-\infty \le X \le \infty) = \int_{-\infty}^{\infty} f(x)\,dx = 1 \qquad (9.2)$$

NOTE If Eq. (9.2) is violated, then the probabilities will not be properly restricted to the interval [0, 1].

Example 9.1 A probability density function is defined as:

$$f(x) = a/x\,,\ 1 < x < 12$$

For $f(x)$ to be a valid probability density function, what is the value of a?

Solution To be a valid probability density function, all values of $f(x)$ must be positive, and the area beneath $f(x)$ must equal one. The first condition is met by restricting a

and x to positive numbers. To meet the second condition, the integral of $f(x)$ from 1 to 12 must equal 1.

$$\int_1^{12} \frac{a}{x} dx = 1 = aln(12) - aln(1) = aln(12)$$

$$a = \frac{1}{\ln(12)} = 0.40243$$

The *cumulative distribution function F(X)* is defined as the probability that the random variable X assumes a value less than or equal to a given x.

This is calculated from the probability density function, as follows:

$$F(X) = P(X \leq x) = \int_{-\infty}^{x} f(u)\,du \tag{9.3}$$

9.2 Typical Distribution Functions

For a typical distribution function, in addition to its probability (or sometimes called point) density function (pdf), the distribution is further characterized by its mean and variance, which respectively reflect, in general, the height and width of the curve.

9.2.1 Uniform Distribution

The uniform distribution on the interval [0, 1] has the probability density $f(x) = 1$ for $0 \leq x \leq 1$ and $f(x) = 0$ elsewhere. The distribution is defined by the two parameters, a and b, which are its minimum and maximum values. The distribution is often abbreviated *U(a, b)*. See Fig. 9.2.

$$f(x) = \begin{cases} \dfrac{1}{b-1} & for\ a \leq x \leq b \\ 0 & for\ x < a\ or\ x > b \end{cases} \tag{9.4}$$

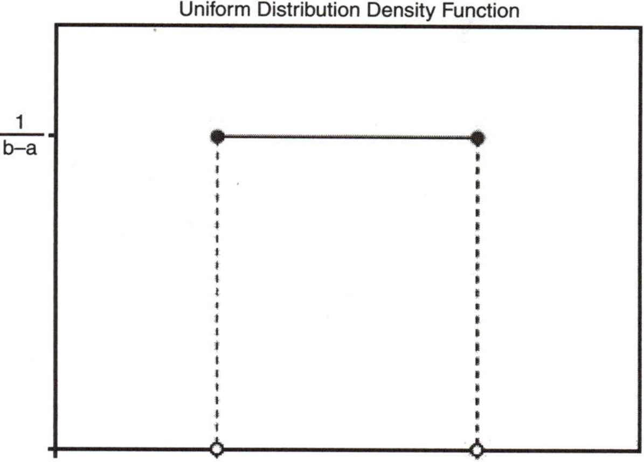

Figure 9.2 Representation of the uniform distribution function.

The mean and variance of the uniform distribution are given by:

$$E(X) = \frac{1}{2}(a+b) \tag{9.5}$$

$$V(X) = \frac{1}{12}(b-a)^2 \tag{9.6}$$

The MATLAB Statistics Toolbox functions **unifpdf(x, a,b)** and **unifcdf(x,a,b)** are used to obtain the uniform probability density function and the cumulative distribution function, respectively, over the interval [a,b].

9.2.2 Normal Distribution

The normal (or Gaussian) distribution is a continuous probability distribution that is often used as a first approximation to describe real-valued random variables that tend to cluster around a single mean value. The graph of the associated probability density function, which is "bell" shaped, is known as the Gaussian function or bell curve. See Fig. 9.3.

$$f(x) = \frac{1}{\sqrt{2\pi\sigma^2}} e^{-\frac{(x-\mu)^2}{2\sigma^2}} \tag{9.7}$$

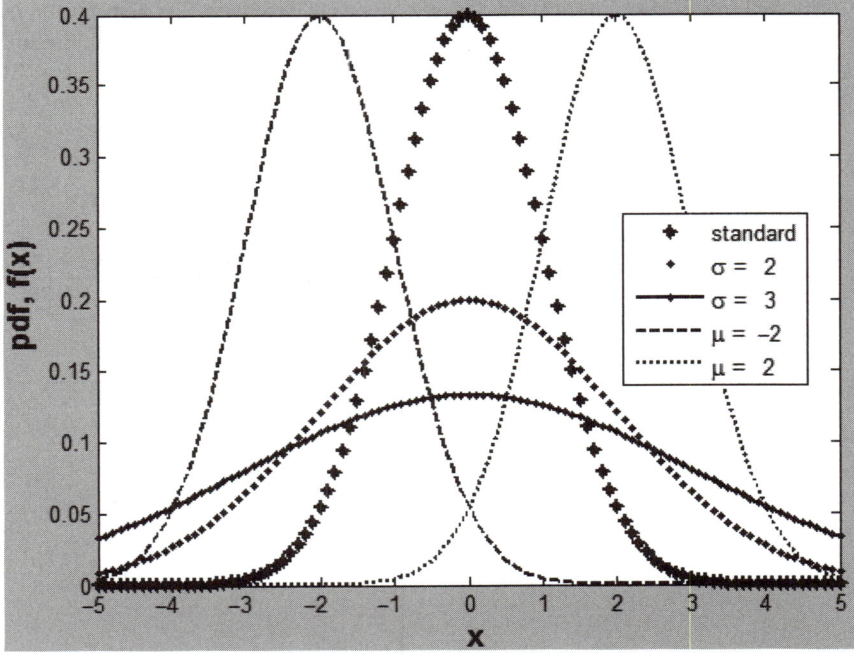

Figure 9.3 The plot of the normal distribution function as a function of x. The "*" curve represents the standard ($\mu = 0$; $\sigma^2 = 1$) normal distribution. Others differ by μ *or* σ.

$$E(X) = \mu \tag{9.8}$$

$$V(X) = \sigma^2 \tag{9.9}$$

The functions **normpdf(x,mu,sigma)** and **normcdf(x,mu,sigma)** are available in the MATLAB Statistics Toolbox for calculating the normal probability density function and cumulative normal distribution function, respectively. There is another special function called **normspec** that determines the probability that a random variable X assumes a value between two limits, where X is normally distributed with mean μ and standard deviation σ. This function also plots the normal density, where the area between the specified limits is shaded. The syntax is:

```
>> mu=4.0;
>> sigma=3;
>> spec=[1 8];
>> prob = normspec(spec, mu, sigma);
```

Figure 9.4 shows the plot generated by the previous set of commands.

9.2.3 *Binomial Distribution*

The binomial distribution is the discrete probability distribution of the number of successes in a sequence of n independent yes/no experiments, each of which

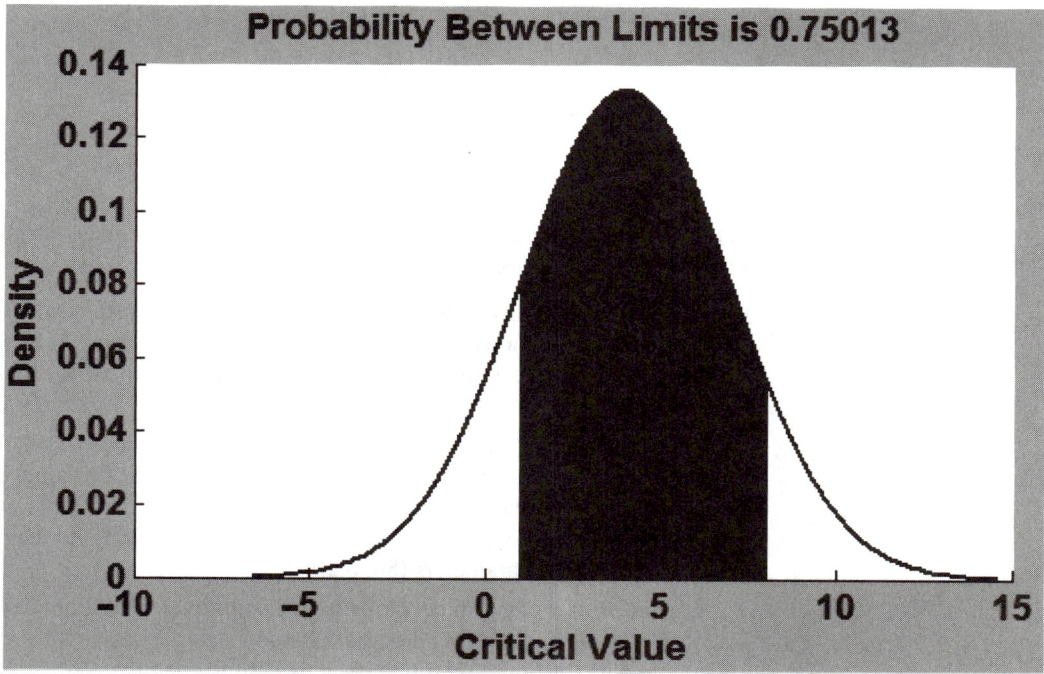

Figure 9.4 The **normspec** function that plots the normal distribution curve and evaluates the area that lies under the curve, between the two limits given by the **spec** vector.

yields success with probability p. Such a success/failure experiment is also called a Bernoulli experiment or Bernoulli trial; when $n = 1$, the binomial distribution is a Bernoulli distribution. The binomial distribution is $n \times$ repeated Bernoulli trial. The binomial distribution is the basis for the popular binomial test of statistical significance.

The binomial distribution is frequently used to model the number of successes in a sample of size n drawn *with replacement* from a population of size N. If the sampling is carried out *without replacement*, the draws are not independent and so the resulting distribution is hyper-geometric, not binomial. However, for N much larger than n, the binomial distribution is a good approximation and is widely used.

In general, if the random variable K follows the binomial distribution with parameters n and p, we write $K \sim B(n, p)$. The probability of getting exactly k successes in n trials is given by the *probability mass function*:

$$f(k; n, p) = \binom{n}{k} p^k (1-p)^{n-k} \tag{9.10}$$

for $k = 0, 1, 2, \ldots, n$

$$\binom{n}{k} = \frac{n!}{(n-k)!\,k!} \tag{9.11}$$

The mean and variance of a binomial distribution are given by:

$$E(K) = np \tag{9.12}$$

$$V(K) = np(1-p) \tag{9.13}$$

See Fig. 9.5.

Here are some examples in which the results of an experiment can be modeled by a binomial random variable:

1. A drug has probability 0.95 of curing a disease. It is administered to 200 patients, and the outcome for each patient is either cured or not cured. If X is the number of patients cured, then X is a binomial random variable with parameters (200, 0.95).
2. The National Institute of Mental Health estimates that there is a 20% chance that an adult American suffers from a psychiatric disorder. 80 adult Americans are randomly selected. If we let X represent the number of adults who have a psychiatric disorder, then X takes on values according to the binomial distribution with parameters (80, 0.20).
3. A manufacturer of computer chips finds that on average 6% are defective. To monitor the manufacturing process, a random sample of size 80 is used. If the sample contains more than five defective chips, then the process is stopped. The binomial distribution with parameters (80, 0.06) can be used to model the random variable X, where X represents the number of defective chips.

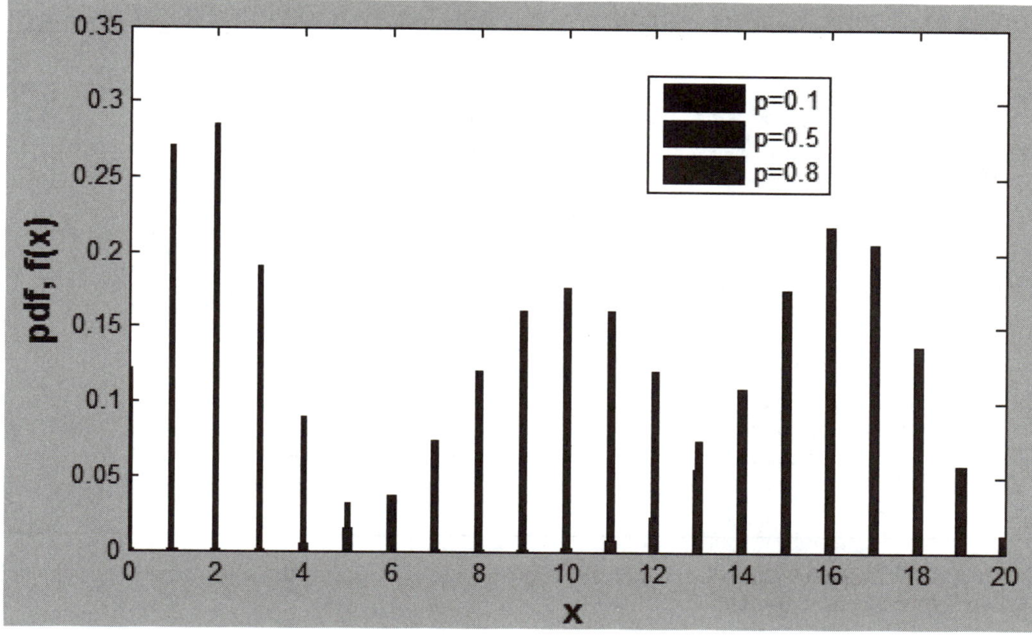

Figure 9.5 A binomial distribution for $n = 20$; $p = 0.1$ (*left*); $p = 0.5$ (*middle*); and $p = 0.8$ (*right*).

Example 9.2 Suppose there is a 20% chance that an adult American suffers from a psychiatric disorder. We randomly sample 30 adult Americans. If we let X represent the number of people who have a psychiatric disorder, then X is a binomial random variable with parameters (30, 0.2). We are interested in the probability that at most four of the selected people have such a disorder. We can use MATLAB Statistics Toolbox function **binocdf** (the binomial cumulative mass function) to determine $P(X \leq 4)$, as follows:

```
>> Prob_Val=binocdf(4,30,0.2)
Prob_Val =
    0.2552
```

Alternatively, we may also sum up the individual values of the binomial probability mass function (**binopdf**):

```
>> Prob_Val2=sum(binopdf(0:4,30,0.2))
Prob_Val2 =
    0.2552
```

In words, the probability that at most 4 out of 30 people have psychiatric disorder is about 25%. Figure 9.6 shows the M-file needed to graphically show both the binomial (point) mass and binomial cumulative mass function describing such a case (see Fig. 9.7).

```
% Get the values for the domain, x.
x = 0:4;
% Get the values of the probability mass function.
% For n = 30, p = 0.2:
pdf = binopdf(x,30,0.2);
pdf2=binocdf(x,30,0.2);
% Do the plots.
subplot(1,2,1), bar(x,pdf,1,'k')
title(' n = 30, p = 0.2')
xlabel('\bf \fontsize{14} X'),ylabel('\bf \fontsize{14} f(X)')
axis([0 5 0 0.26]);
axis square;
subplot(1,2,2), bar(x,pdf2,1,'k')
title(' n = 30, p = 0.2')
xlabel('\bf \fontsize{14} X'),ylabel('\bf \fontsize{14} F(X)')
axis([0 5 0 0.26]);
axis square;
```

Figure 9.6 **example9_2.m** file that evaluates and graphically presents both the point mass and cumulative mass function of the binomial distribution.

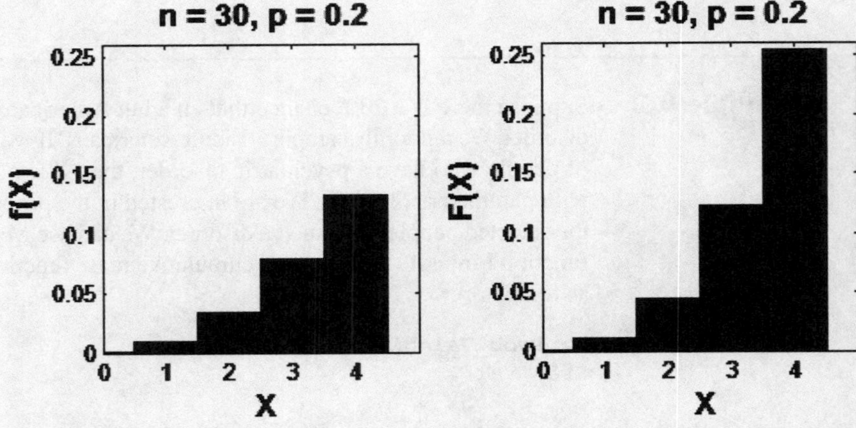

Figure 9.7 The plot of $f(X)$ and $F(X)$ as a function of X, where X takes on the values between 0 and 4. The sample size, n, is 30 with $p = 0.2$.

9.2.4 *Poisson Distribution*

The Poisson distribution (pronounced [pwasɔ̃]) is a discrete probability distribution that expresses the probability of a given number of events occurring in a fixed interval of time and/or space if these events occur with a known average rate and independently of the time since the last event. (Poisson distribution can also be used for the number of events in other specified intervals such as distance, area, or volume.)

If the expected number of occurrences in this interval is λ, then the probability that there are exactly x occurrences (x being a non-negative integer, $x = 0, 1, 2, ...$) is equal to:

$$f(x;\lambda) = \frac{\lambda^x e^{-\lambda}}{x!} \qquad x = 0, 1, 2, 3, \ldots \tag{9.14}$$

where $\boldsymbol{e}$ is the base of the natural logarithm ($e = 2.71828...$), $\boldsymbol{x}$ is the number of occurrences of an event—the probability of which is given by the function, $\boldsymbol{x!}$ is the factorial of x, and $\boldsymbol{\lambda}$ is a positive real number, equals the expected number of occurrences in the given interval.

For instance, if the events occur, on the average, five times per minute, and one is interested in the probability of an event occurring k times in a 6-min interval, one would use a Poisson distribution as the model with $\lambda = 5\frac{events}{min} \times 6\,min = 30\,events.$

The mean and variance of Poisson distribution are given by:

$$E(X) = \lambda \tag{9.15}$$

$$V(X) = \lambda \tag{9.16}$$

Example 9.3 Suppose the number of individual plants of a given species we expect to find in a one meter square follows the Poisson distribution with mean $\mu = 10\frac{plants}{m^2} \times 1m^2 = 10 plants$. Find the probability of finding exactly 12 individuals.

$$f(x;\lambda) = f(12;10) = \frac{\lambda^x e^{-\lambda}}{x!} = \frac{10^{12} e^{-10}}{12!} = 0.0948$$

Using MATLAB:

```
>> Prob=poisspdf(12,10)
Prob =
   0.0948
```

In words, the probability of finding 12 plants in one square meter will be about 10%.

Example 9.4 Suppose that we assume that at a location, a particular species of plant is distributed according to a Poisson process with expected density $0.3\frac{individuals}{m^2}$. In a 9-m square, what is the probability of no individuals?

$$\lambda = 0.3\frac{individuals}{m^2} \times 9m^2 = 2.7\ individuals$$

$$f(x;\lambda) = f(0;2.7) = \frac{2.7^0 e^{-2.7}}{0!} = e^{-2.7} = 0.0672$$

```
>>Prob=poisspdf(0,2.7)
Prob =
   0.0672
```

In words, the probability of finding no single individual in a 9 m^2 area will be about 7%.

Example 9.5 Suppose that a metal wire producing company would like to inspect its product quality. The number of defects in a metal wire follows the Poisson process with expected density of 1 defect per 100 longitudinal meters. In a 100-meter wire length, what is the probability that such a wire has three defects or more? The company will reject the metal wire if it has three defects or more.

$$\lambda = 1\frac{defect}{100m} \times 100m = 1\ defect$$

$$P(X \geq 3) = 1 - P(X < 3) = 1 - P(X = 2) - P(X = 1) - P(X = 0)$$

$$P(X \geq 3) = 1 - \left[\frac{1^2 e^{-1}}{2!} + \frac{1^1 e^{-1}}{1!} + \frac{1^0 e^{-1}}{0!}\right] = 1 - [0.1839 + 0.3679 + 03679] = 0.0803$$

This means that out of 100 samples being probed, wherein each sample is 100 meters long, 8.0% will be rejected.

Now, using MATLAB:

```
>> Prob=1-poisscdf(2,1)
Prob =
         0.0803
```

9.2.5 *Exponential Distribution Function*

$$f(x;\mu) = \frac{1}{\mu} e^{-\frac{x}{\mu}} \qquad x \geq 0;\ \ \mu > 0 \tag{9.17}$$

The mean and variance of the exponential distribution are given by:

$$E(X) = \mu \tag{9.18}$$

$$V(X) = \mu^2 \tag{9.19}$$

The functions **exppdf(x,mu)** and **expcdf(x,mu)** are available in the MATLAB Statistics Toolbox for calculating the exponential probability density and exponential cumulative distribution function, respectively.

Example 9.6

```
>> y = exppdf(5,1:5)
y =
    0.0067    0.0410    0.0630    0.0716    0.0736
>> y = exppdf(1:5,1:5)
y =
    0.3679    0.1839    0.1226    0.0920    0.0736
>> expcdf(0:3,2)
ans =
         0    0.3935    0.6321    0.7769
>> expcdf(0:3,2:5)
ans =
         0    0.2835    0.3935    0.4512
```

Example 9.7 The time between arrivals of vehicles at an intersection follows an exponential distribution with a mean of 12 seconds. What is the probability that the time between arrivals is 10 seconds or less? We are given the average interarrival time, so $\mu = 12$ seconds. The required probability is obtained as follows:

$$P(X \leq 10) = \int_0^{10} \frac{1}{\mu} e^{-\frac{x}{\mu}} dx = -e^{-\frac{x}{\mu}} \Big|_0^{10} = e^{-\frac{x}{\mu}} \Big|_{10}^{0} = 1 - e^{-\frac{10}{12}} = 0.5654$$

Using MATLAB:

```
>> expcdf(10,12)
ans =
   0.5654
```

In words, the probability for the elapsing time between two vehicle arrivals to be 10 seconds or less will be about 56%.

9.2.6 *Gamma Distribution Function*

The gamma probability density function with parameters $a > 0$ and $b > 0$:

$$f(x;\, a,\, b) = \frac{1}{b^a \Gamma(a)} x^{a-1} e^{-\frac{x}{b}}; \qquad x \geq 0 \tag{9.20}$$

The gamma function $\Gamma(a)$ is defined as:

$$\Gamma(a) = \int_0^{\infty} e^{-u} u^{a-1} du \tag{9.21}$$

For integer arguments of the gamma function:

$$\Gamma(a) = (a-1)! \tag{9.22}$$

The mean and variance of the gamma distribution are given by:

$$E(X) = ab \tag{9.23}$$

$$V(X) = ab^2 \tag{9.24}$$

The gamma cdf is:

$$ProbSum = F(x;\, a,\, b) = \frac{1}{b^a \Gamma(a)} \int_{-\infty}^{x} t^{a-1} e^{-\frac{t}{b}} dt \tag{9.25}$$

The result, *ProbSum*, is the sum of probabilities for all possible observations of x that fall in the interval $[0\ x]$ with a probability given by a gamma distribution with parameters a and b.

The functions **gampdf(x,a,b)** and **gamcdf(x,a,b)** are available in the MATLAB Statistics Toolbox for calculating gamma probability density and gamma cumulative distribution function. The gamma probability density function is useful in reliability models of lifetimes. The gamma distribution is frequently a probability model for waiting times; for instance, in life testing, the waiting time until death is a random variable that is frequently modeled with a gamma distribution. The gamma distribution is more flexible than the exponential distribution in that the probability of a product surviving an additional period may depend on its current age. The exponential and chi square (χ^2) functions are special cases of the gamma function.

Example 9.8 Engineers designing the next generation of space shuttles plan to include two fuel pumps—one active, the other standby. If the primary pump malfunctions, the second is immediately brought into service. Suppose a typical mission is expected to require that fuel be pumped for at most 50 hours. According to the manufacturer's specifications, pumps are expected to fail once every 100 hours. What are the chances that such a fuel pump system would completely fail for the next 50 hours?

Let X denote the elapsing time until $a = 2$ (i.e., the second pump breaks down) and $b = 100$ hours (lifetime).

$$P(X < 50) = gamcdf(x, a, b) = gamcdf(50, 2, 100)$$

```
>> Fail_Prop2=gamcdf(50,2,100)
Fail_Prop2 =
    0.0902
```

In words, the probability that both pumps fail in the next 50 hours will be about 9%.

If you run:

```
>> Fail_Prop1=gamcdf(50,1,100)
Fail_Prop1 =
    0.3935
```

In words, the probability that the first pump fails in the next 50 hours will be about 39%.

Example 9.9 The time between major stock market crashes in the years 2009–2015 can be described by gamma distribution. The time in days for one major stock market to crash will occur within 360. How many days can one expect to pass until the fourth stock market crash?

The time x until the fourth stock market crash will follow a gamma distribution with $a = 4$ and $b = 360$, and so $E(X) = ab = 4 \times 360 = 1440$ days.

9.2.7 *Chi-Square Distribution*

A gamma distribution where $b = 2$ and $a = \frac{\nu}{2}$ with ν a positive integer is called a chi-square distribution (denoted as χ_ν^2) with ν degrees of freedom. The chi-square distribution is used to derive the distribution of the sample variance and is important for goodness-of-fit tests in statistical analysis (see Example 9.11). The probability density function for a chi-square random variable with ν degrees of freedom is:

$$f(x;\, \nu) = \frac{1}{2^{\nu/2}\Gamma(\nu/2)} x^{\frac{\nu}{2}-1} e^{-\frac{x}{2}}; \qquad x \geq 0$$

The mean and variance of chi square are given by:

$$E(X) = \nu \tag{9.26}$$

$$V(X) = 2\nu \tag{9.27}$$

The functions **chi2pdf(x,ν)** and **chi2cdf(x,ν)** are available in the MATLAB Statistics Toolbox for calculating the probability density function and cumulative distribution function for chi-square function. The following examples demonstrate the application of chi-square distribution as a tool to analyze studies, tests, or surveys.

Example 9.10 A battery company has developed a new cell phone battery. On average, the battery lasts 90 minutes on a single charge. The standard deviation is 6 minutes.

Suppose the manufacturing department runs a quality control test. They randomly select nine batteries. The standard deviation of the selected batteries is 7 minutes. What would be the chi-square statistic represented by this test?

Suppose the test was repeated with a new random sample of nine batteries. What is the probability that the standard deviation in the new test would be greater than 7 minutes?

$$\chi_\nu^2 = \frac{\{(n-1)s^2\}}{\sigma^2}$$

where n is the sample size; s is the sample standard deviation; and σ is the population standard deviation.

NOTE The chi-square statistic (χ_ν^2) increases with s and decreases with σ, for the same sample size, n.

$$\chi_8^2 = \frac{\{(9-1)7^2\}}{6^2} = 10.9$$

$$P(X > 10.9) = 1 - P(X \leq 10.9) = 1 - chi2cdf(10.9, 8) = 1 - 0.793 = 0.207$$

In words, the probability that the new samples have a standard deviation greater than 7 (i.e., $\chi_8^2 > 10.9$) will be about 21%.

Example 9.11 A company hired you as a researcher to determine which kind of computer system it should use. The systems being considered are Apple, IBM, Mac, and Zenith.

The company is eager to know if these brands significantly differ in quality. You select 60 computer experts as your test group. You ask them to use each of the computers for a month and then to choose one computer that they think is the best. After you have tabulated the results, you find that the selections are distributed in the following way:

Apple	IBM	Mac	Zenith
18	12	12	18

Let us analyze the data using the chi-square statistic. Chi-square is a statistical test commonly used to compare observed data with data we would expect to obtain according to a specific hypothesis. Here, we expect that 15 PCs (out of 60) will be dedicated for each of the four brands. The "goodness to fit" between the observed and expected will be formulated as chi-square distribution. Then you have to answer if the deviations (differences between observed and expected) are the result of chance (i.e., the observed values are randomly distributed around their mean value), or if they were due to other factors. What is the maximum tolerated deviation that can occur before you reach a conclusion that something other than chance is the cause behind the significant difference between the observed and the expected value?

The chi-square test is always testing what scientists call the *null hypothesis*, which states that there is no significant difference between the expected and observed results.

Table 9.1 shows how the chi-square statistic is calculated, based on both the observed and expected values.

Table 9.1 Chi-square statistic calculation based on both the observed and expected values.

	Apple	IBM	Mac	Zenith
Observed (O_i)	18	12	12	18
Expected (E_i)	15	15	15	15
Deviation, $D_i = (O_i - E_i)$	3	−3	−3	3
$D_i^2 = (O_i - E_i)^2$	9	9	9	9
$\chi_i^2 = \frac{D_i^2}{E_i}$	0.6	0.6	0.6	0.6

$$\chi_3^2 = \sum_{i=1}^{Group\ \#} \chi_i^2 = \frac{D_i^2}{E_i} = 0.6 + 0.6 + 0.6 + 0.6 = 2.4$$

If the p value for the calculated χ_3^2 is > 0.05 (this is the level of significance, or the relative standard commonly used in different studies), *accept your null hypothesis*. The deviation is small enough that chance alone accounts for it. A p value of

Table 9.2 p value as a function of the chi-square statistic and its associated degree of freedom, ν.

ν	Probability (p)										
	0.95	**0.90**	**0.80**	**0.70**	**0.50**	**0.30**	**0.20**	**0.10**	**0.05**	**0.01**	**0.001**
1	0.004	0.02	0.06	0.15	0.46	1.07	1.64	2.71	3.84	6.64	10.83
2	0.10	0.21	0.45	0.71	1.39	2.41	3.22	4.60	5.99	9.21	13.82
3	0.35	0.58	1.01	1.42	2.37	3.66	4.64	6.25	7.82	11.34	16.27
4	0.71	1.06	1.65	2.20	3.36	4.88	5.99	7.78	9.49	13.28	18.47
5	1.14	1.61	2.34	3.00	4.35	6.06	7.29	9.24	11.07	15.09	20.52

0.8, for example, means that there is 80% probability that any deviation from the expected is due to chance only. This is within the range of acceptable deviation.

On the other hand, if the p value for the calculated χ_3^2 is < 0.05, *reject your null hypothesis*, and conclude that some factor other than chance is the cause behind the deviation being so great. For example, a p value of 0.01 means that there is only a 1% chance that this deviation is due to chance alone. Therefore, other factors must be involved.

Table 9.2 shows the p value of the chi-square statistic that has a degree of freedom ν. From Table 9.2, the p value for the calculated chi-square statistic, $\chi_3^2 = 2.4$, with $v = 3$ is about 0.5 (exactly 0.4936, see below). That is greater than 0.05. Hence, the conclusion is to *accept the null hypothesis*; that is, the deviation is small enough that chance (monitored values randomly fluctuate around their mean) alone accounts for it. Another way of looking at the result is simply realizing that a chi-square value of 2.4, which represents the sum of normalized deviations for all four groups, is less than the critical value that is equal to 7.82, for a significance level, α, set at 0.05. If the p value had been less than 0.05 (alternatively, the sum of normalized deviations > 7.82) then the null hypothesis would have been rejected.

Using MATLAB:

```
>> p_Val=1-chi2cdf(2.4,3)
p_Val =
    0.4936
```

Evaluated at $\chi_3^2 = 2.4$, and $v = 3$, this represents the p value $= 0.4936 > 0.05$.

```
>> p_Val_alpha05=1-chi2cdf(7.82,3)
p_Val_alpha05 =
                0.0499
```

This is the p value $= 0.0499 \cong 0.05$, which represents the level of significance $\alpha = 0.05$, commonly used in studies and surveys.

9.2.8 Weibull Distribution

Waloddi Weibull (1887–1979) offered the distribution that bears his name as an appropriate analytical tool for modeling the breaking strength of materials. Current usage also includes reliability and lifetime modeling. The primary advantage of

Weibull analysis is the ability to provide reasonably accurate failure analysis and failure forecasts with extremely small samples. Solutions are possible at the earliest indications of a problem without having to "crash a few more."

The Weibull distribution is more flexible than the exponential for these purposes. The probability density function for Weibull random variable is:

$$f(x;\, a,b) = ba^{-b}x^{b-1}e^{-\left(\frac{x}{a}\right)^b}; \quad x \geq 0 \tag{9.28}$$

where $b > 0$ is the shape parameter and $a > 0$ is the aging time/scale parameter of the distribution.

The characteristic life, $a > 0$, is the typical aging time/factor until failure occurs in Weibull analysis and it is related to the mean time/factor to failure. The units of ageing time/factor depend on the part usage and the failure mode. For example, low cycle and high cycle fatigue may produce cracks leading to rupture. The age units would be fatigue cycles. The age of a starter may be the number of starts. Burner and turbine parts may fail as a function of time at high temperature, or as the number of cold to hot to cold cycles. In this regard, knowledge of the physics-of-failure will provide the age scale/factor.

On the other hand, parameter b provides a clue to the physics of the failure. In general, parameter b indicates which class of failures is present:

- $b > 1.0$ indicates infant mortality (i.e., decay or aging constant is high)
- $b = 1.0$ means random failures (independent of age)
- $b < 1.0$ indicates wear out failures (i.e., normal decay or aging constant)

The mean and variance of Weibull are given by:

$$E(X) = a\Gamma\left(1 + \frac{1}{b}\right) \tag{9.29}$$

$$V(X) = a^2\left[\Gamma\left(1 + \frac{2}{b}\right) - \Gamma\left(1 + \frac{1}{b}\right)^2\right] \tag{9.30}$$

NOTE There are different versions, other than Eq. (9.28), for defining the pdf of Weibull distribution. Here, MATLAB notation is used.

The functions **wblpdf(x,a,b)** and **wblcdf(x,a,b)** are available in the MATLAB Statistics Toolbox for calculating the probability density function and cumulative distribution function of Weibull.

Example 9.12 To demonstrate the effect of parameter b on the lifetime of a working part/machine, consider the three possible cases for b:

```
>> fail_prob_of_time_lt_300h_beta_gt_1=wblcdf(300,100,2)
fail_prob_of_time_lt_300h_beta_gt_1 =
    .9999
```

```
>> fail_prob_of_time_lt_300h_beta1=wblcdf(300,100,1)
fail_prob_of_time_lt_300h_beta1 =
    0.9502
>> fail_prob_of_time_lt_300h_beta_lt_1=wblcdf(300,100,0.5)
fail_prob_of_time_lt_300h_beta_lt_1 =
    0.8231
```

Obviously, the probability that a given working part/machine to fail before 300 units of aging (e.g., hours of operation) is highest for $b = 2$ (i.e., infant mortality case).

Example 9.13 Suppose the time to failure of a metal wire can be modeled by the Weibull distribution with $b = 1/5$ and $a = 300$ h. To find the mean time to failure using the expected value of a Weibull random variable:

$$E(X) = 300 \times \Gamma\left(1 + \frac{1}{0.2}\right) = 300 \times 5! = 300 \times 120 = 36000 \text{ h.}$$

Let us say we want to know the probability that the wire will fail before 10,000 hours. We can calculate this probability using MATLAB **wblcdf(x,a,b)**:

```
>> Time_2_Fail=wblcdf(10000,300,1/5)
Time_2_Fail =
    0.8669
```

In words, the probability that the wire will fail before 10,000 hours is about 87%.

Example 9.14 Suppose that the lifetime of a certain kind of emergency backup battery (in hours) is a random variable X having a Weibull distribution with $a = 100$ hours and $b = 0.5$. Find (1) the probability that such a battery will last more than 300 hours

```
>> prob_of_time_gt_300h=1-wblcdf(300,100,0.5)
prob_of_time_gt_300h =
    0.1769
```

and (2) the mean lifetime of these batteries

$$E(X) = a\Gamma\left(1 + \frac{1}{b}\right) = 100 \times \Gamma\left(1 + \frac{1}{0.5}\right) = 100 \times 2! = 200 \text{ h}$$

9.2.9 *Beta Distribution*

The beta distribution is a family of continuous probability distributions defined on the interval [0, 1] parameterized by two positive shape parameters, typically denoted by a and b. The beta distribution can be suited to the statistical modeling of proportions in applications in which values of proportions equal to 0 or 1 do not occur. There are many random variables expressed as percentages, or fractions. These variables have density functions defined on [0, 1]. A large class of random variables, such as the percentage of new businesses that turn a profit in their first year, the percentage of banks that default in a given year, and the percentage of time a plant's machinery is inactive, can be modeled by a **beta density function**. The probability density function for the beta random variable is:

$$f(x;\, a,b)=\frac{\Gamma(\mathrm{a+b})}{\Gamma(\mathrm{a})\times\Gamma(\mathrm{b})}(x)^{a-1}(1-x)^{b-1};\quad 0\le x\le 1 \tag{9.31}$$

where $a > 0$ and $b > 0$ are two positive shape parameters for the distribution.

The mean and variance of beta are given by:

$$E(X)=\frac{a}{a+b} \tag{9.32}$$

$$V(X)=\frac{ab}{[(\mathrm{a+b})^2\times(\mathrm{a+b+1})]} \tag{9.33}$$

The functions **betapdf(x,a,b)** and **betacdf(x,a,b)** are available in the MATLAB Statistics Toolbox for calculating beta probability density function and beta cumulative distribution function.

Example 9.15 A utilities industry consultant predicts a cutback in the Canadian utilities industry during 2010–2015 by a percentage specified by a beta distribution with $a = 1.5$ and $b = 2$. Calculate the probability that Ontario Hydro will downsize by between 20% and 40% during the given 5-year period.

Using MATLAB **betacdf(x,a,b)**, we can calculate the probability that x lies between 20% and 40% during the given 5-year period.

$$P(0.2\le X\le 0.4)=betacdf(0.4,1.5,2)-betacdf(0.2,1.5,2)$$

```
>> Down_Size_Prob=betacdf(0.4,1.5,2)-betacdf(0.2,1.5,2)
Down_Size_Prob =
0.2839
```

In words, the probability that Ontario Hydro will downsize by between 20% and 40% is about 28%.

Example 9.16 The relative humidity, Y, when measured at a certain location, can be described by the beta probability density function with parameters $a = 4$ and $b = 3$. Find the probability that humidity lies between 40% and 60%.

```
>> Relative_Humidity_Prob=betacdf(0.6,4,3)-betacdf(0.4,4,3)
Relative_Humidity_Prob =
          0.3651
```

In words, the probability that the relative humidity lies between 40% and 60% is about 36%.

Example 9.17 The percentage of impurities per batch in a certain chemical product is a random variable Y having beta density function with parameters $a = 0.5$ and $b = 10$. A batch with more than 10% impurities will be rejected. What is the probability that a randomly selected batch will be rejected because of excessive impurities?

```
>> Impurity_GT_10_Prob=1-betacdf(0.1,0.5,10)
Impurity_GT_10_Prob =
        0.1516
```

In words, the probability that the selected batch will be rejected because it contains more than 10% impurities is about 15%.

9.3 Maximum Likelihood Estimates (MLE)

MATLAB has an additional tool that allows the user to estimate the maximum likelihood estimates (MLEs) for the parameters of different distributions, using the sample data in the vector **data**. The command is as follows:

```
>> phat = mle(data)
```

Or,

```
>>[phat,pci] = mle(data,'distribution',dist)
```

where **phat** is a vector that represents the estimated parameters, **pci** is the parameter's 95% confidence interval bracketing the estimated value, and ***dist*** is replaced by the specific name of the distribution function under concern, which can be one of the following list:

- `'beta'`
- `'bernoulli'`
- `'binomial'`
- `'birnbaumsaunders'`
- `'discrete uniform'` or `'unid'`
- `'exponential'`
- `'extreme value'` or `'ev'`
- `'gamma'`
- `'generalized extreme value'` or `'gev'`
- `'generalized pareto'` or `'gp'`
- `'geometric'`
- `'inversegaussian'`
- `'logistic'`
- `'loglogistic'`
- `'lognormal'`
- `'nakagami'`
- `'negative binomial'` or `'nbin'`

- 'normal'
- 'poisson'
- 'rayleigh'
- 'rician'
- 'tlocationscale'
- 'uniform'
- 'weibull' or 'wbl'

NOTE For the 'binomial' choice, the user has to add an additional parameter 'ntrials,' which stands for the total number of trials followed by its numerical value.

Example 9.18 First, the MATLAB built-in **binornd** will synthesize a (100 × 1) column vector that follows a random binomial distribution with sample size $n = 20$ and $p = 0.85$.

```
>> data = binornd(20,0.85,100,1); % Simulated data, p = 0.85
```

Second, using the MATLAB **mle** function:

```
>>[phat,pci] = mle(data,'distribution','binomial','alpha',
.05,'ntrials',20)
phat =
    0.8625
pci =
   0.8466
   0.8773
```

phat is the curve-fitted value for the p value that was assigned a value of 0.85 in the simulated data (i.e. 100 × 1 column vector). Notice how close the curve-fitted to the assigned value is. **pci** represents the 95% confidence interval for the estimated value.

Example 9.19 First, we synthesize random data (**n1:** 100 × 1 column vector) that follow the gamma distribution with parameters $a = 1$ and $b = 2$ (see Sec. 9.2.6):

```
>> n1=gamrnd(1,2,100,1);
```

Second, using the MATLAB **mle** function:

```
>> [phat,pci]=mle(n1,'distribution','gamma')
phat =
    0.8817 2.4664
pci =
   0.6927 1.7927
   1.1223 3.3934
```

Here, **mle** estimates the gamma distribution function parameters (a & b) and their 95% C.I.

Notice that the estimated parameters a and b of the gamma distribution function are 0.8817 and 2.4664, respectively.

9.4 A Full-Fledged Example Showing Statistical Data Acquisition, Analysis, and Visualization by MATLAB

A prototype aircraft was tested in a wind tunnel. The test focused on understanding the wing stresses during the climbing phases of the flight, when stresses are typically high. The wind tunnel's gust generator induces a controlled gust field to simulate severe wind conditions. The stress was measured at seven locations on each wing (see Fig. 9.8).

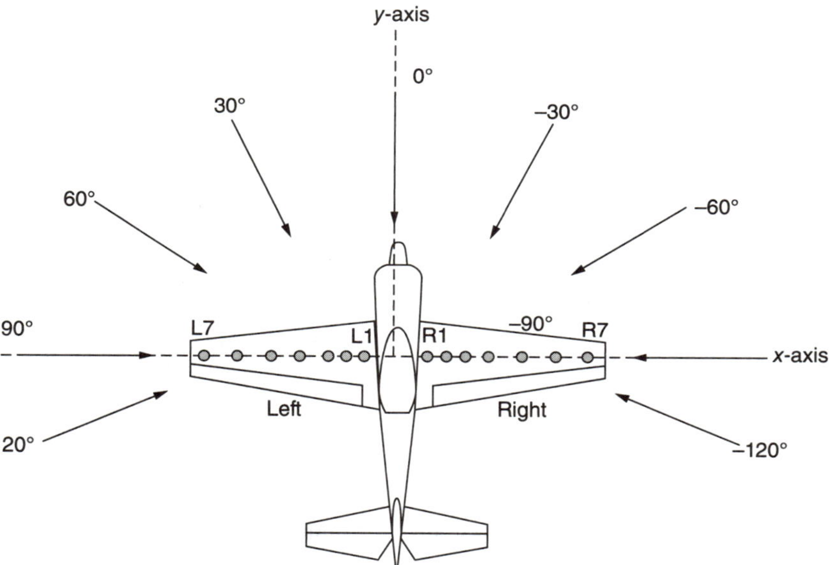

Figure 9.8 The location of test spots and direction (i.e., angle) of wind flowing on the body of the aircraft, which makes a certain angle with the zero-direction (up-front) facing the plane.

The data from testing are stored in WingStress.xls, a Microsoft Excel spreadsheet available at www.mhprofessional.com/MATLABnumericalmethods. The data include measurements of wind velocity (*m/s*), wind angle (degrees), and stress measurements (*kPa*) for the left (L) and right (R) wings, where 1 denotes the sensor location near the base of the wing and 7 denotes the sensor near the tip of the wing.

The way to import a Microsoft Excel spreadsheet into MATLAB environment (i.e., Workspace) is explained in detail in Sec. 5.2. Here, the data will be brought to the MATLAB environment via creating a dataset from an Excel spreadsheet.

9.4.1 Load In the Data Using the Dataset Array

Take advantage of the dataset array provided in the Statistics Toolbox for manipulating and storing the experimental data.

NOTE You need to have the Statistics Toolbox version 6.0 or higher to use dataset arrays.

```
>>clear all; close all % clean up previous work
```

% Load in data to a dataset array (Statistics Toolbox data type):

```
>> myds = dataset('XLSFile','WingStress.xls');
```

Figure 9.9 shows that MATLAB created **myds** which is a dataset that contains 72 entries (i.e., rows) by 16 columns: the first column for velocity, the second column for angle, and the other 14 columns for stress data for both right and left wings.

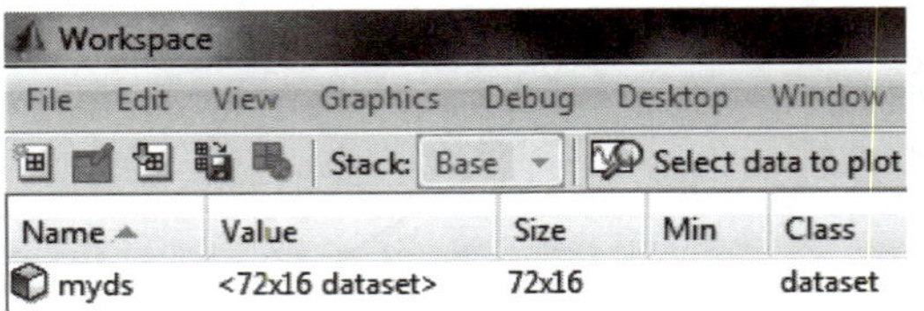

Figure 9.9 MATLAB's Workspace showing the created dataset named **myds**, which has 16 columns that represent the velocity, angle, and 14 stress datasets for right and left wing.

The command **repmat** in MATLAB:

```
>> B = repmat(A,m,n)
```

creates a large matrix *B* consisting of an *m*-by-*n* tiling of copies of A. The size of B is [size(A,1)*m, (size(A,2)*n].

```
>> StrainUnits = repmat({'kPa = kN/m^2'},1,14); % creates a cell structure which is 1x14 row vector, where 'kPa = kN/m^2' is coined 14 times.
```

The command:

```
>> myds.Properties.Units = {'m/s','degrees',StrainUnits{:}};
```

will assign the units for each column: the first is the velocity, the second the angle, and the other 14 columns are stress values that have **'kPa = kN/m^2'** units.

```
>> summary(myds) % Shows summary
```

A portion of the summary is shown in Fig. 9.10. It shows the minimum, maximum, median, first quartile, and third quartile values for each vector (i.e., column in this case).

```
Velocity: [72x1 double, Units = m/s]
     min         1st Qrtl     median     3rd Qrtl     max
     4.4538      7.8577       12.001     15.94        20.598
Angle: [72x1 double, Units = degrees]
     min      1st Qrtl     median     3rd Qrtl     max
     -120     -60          0          60           120
L1: [72x1 double, Units = kPa = kN/m^2]
     min        1st Qrtl     median     3rd Qrtl     max
     65.275     140.88       199.79     252.53       370.11
L2: [72x1 double, Units = kPa = kN/m^2]
     min        1st Qrtl     median     3rd Qrtl     max
     60.963     154.72       201.61     252.22       341.56
L3: [72x1 double, Units = kPa = kN/m^2]
     min        1st Qrtl     median     3rd Qrtl     max
     27.762     75.422       95.742     128.49       198.73
```

Figure 9.10 A portion of the summary for the dataset **myds**, showing the minimum, maximum, median, first quartile, and third quartile values for each vector (i.e., column in this case).

One can see how the dataset loaded the data from the spreadsheet file and automatically used the column names as variable names within the dataset. The summary command lists the dataset contents, including the type, size, assigned units, and summary of statistics for each variable. You can directly access the variables using the "dot" notation, for example, if you type at the command prompt:

```
>> myds.L2(1:6)
ans =
   146.5684
   161.5095
   192.3588
   203.5443
   200.2862
   204.9604
```

9.4.2 *Data Management and Manipulation*

Datasets make it easy to access the data by name and operate on the data within the dataset as a group. For example, to access only 90° data for L1 and R1:

```
>> L1R190 = myds(myds.Angle == 90,{'Angle','L1','R1'})
L1R190 =
    Angle     L1         R1
    90        159.36     156.15
    90        160.8      163.92
    90        150.6      137.59
    90        173.62     131.17
    90        183.5      160.35
    90        151.1      164.99
    90        166.01     123.94
    90        155.57     137.87
```

To access only entries with velocity less than 5.0 for the three columns L1, R1, and R3:

```
>> vel_lt_5=myds(myds.Velocity<5.0,{'Velocity','L1','R1',
'R3'})
vel_lt_5 =
    Velocity    L1        R1        R3
    4.8292      189.17    170.66    187.6
    4.9279      202.52    197.59    180.25
    4.881       193.97    181.67    216.77
    4.9362      201.49    156.98    200.61
    4.4538      159.3     162.61    138.12
```

Using MATLAB **grpstats**, one can also calculate descriptive statistics by **Angle** for **L1**:

```
>>ByAngle=grpstats(myds,{'Angle'},{@mean,@std},
'DataVars',{'L1'})
ByAngle =
            Angle    GroupCount    mean_L1    std_L1
    -120    -120         8         100.16     24.16
    -90     -90          8         149.31     26.541
    -60     -60          8         230.35     27.332
    -30     -30          8         262.22     44.05
    0        0           8         275.32     51.681
    30       30          8         260.96     60.526
    60       60          8         216.21     12.654
    90       90          8         162.57     11.398
    120      120         8         102.59     29.677
```

GroupCount indicates that 8 entries are involved in the calculation of statistical parameters for each angle. On the other hand, typing:

```
>>ByAngle=grpstats(myds,{'Angle'},{@mean,@std},
'DataVars',{'Velocity'});
```

will present the mean and standard deviation for velocity at each angle. Try it!

9.4.3 *Show Plot by a Category*

Let us generate a scatter plot to visualize the data using the angle as a group/category indicator.

```
>> gscatter(myds.L1, myds.R1, myds.Angle)
>> xlabel('L1 data'),ylabel('R1 data')
```

Figure 9.11 shows the output of the previous command, with some modifications to fit the black and white version.

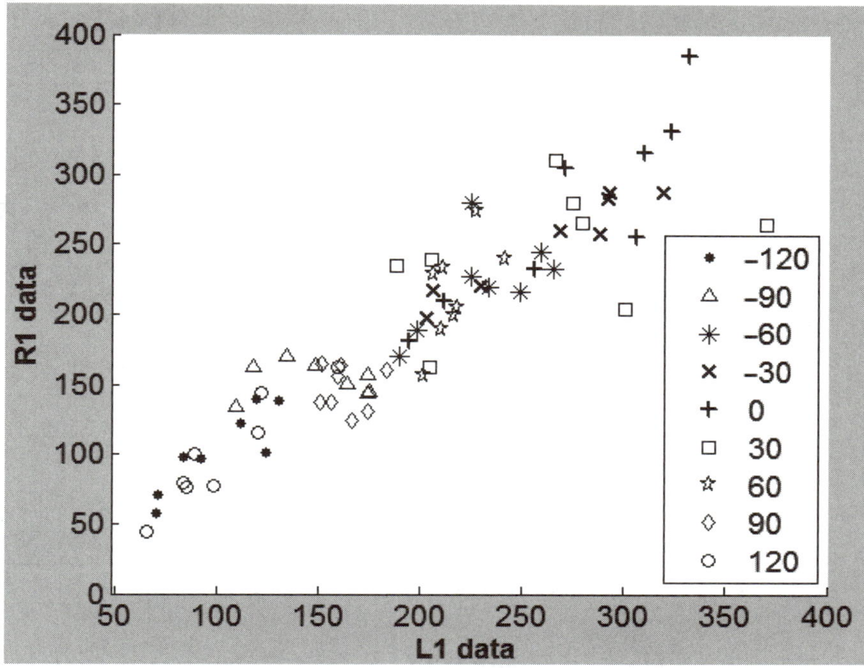

Figure 9.11 A plot of R1 vs. L1 shows a linear correlation. One may also notice how angle affects the stress level. It appears that at a −120°/120° angle, stress is the lowest and at a 0° angle, it is the highest. Therefore, flying directly into the wind (upfront) causes the highest stress on the wings.

9.4.4 *Customize the Plot*

One can create a custom plot, for example, the data will follow the 45° (i.e., $y = x$) line if the left and right sensors are behaving the same (i.e., correlated).

The following set of commands will create a plot of *R1* vs. *L1* while showing the 45° (diagonal) line on the same plot (Fig. 9.12):

```
>> plot(myds.L1,myds.R1,'*')
>> hold on;
>> x=0:10:400;
>> y=0:10:400;
>> plot(x,y,'-')
>> xlabel('L1 data'),ylabel('R1 data');
>> title('R1 vs. L1');
```

We can see from the plot of *R1* vs. *L1* that the data on the left and right wings are similar and do follow a 45° line. On the other hand, a plot of *R2* vs. *L2* (Fig. 9.13) will reveal some bad data in *R2*, as can be seen by the zero values in the plot. It is reported from testing that *R2* failed during the latter part of testing. Hence, we need to remove the bad data from *R2* and keep the good data points that are at or

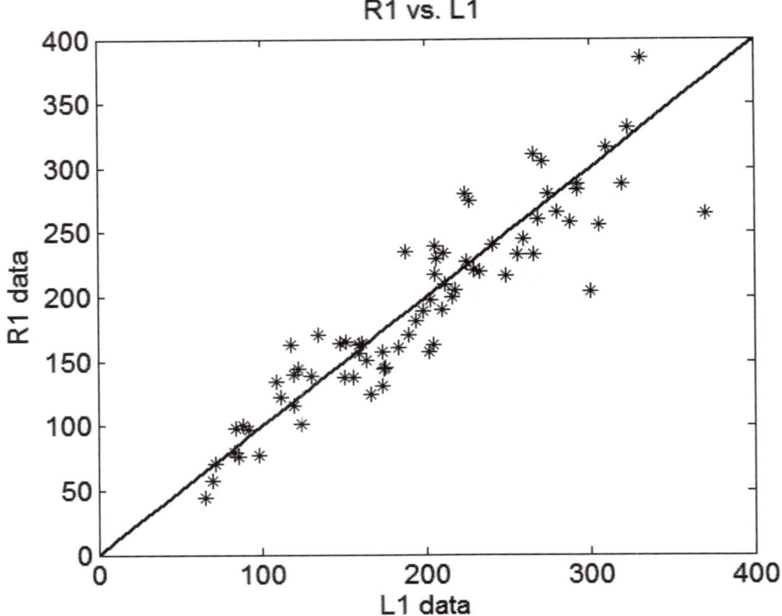

Figure 9.12 A plot of *R1* vs. *L1* accompanied by the diagonal line $y = x$.

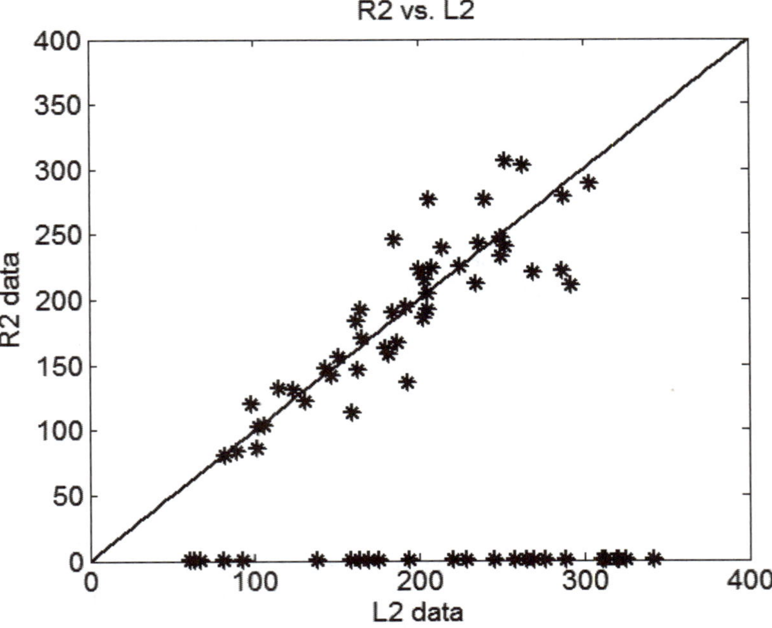

Figure 9.13 A plot of *R2* vs. *L2* accompanied by the diagonal line $y = x$. There were bad data points where *R2* measurement failed (i.e., $R2 = 0$).

close to the 45° line. Moreover, one can define the *R1/L1* ratio and take the mean to see that it is very close to 1.00.

```
>> R1L1ratio=myds.R1./myds.L1;
>> mean(R1L1ratio)
ans =
     0.9831
```

On the other hand, if we define the *R2/L2* ratio and take the mean to see that 0.63823 is not close to one as is the case with *R1/L1* ratio.

```
>> R2L2ratio=myds.R2./myds.L2;
>> mean(R2L2ratio)
ans =
     0.6383
```

9.4.5 Hypothesis Testing

Let us test the hypothesis that *L1* = *R1* and *L2* = *R2*. This will perform the same test as the customized plot did, but does not require us to visually inspect the data. The hypothesis testing is done using the MATLAB function **kstest2** from the Statistics Toolbox.

`h = kstest2(x1,x2,alpha)` performs a two-sample Kolmogorov-Smirnov test to compare the distributions of the values in the two data vectors **x1** and **x2**. The null hypothesis is that **x1** and **x2** are from the same continuous distribution. The alternative hypothesis is that they are from different continuous distributions. The result *h* will be one if the test rejects the null hypothesis at the 5% significance level; it will be zero if the null hypothesis is accepted.

```
>> h1 = kstest2(myds.L1,myds.R1,0.05)
h1 =
      0
```

Note that the hypothesis test (h1) returns 0 for the assumption that *L1* = *R1*. This means that our assumption *L1* = *R1* is valid, or in other words, the differences observed in *L1* and *R1* are due to normal and expected variation.

On the other hand, the hypothesis test (h2) returns 1 for the assumption that *L2* = *R2*. This tells us that *L2* is not equal to *R2*, which is mainly due to the presence of bad data (i.e., *R2* = 0).

```
>> h2 = kstest2(myds.L2,myds.R2,0.05)
h2 =
      1
```

Hypothesis testing thus allows us to systematically test two sets of data for differences in measured values where we normally would only detect these differences by visual inspection.

9.4.6 Screening Data Using Hypothesis Testing

Figure 9.14 shows the **ch9_4_step6.m** file that we will test for the null hypothesis that $h: L_{(i)} = R_{(i)}$. If the answer is one, then it means that $L_{(i)} \neq R_{(i)}$. In other words, the null hypothesis is rejcted.

```
clear all; close all;
myds = dataset('XLSFile','WingStress.xls');
StrainUnits = repmat({'kPa = kN/m^2'},1,14);
myds.Properties.Units = {'m/s','degrees',StrainUnits{:}};
Left = {'L1','L2','L3','L4','L5','L6','L7'};
Right = {'R1','R2','R3','R4','R5','R6','R7'};
for i = 1:length(Left)
 %Lx = double(myds,vars) returns the contents of the dataset variables
 %specified by vars. vars can be a positive integer; a vector of positive
 %integers; a variable name; a cell array containing one or more
 %variable names; or a logical vector.
 Lx = double(myds,Left{i});
 Rx = double(myds,Right{i});
 h(i) = kstest2(Lx,Rx,0.05); % test Lx = Rx at 5% level of significance
 if h(i) > 0 % Plot potentially bad data
 figure;
 plot(Lx,Rx,'*');
 hold on;
 x=0:10:400;
 y=0:10:400;
 plot(x,y,'-')
 xlabel(Left{i}),ylabel(Right{i}
 label = [Right{i},' vs. ',Left{i}];
 title(label);
 fprintf('%s failed test, h = %i\n',label,h(i))
 end
end
```

Figure 9.14 **ch9_4_step6.m** file that tests for the null hypothesis, that is, $h: L_{(i)} = R_{(i)}$. It also plots the case where the null hypothesis is rejected.

If **ch9_4-step6** is executed at the command prompt, the following results will show up:

```
>> ch9_4_step6
R2 vs. L2 failed test, h = 1
R3 vs. L3 failed test, h = 1
```

Figure 9.15 shows the two separate plots (combined in one figure) for the two cases where the null hypothesis is rejected ($L_{(i)} \neq R_{(i)}$). It can be seen that *R3* appears to have a higher stress measurement than *L3*. However, which one is correct? See the answer in Sec. 9.4.8.

9.4.7 Flagging or Labeling Data: Alive vs. Dead Sensor

At this point, it is desired to start removing data that we know is bad from further analysis. We'll use categorical arrays from the Statistics Toolbox to categorize each sensor as Alive, Dead, or Suspect. This will allow us to filter the data based

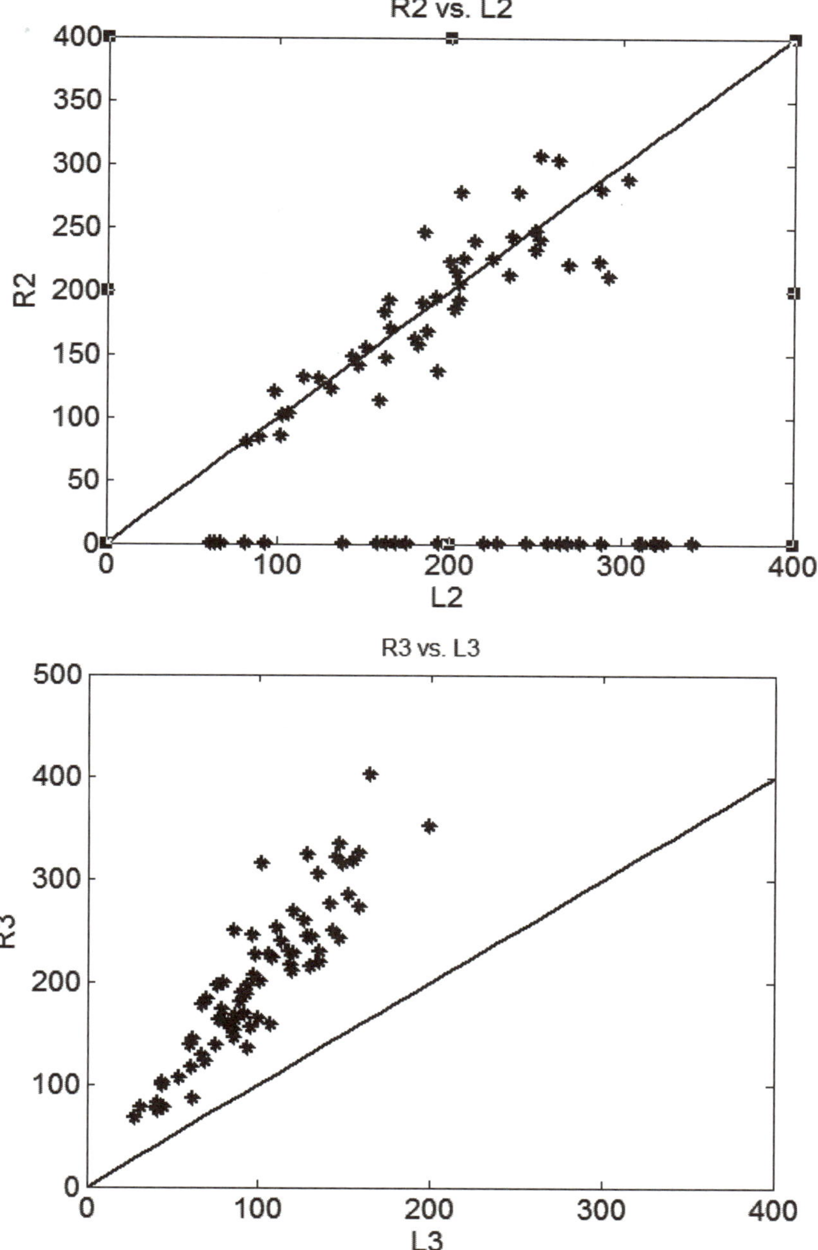

Figure 9.15 The plot of *R2* vs. *L2* (*top*) and *R3* vs. *R3* (*bottom*) for the cases where the null hypothesis is rejected (i.e., $L_{(i)} \neq R_{(i)}$).

```
clear all; close all;
myds = dataset('XLSFile','WingStress.xls');
StrainUnits = repmat({'kPa = kN/m^2'},1,14);
myds.Properties.Units = {'m/s','degrees',StrainUnits{:}};
% Construct nominal categorical 1x1 array named Health
Health = nominal('Alive');
% Health array will be 72x14
Health = repmat(Health,length(myds),14);
% Additional 14 columns to be created for the new dataset myds2
% where all will be initially labeled as 'Alive'
HLeft = {'HL1','HL2','HL3','HL4','HL5','HL6','HL7'};
HRight = {'HR1','HR2','HR3','HR4','HR5','HR6','HR7'};
% myds2 will be 72x30 matrix with the 14 additional columns that are
% labeled by: HL1,..., HL7, HR1, ..., HR7. All new 72x14 entries are
% labeled as "Alive"
myds2 = [myds dataset({Health,HLeft{:},HRight{:}})];
HealthUnits = repmat({'[-]'},1,14);
myds2.Properties.Units= [myds.Properties.Units HealthUnits];
R2L2 = myds2.R2./myds2.L2; % ratio
% indx is 72x1 logical array which has either 0 or 1 entry.
indx = R2L2 < 0.5;
% If indx is one (i.e., (R2/L2)<0.5) then label R2 entry as "Dead" in myds2
% matrix
myds2.HR2(indx) = 'Dead';
%subscript indices of any matrix must either
%be real positive integers or logicals. In this case, index is logical
summary (myds2)
figure;
    % plot while excluding bad data points in R2 column and the
    % corresponding column L2.tilde (~) negates the logical array (i.e., 0 is
replaced by 1 and vice versa
    plot(myds2.L2(~indx),myds2.R2(~indx),'*');
    hold on;
    x=0:10:400;
    y=0:10:400;
    plot(x,y,'-')
    xlabel('L2'),ylabel('R2')
    label = ['R2',' vs. ','L2'];
    title(label);
```

Figure 9.16 **ch9_4_step7.m** file that declares at the start that all sensors are "Alive" and then excludes data points having *R2/L2* less than 0.5 while labeling them as "Dead."

upon these categories. Figure 9.16 shows the **ch9_4_step7.m** file in which at the beginning all sensors are declared "Alive." If the ratio *R2/L2* is less than 0.5, then the sensor will be declared "Dead." At the end, a plot of *R2* vs. *L2* will be generated with no bad data.

The following is a portion of the output (i.e., via executing the **ch9_4_step7.m** file at the command prompt) where it shows that 26 out of 72 data points of the *R2* vector are labeled "Dead" simply because their *R2/L2* value is less than 0.5:

HR1: [72x1 nominal, Units = [-]]
  Alive
  72
HR2: [72x1 nominal, Units = [-]]
  Alive Dead
  46 26
HR3: [72x1 nominal, Units = [-]]
  Alive
  72

Figure 9.17 shows a plot of *R2* vs. *L2* but without the bad data.

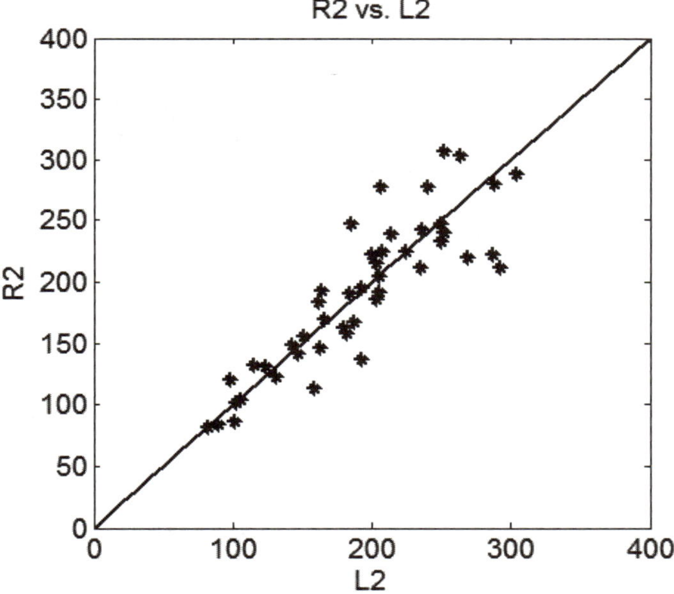

Figure 9.17 A plot of *R2* vs. *L2* while excluding the "bad" (those having *R2/L2* value less than 0.5) data.

9.4.8 *Analysis of Variance (ANOVA)*

We can use ANOVA techniques to test the assumption that all of the sensors have the same mean (i.e., *L1* = *R1* = *R2* = ...*L7* = *R7*). This test will tell us if the differences in stress measurements are different or are a result of the normal and expected variation in the data.

Figure 9.18 shows the code (**ch9_4_step8.m**); after deducting the bad data, the analysis was carried out on the remaining good data. It should be pointed out that **p = anova1(X)** performs balanced one-way ANOVA for comparing the means of

```
clear all; close all;
myds = dataset('XLSFile','WingStress.xls');
StrainUnits = repmat({'kPa = kN/m^2'},1,14);
myds.Properties.Units = {'m/s','degrees',StrainUnits{:}};
Left = {'L1','L2','L3','L4','L5','L6','L7'};
Right = {'R1','R2','R3','R4','R5','R6','R7'};
R2L2 = myds.R2./myds.L2; % ratio and remove values below 50% (0.5)
indx = R2L2 < 0.5;
% Remove bad data points from R2 vector
myds.R2(indx) = NaN;
% stress is 72x14 Double-type matrix that contains stress values evaluated
% at 14 locations (L1, ..., L7, R1, ..., R7) for nine different degrees.
stress = double(myds,[Left,Right]);% stress contains NaN values for R2 if
R2/L2 < 0.5.
[p,t,stats] = anova1(stress,[Left,Right]);
```

Figure 9.18 Analysis of variance (ANOVA) for all stress data except the bad ones (**ch9_4_step8.m**).

two or more columns of data in the matrix X, where each column represents an independent sample containing mutually independent observations. The function returns the p value under the null hypothesis that all samples in X are drawn from populations with the same mean.

If p is near zero, it casts doubt on the null hypothesis and suggests that at least one sample mean is significantly different from the other sample means. Common significance levels are 0.05 or 0.01.

The results of ANOVA are shown in Figs. 9.19 and 9.20.

Figure 1: One-way ANOVA

File Edit View Insert Tools Desktop Window Help

ANOVA Table

Source	SS	df	MS	F	Prob>F
Groups	2.7084e+006	13	208338.7	54.8	2.18739e-106
Error	3.67996e+006	968	3801.6		
Total	6.38836e+006	981			

Figure 9.19 ANOVA results for all 14(*L1…L7, R1…R7*) spot stress values telling us that with the p value being essentially zero, at least one sample mean is significantly different from the other sample means.

Notice that from Fig. 9.19, the p value ($2.18739e^{-106}$) is very close to zero, which suggests that at least one sample mean is significantly different from the other sample means.

One can see from the box plot (Fig. 9.20) the amount of variation each sensor has in the data. The y-axis shows the median (the neck), the quartile range of the data (end of boxes), the range of the data (⊥ shape), and any outliers that might be in the data (+ symbol). We can see a trend for each wing (looking left to right) that

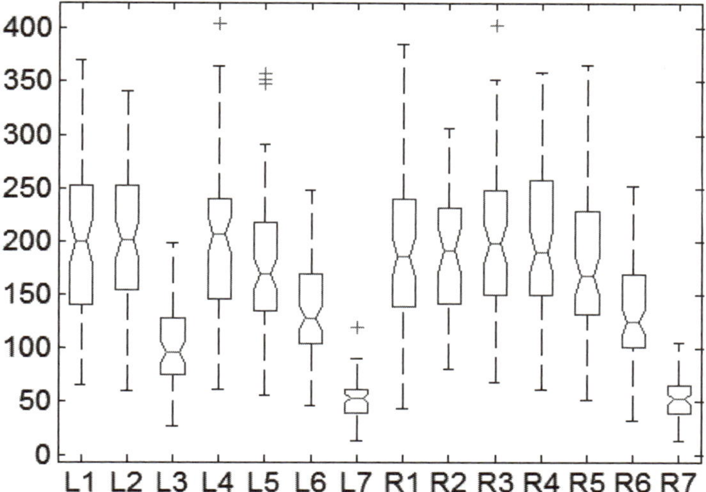

Figure 9.20 The box plot for the 14 stress groups showing the amount of variation each group has. Each group is characterized by its median, quartiles, range, and the outliers.

shows decreasing (median) values as we go from the wing base (*L1* or *R1*) to the wing tip (*L7* or *R7*). Also, note that *L3* and *R3* do not behave similarly. We can conclude from this plot that *L3* is the sensor that has weird behavior. Moreover, if we also want to test for differences across sensors individually, we can do this using the MATLAB **multcompare** function using stats as the input argument. The **stats** structure is the output of the MATLAB **anova1** function (see Fig. 9.18).

```
>> multcompare(stats);
```

Figure 9.21 is the output. The **multcompare** function provides an *interactive plot* that we can use to test for differences across groups. For example, if you click on *L3*, you can see that it is different from 12 sensors. Therefore, *L3* must be the sensor that is wrong (i.e., odd in the sense that it is so different from all other sensors). On the other hand, *R3* is only different from five other sensors. We do not know the cause of *L3*'s oddness. It could be bad data acquisition hardware, improper installation or placement of the *L3* sensor, or some other reason. The point is that *L3*'s values are suspicious and will be removed from further analysis. The command:

```
>> stress(:,3) = [];
```

will delete column *L3* from the stress matrix. It is worth mentioning here that the odd behavior may sometimes unravel or uncover a new phenomenon, event, or behavior for the examined physical system. The odd data (i.e., behavior) ought to be excluded if there is prior information that something went wrong, such as the experimental environment in which the experiment being carried out has changed, or the assessment tool(s) and procedure may have changed.

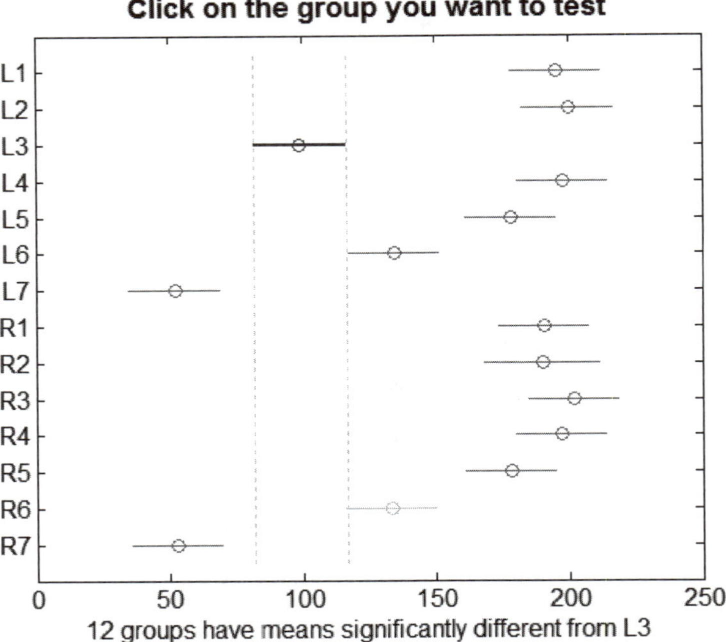

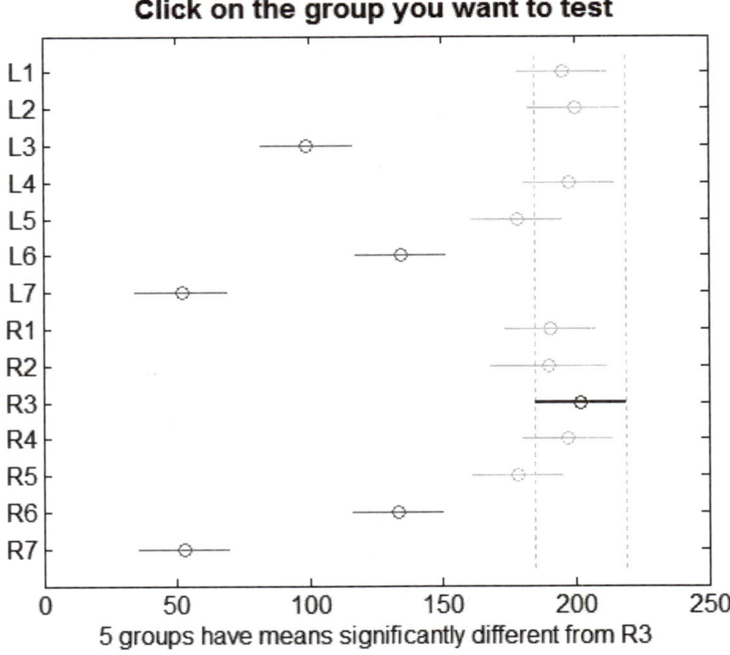

Figure 9.21 *L3* shows that it is different from (or, odd from) 12 other sensors (*top*). On the other hand, *R3* is different from 5 other sensors only (*bottom*).

9.4.9 *Clustering (Showing Groups and Outliers)*

We looked at our data by sensor location and found that *R2* contains bad data. We also found that *L3* to be suspect, and it should be removed from analysis based upon not behaving as expected. Let us look at angle effects on our data. We will use cluster analysis here instead of hypothesis testing, or ANOVA, to show a different method from the Statistics Toolbox that can be used to identify differences in the data.

Let us first define two MATLAB Statistics Toolbox functions. The first function:

```
>>d = manova1(X,group)
```

performs one-way Multivariate Analysis of Variance (MANOVA) for comparing the multivariate means of the columns of **X**, grouped by **group**. **X** is an m-by-n matrix of data values, and each row is a vector of measurements on n variables for a single observation. **group** is a grouping variable defined as a *categorical variable*, vector, string array, or cell array of strings. Two observations are in the same group if they have the same value in the group array. The function returns ***d***, an estimate of the dimension of the space containing the group means. **manova1** tests the null hypothesis that the group means are all the same for the n-dimensional multivariate vector, and that any difference observed in the sample **X** is due to random chance. If $\boldsymbol{d} = 0$, there is no evidence to reject that hypothesis. If $\boldsymbol{d} = 1$, then you can reject the null hypothesis at the 5% level, but you cannot reject the hypothesis that the multivariate means lie on the same line. Similarly, if $\boldsymbol{d} = 2$ the multivariate means may lie on the same plane in n-dimensional space, but not on the same line.

The second function:

```
>>manovacluster(stats)
```

generates a dendrogram plot of the group means after a multivariate analysis of variance (MANOVA). **stats** is the output stats structure resulting from executing **manova1**.

Figure 9.22 shows the **ch9_4_step9.m** file that contains the code for Multivariate Analysis of Variance, MANOVA (i.e., testing the null hypothesis that the group means are all the same for the n-dimensional multivariate vector, and that any difference observed in the sample stress is due to random chance). The group means must lie in a maximum of **dfb**-dimensional space, where **dfb** is the degree of freedom of B matrix. B is the between-groups sum of squares and cross products matrix (see MATLAB online help for **manova1**). MATLAB **manova1** will take care of this maximum dimension, **dfb**. Because $d = 1$ as calculated by **manova1**, we cannot reject the hypothesis that the means lie in a *1-D* subspace.

Notice that the first p value is near zero; this casts doubt on the null hypothesis that the group means lie on a *space of zero dimension* (i.e., all means are equal). The choice of a critical p value to say whether the result is statistically significant is governed by the value of the input argument alpha, which is set to 0.05 here. Groups are categorized by the angle and with $d = 1$, one can say that there is a significant difference in the mean value between one cluster and another. The cluster may contain at least one *degree-categorized group*.

Figure 9.23 shows the MANOVA cluster (dendrogram plot) for stress data.

```
clear all; close all;
myds = dataset('XLSFile','WingStress.xls');
StrainUnits = repmat({'kPa = kN/m^2'},1,14);
myds.Properties.Units = {'m/s','degrees',StrainUnits{:}};
R2L2 = myds.R2./myds.L2; % ratio and remove values below 50% (0.5)
for indx=1:length(myds)
if R2L2(indx)<0.5
% Assign Nan value for the entry in R2 column in case R2/L2 <0.5.
myds.R2(indx) = NaN;
end
end
% The dataset will exclude L3, as was explained earlier that it is odd in
% behavior
myds2=[myds.L1 myds.L2 myds.L4 myds.L5 myds.L6 myds.L7];
myds3=[myds2 myds.R1 myds.R2 myds.R3 myds.R4 myds.R5 myds.R6 myds.R7];
[d,p,stats] = manova1(myds3,myds.Angle,0.05)
manovacluster(stats)
xlabel('Wind Gust Angle (deg)')
```

Figure 9.22 **ch9_4_step9.m** file that contains the code for Multivariate Analysis of Variance (MANOVA) for testing the null hypothesis. $d = 1$; this means that the null hypothesis is rejected, but the group means lie in 1D sub-space (multivariate means lie on the same line).

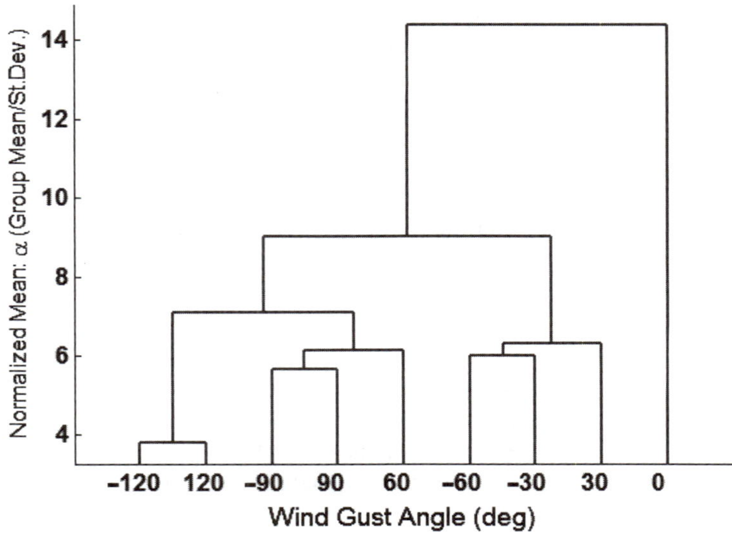

Figure 9.23 The dendrogram plot where groups are paired based upon their normalized mean values $\left(\propto\left(\frac{mean}{st.dev.}\right)_{group}\right)$.

The dendrogram plot (Fig. 9.23) shows that –120/120° is a group. We can see the same for –90/90. The height of the bars indicates how closely related the different clusters are to one another. For example, if we cut the bars at around a value of 6 (ordinate value), we would get the grouping of –120/120, –90/90, and 60 for the left half of the plot. The 60 cluster is interesting in that it is more like –90 and 90 than it is like –60. This is surprising in that we should have a –60/60 grouping, based upon symmetry of the airplane. This requires further investigation why this might be so.

NOTE 0-Degree group is in a cluster all by itself. This is expected, since it is the only measurement that does not have a symmetric counterpart.

9.4.10 *Curve-Fitting*

If the user is interested in curve-fitting $y = f(x)$ for any of the aforementioned variables, the procedure for doing that is explained in detail in Sec. 5.6. If the user is interested in surface-fitting $z = f(x, y)$, the procedure for doing that is explained in detail in Sec. 5.7.

9.5 Problems

9.1 Let's say the probability of having a boy is 0.50. Using the Multiplication Rule, find the probability that a family's first and second children are boys. What is the probability that the first child is a boy and the second child is a girl?

Binomial Distribution

9.2 Military radar and missile detection systems are designed to warn a country of enemy attacks. A reliability question deals with the ability of the detection system to identify an attack and issue a warning. Assume that a particular detection system has a 0.90 probability of detecting a missile attack. Answer the following questions using the *binomial* probability distribution.

a) What is the probability that one detection system will detect an attack?
b) If two detection systems are installed in the same area and operate independently, what is the probability that at least one of the systems will detect the attack?
c) If three systems are installed, what is the probability that at least one of the systems will detect an attack?
d) Would you recommend that multiple detection systems be operated? Explain.

9.3 A company owns 500 laptops. Each laptop has a 10% probability of not working. You randomly select 30 laptops for your salespeople.

a) What is the likelihood that 8 will be broken?
b) What is the likelihood that they will all work?
c) What is the likelihood that they will all be broken?

9.4 A study indicates that 5% of American teenagers have tattoos. You randomly sample 40 teenagers. What is the likelihood that exactly 7 will have a tattoo?

9.5 A cell phone is made from 45 components. Each component has a .003 probability of being defective. What is the probability that such a cell phone will not work perfectly?

9.6 A toy company manufactures toy robots. About 1 toy robot per 100 does not work. You purchase 40 toy robots. What is the probability that exactly 6 do not work?

9.7 A tire company manufactures tires. They claim that only .008 of their tires are defective. What is the probability of finding 4 defective tires in a random sample of 60 tires?

9.8 A TV set is made from 80 components. Each component has a .002 probability of being defective. What is the probability that a TV set will not work perfectly?

Normal Probability Distribution

9.9 Family incomes in two particular neighborhoods follow a *normal distribution*. For neighborhood 1 the mean is $26,500, with a standard deviation of $15,000. For neighborhood 2, the mean is $30,000 with a standard deviation of $18,000.

1. What proportion of families in neighborhood 1 has incomes below the mean family income in neighborhood 2?
2. What income level in neighborhood 1 represents the top 10% for that neighborhood?
3. What income level for neighborhood 2 represents the lowest 5% of incomes for that neighborhood?
4. For neighborhood 1, what percentage of families has incomes that fall between the mean incomes of neighborhood 1 and neighborhood 2?

9.10 One of the jobs that a tire manufacturer does is to determine the warranty policy for the tires the company produces. From the historical data, it was found that the lifttime of a special brand tire can be approximated by a *normal distribution* with an average lifetime of 50,000 miles with a standard deviation of 12,000 miles. Currently, there are two different warranty policies for this brand name tire:

1. Basic Warranty: 40,000 miles;
2. Extended Warranty: 60,000 miles.

The company would like to find the following:

1. What proportion of the tires (probability of a tire) will fail before the basic warranty expires?
2. What proportion of the tires will fail after the basic warranty expires but before the extended warranty expires?
3. If the company would like to change the mileage of the basic warranty policy so that less than 5% of the tires will fail before the basic warranty expires, what should be the mileage for the Basic Warranty?

9.11 Pressure gauges manufactured by Precision Corp. must be checked for accuracy before being placed on the market. To test a pressure gauge, a worker uses it to measure the pressure of a sample of compressed air known to be at a pressure of exactly 50 pounds per square inch. If the gauge reading is off by more than 1% (0.5 psi), the gauge will be rejected. Assuming that the reading of a pressure gauge under these circumstances is a *normal random variable* with mean 50 psi and standard deviation 0.5 psi, find the percentage of gauges rejected.

Exponential Density Function

9.12 You are an investment analyst, and you have information that mortgage lenders are failing continuously at 6% per year.

1. What is the probability density function (pdf) for the mortgage lenders failing with time x in years?

2. What is the probability that a given mortgage lender will fail between 3 and 5 years from now?
3. What is the probability that a mortage lender will last 6 or more years?

9.13 The lifetime in years of liquid crystal display (LCD) is a random variable with the exponential probability density function given by:

$$f(x;10) = \frac{1}{10}e^{-\frac{x}{10}} = 0.1e^{-0.1x}$$

1. What is the mean lifetime of the LCD?
2. What is the probability that the display fails before 2 years?
3. What is the probability that the display fails between 2 and 5 years?

9.14 Plutonium 239 decays continuously at a rate of 0.00284% per year. If X is the time a randomly chosen plutonium atom will decay, use the associated probability density function to compute the probability that a plutonium atom will decay between 100 and 500 years from now.

Poisson Distribution

9.15 A small life insurance company has determined that on average it receives 5 death claims per day. Find the probability that the company receives at least 7 death claims on a randomly selected day.

9.16 The number of road construction projects that take place at any one time in a certain city follows a Poisson distribution with a mean of 4. Find the probability that exactly 6 road construction projects are currently taking place in this city.

9.17 The number of traffic accidents that occur on a particular stretch of road during a month follows a Poisson distribution with a mean of 8.7. Find the probability that less than 4 accidents will occur next month on this stretch of road.

9.18 Suppose the number of babies born during an 8-hour shift at a hospital's maternity wing follows a Poisson distribution with a mean of 8 an hour. Find the probability that 6 babies are born during a particular 1-hour period in this maternity wing.

9.19 During a typical football game, a coach can expect 4.3 injuries. Find the probability that the team will have at most 1 injury in this game.

Gamma Distribution

9.20 Suppose that the lifetime (in years) of an electronic device is 4 years. The device is sold with a one-year warranty. The manufacturer is considering offering an extended warranty for an additional one year. What proportion of all devices that are found to be working at the expiration of the regular warranty (i.e., one year) will be working at the end of the extended warranty (i.e., 2 years)?

Chi-Square Distribution

9.21 Jury Selection: One study of grand juries in Alameda County, California, compared the demographic characteristics of jurors with the general population, to see if jury panels were representative. The results for age are shown below. The investigators wanted to know if the 66 jurors were selected at random from the population of Alameda County. (Only persons over 21 are considered; the county age distribution is known from Public Health Department data.) The study was published in the UCLA Law Review.

Age	Count-wide %	# of jurors observed (O)	# of jurors expected (E)	(O-E)	$(O-E)^2/E$
21–40	42%	5			
41–50	23%	9			
51–60	16%	19			
Over 60	19%	33			
Total	100%	66			

Carry out statistical analysis using chi-square statistic to see if there is evidence that grand juries are selected at random for the population of Alameda County.

Weibull Distribution

9.22 The time to failure for a widget follows a Weibull distribution, with $b = 1/2$, and $a = 750$ hours.

1. What is the mean time to failure of the widget?
2. What percentage of the widgets will fail by 2500 hours of operation?

9.23 Suppose that the service life (in hours) of a semiconductor is a random variable having the Weibull distribution with $a = 1600$ and $b = 0.5$. What is the probability that such a semiconductor will still be in operating condition after 4000 hours?

Beta Distribution

9.24 During 8-hour shift, the proportion of time, X, that a sheet metal stamping machine is down for maintenance or repairs has a beta distribution with $a = 1$, and $b = 2$. What is the probability that during one shift it is off by 1 h of the total time (8-hour shift)?

What is the mean time, expressed in hours, for such a machine when it is off service?

9.25 Errors in measuring the time (microseconds) of arrival of a wave front from an acoustic source can be approximated by a beta distribution with $a = 1$, and $b = 2$.

What is the probability that a measurement error is less than 0.4 microseconds?

What are the mean and standard deviation of the measurement errors?

MATLAB Statistics Toolbox Applications

9.26 For the sake of dealing with datasets, create a dataset using the following M-code:

```
clear all;
myzero=zeros(100,6);
myzero(:,1)=0:0.01:0.99;
myzero(:,2)=betarnd(1,3,100,1);
myzero(:,3)=gamrnd(1,2,100,1);
myzero(:,4)=normrnd(5,1,100,1);
myzero(:,5)=gevrnd(0.2,5,0.1,100,1);
myzero(:,6)=wblrnd(1,2,100,1);
YourDS = dataset({myzero, 'Feature1','Feature2', 'Feature3',
'Feature4', 'Feature5','Feature6'});
MyColor=nominal({'Red','Green','Blue','Black'});
mycell=cell(100,1);
mycell(1:25,1)=cellstr(repmat([MyColor(1,1)],25,1));
mycell(26:50,1)=cellstr(repmat([MyColor(1,2)],25,1));
```

```
mycell(51:75,1)=cellstr(repmat([MyColor(1,3)],25,1));
mycell(76:100,1)=cellstr(repmat([MyColor(1,4)],25,1));
TempDS=dataset({mycell,'Color'});
DS2=[TempDS YourDS];
```

The generated dataset **DS2** will have 100 entries and 7 columns; the first column is the color category, which has four colors; the second column is the [0, 1] interval; and the other five columns each represent a randomly generated point density function. Try the following MATLAB Statistics Toolbox functions:

1. **mle** using beta, gamma, normal, generalized extreme value, and Weibull distribution.
2. **kstest2** with $\alpha = 0.05$ to compare between columns of DS2 (columns 3–7).
3. Convert **YourDS** dataset into double type and apply **[p,t,stats] = anova1(DDS)**, where DDS is the double version of **YourDS** dataset.
4. Use **multcompare(stats)**; where **stats** structure is the output of MATLAB **anova1** function, used in Step 3. **multcompare** generates an interactive plot where you can tell how many groups are compared with the group of interest.
5. Use **[d,p,stats2]=manova1(DDS,DS2.Color,0.05)**, where **DS2.Color** is the categorical group. Do all groups have the same mean?
6. Use **manovacluster(stats2)** to generate the dendrogram, where **stats2** structure is the output of MATLAB **manova1** function, used in Step 5.

10 Chemical Engineering Applications

Chapter Outline

This chapter is intended to demonstrate full-fledged examples borrowed from the chemical engineering domain. There are two steps to consider.

The first step deals with the mathematical modeling of a given physical (nonreactive) or chemical (reactive) system. This means writing down equations in the algebraic, differential, or both forms. In general, such equations stem from:

1. Conservation laws: Mass, component, energy, and so on.
2. Thermodynamics: Equations of state for density, enthalpy, or any other property for a single- and multi-component system; and phase equilibrium that connects a thermodynamic property in one phase with that in another adjacent phase.
3. Transport: Newton's viscosity, Fourier's, and Fick's.
4. Chemical kinetics and chemical equilibrium: Mass action law, Arrhenius law, stoichiometric relations, yield, conversion, and equilibrium constant.
5. Geometrical constraints: This includes initial and boundary value for a monitored property. Another example; in a vessel or storage tank, $V_{Liquid} + V_{Vapor} = V_{total}$.
6. Judicious assumptions: Assumptions made by the modeler regarding how a given property behaves with space (or location) and time.
7. Any other physical constraint that may be imposed upon the system under consideration.

The most important point to spell out is simply that the number of written equations (both algebraic and differential) must match with the number of unknown "to-be-monitored" properties.

The second step deals with the simulation, or solution, of the set of algebraic/differential equations that the modeler comes up with from the first step. All possible means for the solution of such equations should be sought, implemented, and then verified.

The approach used in this chapter considers each separate case study in terms of both modeling and simulation steps. Because this textbook is written for a junior- or senior-level student, simple, but not trivial, cases are presented here, while keeping in mind that there are more complex case studies to tackle in the chemical engineering realm. The theme is to give the user numerous cases from chemical engineering life and present MATLAB as a simulation tool in chemical engineering applications. Hopefully, showing such solved examples will help the user grasp and implement the process of transforming the written equations into MATLAB scripting language. The advantage of using MATLAB over other built-in chemical engineering full-fledged, commercial simulators is simply that MATLAB provides a platform for the user to write his/her own code; gives the user the chance to inspect the code on the microgranular (line-by-line) level; and allows the user to present the results in the form of video, graph, chart, table, and text both on the screen and saved to different format files. In brief, the MATLAB user feels right away what he or she writes and there is no black box or "Jack works in the dark" scenario.

NOTE For any written M-file, the user may impose what is called "flaggers," or stop signs. For each marked line in an M-file, one red circle appears just on the left margin of it. If you make any change in the M-file itself the red circle will turn into a black circle and return to red once you save the file up to its current state. Once the user runs the program (M-code), the MATLAB compiler will execute the M-file line by line down to the first flagger, and the cursor stops there waiting for the user to resume execution. There are different levels of how to resume: starting with one step at a time; step in; step out; and continue. Each level tells the MATLAB compiler how much deeper it should dig in the execution "underground" before it gives control back to the user. While the execution is in pause mode because of a stop sign, you may hover (without really clicking) your mouse over any name of a variable. This will trigger MATLAB to show a pop-up window or highlight box that shows the temporary value(s) assigned to the variable under scrutiny. This excellent means provided by MATLAB enables the user to trace and find logical (semantic) errors inadvertently made by the user. Textual (syntactic) types are either highlighted by corrugated red lines on the spot or detection will be delayed until execution, in which MATLAB tells the user that such a variable name has not been defined to MATLAB before. In Sec. 3.5 more details are given about types of errors involved in writing an M-file. For the sake of clarity on such flaggers or stop signs, refer to Fig. 10.2, in which a snapshot is shown for the M-file itself under execution and in pause mode at line 7. The flaggers are shown inside the left margin of each marked line. The levels of resuming execution are shown at the top right of the toolbar. The value of the highlighted variable L is shown as 1.

10.1 A Simple Electric Circuit

In this section, we will solve the governing equation to examine the dynamics of a single closed-loop electrical circuit. The loop contains a voltage source, V_S (e.g., a battery), a resistor, R (i.e., an energy dissipation device), an inductor, L (i.e., an energy storage device), and a switch that is instantaneously closed at time $t = 0$.

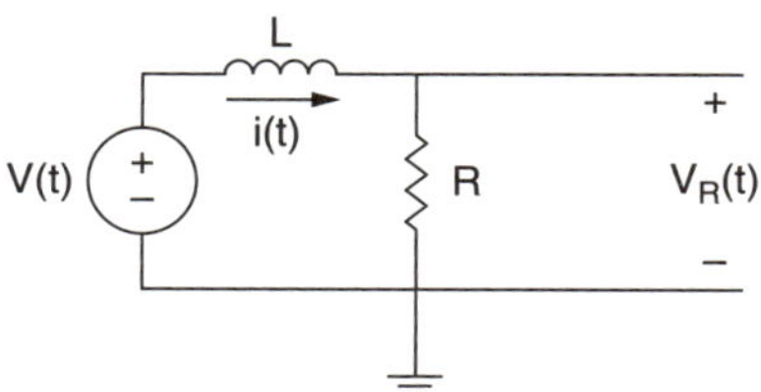

Figure 10.1 RL circuit with a voltage source.

From Kirchhoff's law, the equation describing the response of this system from an initial state of zero current is as follows:

$$L\frac{di}{dt} + Ri = V_S \tag{10.1}$$

where i is the current. At $t = 0$ the switch is on (i.e., a closed-circuit) and the current starts to flow. At this instant of time the voltage is applied to the resistor and inductor (which are connected in series) instantaneously. The equation describes the value of i as a function of time after the switch is on. Hence, for the present purpose, we want to solve this equation to determine i vs. t graphically. Rearranging this equation, we get:

$$\frac{di}{dt}+\frac{R}{L}i=\frac{Vs}{L} \tag{10.2}$$

The solution is:

$$i=\frac{Vs}{R}\left(1-e^{-\frac{R}{L}t}\right) \tag{10.3}$$

The voltage across the resistor is:

$$V_R(t)=Ri(t)=R\times\frac{V_S}{R}\left(1-e^{-\frac{R}{L}t}\right)=V_S\left(1-e^{-\frac{R}{L}t}\right) \tag{10.4}$$

The voltage across the inductor is:

$$V_L(t)=V_S-V_R(t)=V_S-V_S\left(1-e^{-\frac{R}{L}t}\right)=V_Se^{-\frac{R}{L}t} \tag{10.5}$$

Figure 10.2 shows the **SimpleCircuit.m** file that defines the current, the voltage across the resistor, and the voltage across the conductor.

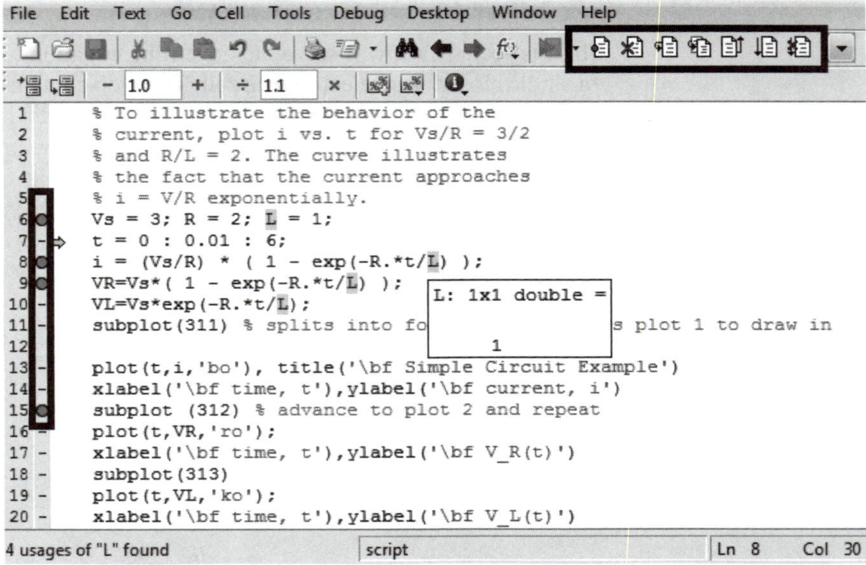

Figure 10.2 The **SimpleCircuit.m** file that defines the current, i, the voltage across the resistor, V_R, and the voltage across the conductor, V_L as a function of time for the simple circuit shown in Fig. 10.1. The image shows the file under execution in pause mode at line 7. The flaggers are shown inside the left margin of each marked line. The levels of execution are shown at the top right of the toolbar. The value of the highlighted variable L is shown as 1.

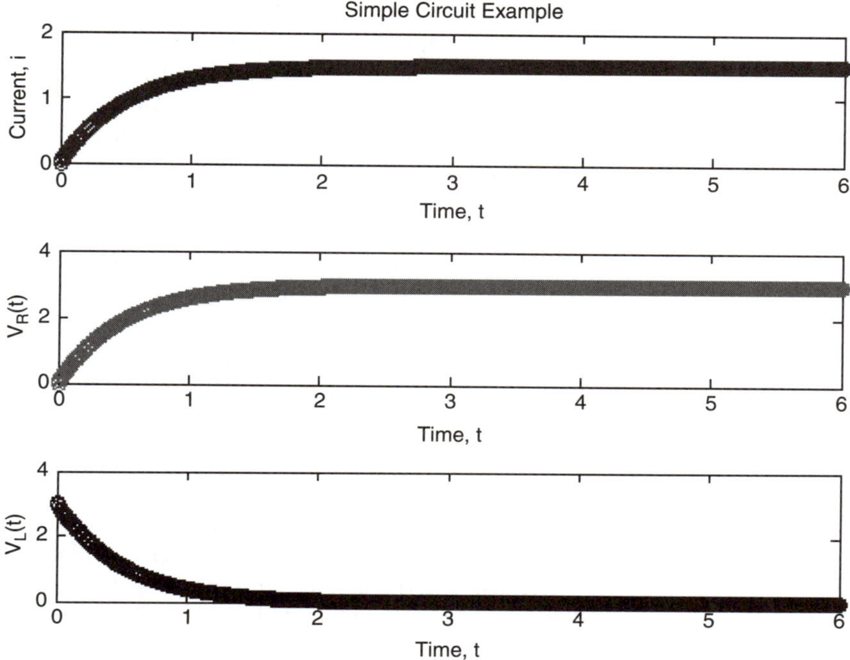

Figure 10.3 The plot of i, V_R, and V_L vs. time for the simple circuit shown in Fig. 10.1.

Figure 10.3 shows the plots of i, V_R, and V_L versus time for the simple circuit shown in Fig. 10.1.

```
>> SimpleCircuit
```

10.2 Van der Waals' Equation of State

An equation of state (EoS) provides a relationship among the three variables: pressure, temperature, and molar volume of a pure substance. Typically, pressure is given as an explicit function of temperature and molar volume. For a pure gas, we have the ideal gas law:

$$P = RT/\tilde{V} \tag{10.6}$$

where R is the universal gas constant, T is the absolute temperature, and $\tilde{V}$ is the molar volume.

Real materials have much more complicated equations of state. In this chapter, we present the equation of state developed by van der Waals. The basic assumption in deriving the ideal gas equation of state is that the interactions between the molecules that compose the gas are negligible. At very low densities, in which the average distance between molecules is extremely large compared with the range of intermolecular interactions (or size of a molecule) and collisions between molecules occur only rarely, this assumption is fairly good. However, for moderately dense gases, this assumption begins to break down as collisions between molecules become more and more frequent. In the case of liquids, this assumption is completely erroneous, as intermolecular interactions dominate the properties of the liquid. To account for the finite size of molecules in the system, van der Waals considered molecules to be of volume b; hence, the actual volume available to the system is approximately $\tilde{V} - b$. Therefore, $\tilde{V}$ in the ideal gas equation of state is replaced by $\tilde{V} - b$. In addition, if the intermolecular distance is sufficiently close, the molecules in a gas exert an attractive interaction between each other because of dispersion forces. Thus, the van der Waals equation of state, which accounts for both effects, is given by:

$$P = \frac{RT}{\tilde{V} - b} - \frac{a}{\tilde{V}^2} \tag{10.7}$$

A schematic of the variation of the pressure with volume, as predicted by the van der Waals equation of state at various temperatures is given in Fig. 10.4. At temperatures above the critical temperature, the pressure-volume variation is monotonic and qualitatively similar to that of an ideal gas (see dotted line). At temperatures below the critical temperature, the pressure-volume curve begins to oscillate, exhibiting a "van der Waals" loop (see dashed line). This behavior is unphysical (i.e., unrealistic), but represents the vapor–liquid transition, and should be replaced by the solid line. The precise location of the solid line is given by the Maxwell construction.

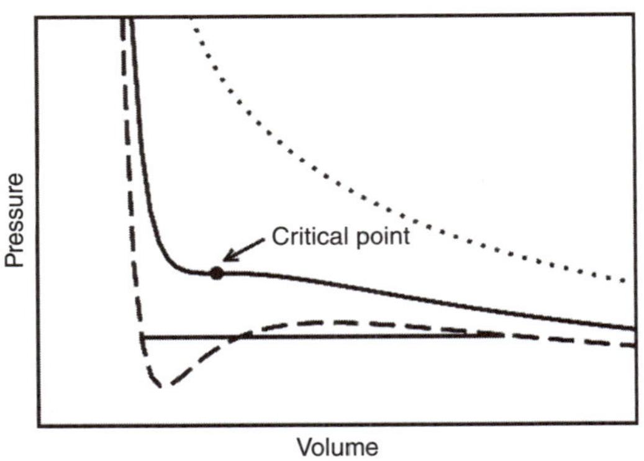

Figure 10.4 Variation of pressure with volume for the van der Waals equation of state for: *a* $T > Tc$ (*dotted line*); *b* $T = Tc$ (*solid line*); and *c* $T < Tc$ (*dashed line*).

When the temperature is equal to the critical temperature, the pressure-volume curve exhibits a point with the inflection (concavity) of the curve changes sign. This occurs at the critical point, where we have:

$$\left(\frac{\partial P}{\partial \tilde{V}}\right)\Bigg|_{T=Tc} = 0 \tag{10.8}$$

$$\left(\frac{\partial^2 P}{\partial \tilde{V}^2}\right)\Bigg|_{T=Tc} = 0 \tag{10.9}$$

Solving simultaneously Eqs. (10.8) and (10.9) allows the estimation of the parameters *a* and *b* appearing in the van der Waals equation.

$$a = \frac{27R^2T_C^2}{64P_C} \tag{10.10}$$

$$b = \frac{RT_C}{8P_C} \tag{10.11}$$

Moreover, the thermal expansion coefficient, α, and the isothermal compressibility, κ, are defined as:

$$\alpha = \frac{1}{\tilde{V}} \times \left(\frac{\partial \tilde{V}}{\partial T} \right) \Bigg|_P \tag{10.12}$$

$$\kappa = -\frac{1}{\tilde{V}} \times \left(\frac{\partial \tilde{V}}{\partial P} \right) \Bigg|_T \tag{10.13}$$

Applying the partial differentiation for the molar volume using the van der Waals equation as the equation of state, we get:

$$\alpha = \frac{1}{\tilde{V}} \times \left(\frac{\partial \tilde{V}}{\partial T} \right) \Bigg|_P = \alpha = \frac{1}{\tilde{V}} \times \frac{\left[R/(\tilde{V} - b) \right]}{\left[\frac{RT}{((\tilde{V} - b)^2)} - \frac{2a}{\tilde{V}^3} \right]} \tag{10.14}$$

$$\kappa = -\frac{1}{\tilde{V}} \times \left(\frac{\partial \tilde{V}}{\partial P} \right) \Bigg|_T = -\frac{1}{\tilde{V}} \times \frac{-1}{\left[\frac{RT}{((\tilde{V} - b)^2)} - \frac{2a}{\tilde{V}^3} \right]} = \frac{1}{\tilde{V}} \times \frac{1}{\left[\frac{RT}{((\tilde{V} - b)^2)} - \frac{2a}{\tilde{V}^3} \right]} \tag{10.15}$$

Given the critical values of pressure and temperature (P_c and T_c) allows the evaluation of *a* and *b* terms for a given substance [Eqs. (10.10) and (10.11)]. Furthermore, if the absolute values of pressure and temperature (*P* and *T*) are given for a pure substance, then Eq. (10.7) will be *a nonlinear algebraic equation* in $\tilde{V}$, which will have, in general, three roots for the molar volume at the given pressure and temperature. MATLAB will be used to find the roots for such an equation that can be put in the form:

$$P = \frac{RT}{\tilde{V} - b} - \frac{a}{\tilde{V}^2} = \frac{RT\tilde{V}^2 - a(\tilde{V} - b)}{(\tilde{V} - b)\tilde{V}^2} = \frac{RT\tilde{V}^2 - a(\tilde{V} - b)}{\tilde{V}^3 - b\tilde{V}^2} \tag{10.16}$$

$$P\tilde{V}^3 - Pb\tilde{V}^2 - RT\tilde{V}^2 + a\tilde{V} - ab = 0 \tag{10.17}$$

Eq. (10.17) is cubic in $\tilde{V}$. A cubic equation, in general, is of the form: $a \times \tilde{V}^3 + b \times \tilde{V}^2 + c \times \tilde{V} + d = 0$.

Figure 10.5 shows the MATLAB (**Van_Der_Waals.m**) code for estimating the molar volume of the gas, $\tilde{V}$, the compressibility factor, *Z*, the thermal expansion

```
clear all;
PureComp = dataset('XLSFile','PureComponentData2.xlsx');
myvar=cellstr(PureComp,'Compound');
myvar2=cellstr(PureComp,'Formula');
SearchKey= str2double(input('Input "1" to search by Compound Name "2" to search by Formula: ', 's'));
if SearchKey==1
CompName= input('Input the name of the Compound, say methane: ', 's');
matchStart = regexpi(myvar, CompName);
elseif SearchKey==2
FormName=input('Input the chemical formula, say CH4: ', 's');
matchStart = regexpi(myvar2, FormName);
else
display('You did not enter 1 or 2');
return
end
tempindex=1;
flag=0;
for idx=1:size(matchStart,1)
    if cell2mat(matchStart(idx,1))>=1
        myindx(tempindex,1)=idx;
        tempindex=tempindex+1;
        flag=1;
    end
end
if flag==0
    display ('No Entry was Found!; Good Luck and Try Again.')
    return
end
for idex2=1:size(myindx,1)
    value=myindx(idex2,1);
    display([value,PureComp.Compound(value),PureComp.Formula(value)])
end
CompNum= str2double(input('Input the number for the chosen compound, say "1" for methane: ', 's'));
Tabs= str2double(input('Input the absolute temperature in Kelvin ', 's'));
Pabs= str2double(input('Input the absolute pressure in atm ', 's'));
Pcr=PureComp.Pcr(CompNum);
Tcr=PureComp.Tcr(CompNum);
Rgas=0.08206;
aterm=(27.0/64.0)*Rgas^2*Tcr^2/Pcr;
bterm=Rgas*Tcr/(8.0*Pcr);
Vols=roots([Pabs, -(Pabs*bterm+Rgas*Tabs), aterm, -aterm*bterm]); % Finds all roots
GasVol=max(Vols(find(imag(Vols) == 0))); % finds largest real root for the gas volume
LiqVol=min(Vols(find(imag(Vols) == 0))); % finds smallest real root for the liquid volume
Zval=Pabs*GasVol/(Rgas*Tabs);
Tr=Tabs/Tcr;
Pr=Pabs/Pcr;
term1=Rgas*Tabs/(GasVol-bterm)^2;
term2=2*aterm/(GasVol^3);
alpha=(1.0/GasVol)*(Rgas/(GasVol-bterm))/(term1-term2);
kappa=(1.0/GasVol)*(1.0/(term1-term2));
display(['The material is:',PureComp.Compound(CompNum)]);
display(['Tr=',num2str(Tr),'    Pr=',num2str(Pr),'    Z=',num2str(Zval)])
display(['Vg=',num2str(GasVol),'  L/mol'])
display(['Vl=',num2str(LiqVol),'  L/mol'])
display(['The Gas Thermal Expansion Coefficient=',num2str(alpha),'    1/K'])
display(['The Gas Isothermal Compressibility Factor=',num2str(kappa),'    1/atm'])
```

Figure 10.5 **Van_Der_Waals.m** file for retrieving critical properties of a substance and estimating the molar volume of both the gas, GasVol, and liquid, LiqVol, the compressibility factor, Zval, the gas thermal expansion coefficient, alpha (α), and the gas isothermal compressibility, kappa (κ), for a selected substance at the given absolute pressure, P, and temperature, T.

coefficient, α, and the isothermal compressibility, κ, for a given substance at the given absolute pressure, *P*, and temperature, *T*. The critical properties for a given substance are read from an Excel sheet (available for students and faculty at www.mhprofessional.com/MATLABnumericalmethods), which are converted into the MATLAB dataset (i.e., the columns are converted into vectors). For more information on dataset handling, please refer to Sec. 9.4.

The user can also search either by the name of the compound or the molecular formula. For example, you can either choose 1 for searching by name (like water) or 2 for searching by formula (like H_2O). Once the user picks up his or her choice, MATLAB will fetch all needed thermodynamic critical properties of the selected substance. The MATLAB built-in root finding **roots** algorithm is used to calculate the three roots of a nonlinear cubic (polynomial) equation. More on the **roots** function is explained in Sec. 2.5. Two roots will be chosen such that the imaginary part is zero, which corresponds to the volume of a gas and liquid. The display statement is used to output the results to the screen for the user's attention. The output is shown in the following for water as the chosen substance:

>>Van_Der_Waals

```
>> Van_Der_Waals
Input "1" to search by Compound Name "2" to search by Formula: 1
Input the name of the Compound, say methane: water
    [188]     'WATER'     'H2O'

Input the number for the chosen compound, say "1" for methane: 188
Input the absolute temperature in Kelvin 300
Input the absolute pressure in atm 1
    'The material is:'     'WATER'

Tr=0.46359   Pr=0.0045923   Z=0.99216
Vg=24.4251  L/mol
Vl=0.036478  L/mol
The Gas Thermal Expansion Coefficient=0.0033906   1/K
The Gas Isothermal Compressibility Factor=1.008   1/atm
```

Exercise 10.1A Rerun **Van_Der_Waals.m** but with different operating conditions in terms of temperature and pressure for the same material. Take, for example, water at room temperature (300 K) and change the pressure from atmospheric maximum up to 32 atm (beyond which the vdW equation predicts one single phase, i.e., the liquid phase) and see how the calculated *Z* is affected by changing the pressure for the same material at a fixed temperature. Draw a conclusion up to what maximum pressure (atm) one can use the ideal gas law for water vapor (i.e., $Z = 1.00 \mp 0.1$). Does it make sense why we use the ideal gas law for an air/water mixture at atmospheric conditions in humidification, air conditioning, and drying/evaporation processes?

Exercise 10.1B Rerun the previous code but with another equation of state, such as Peng-Robinson, Peng-Robinson-Stryjek-Vera, Redlich-Kwong, or Soave-Redlich-Kwong. However, notice that you need to reevaluate α, κ, and the cubic equation shown, respectively, in Eqs. (10.14), (10.15), and (10.17). Do not forget to change the corresponding MATLAB code written for such equations.

10.3 A Simple Cooling Tank

A vertical, cylindrical tank is filled with hot water at 80°C. The tank is insulated at the top and bottom but is exposed on its vertical sides to air at 25°C. The diameter of the tank is 50 cm and its height is 100 cm. The overall heat transfer coefficient, U, is 120 W/(m^2.K). Neglect the metal wall of the tank and assume water in the tank is perfectly mixed (i.e., uniform temperature). Calculate the elapsing time until the temperature drops to $T = 30$°C.

$$\text{DATA: } A = \pi DL = \pi \times 0.5 \times 1 = 1.571\ \text{m}^2$$

The mass of water inside the tank:

$$\text{m}_W = 980\frac{\text{kg}}{\text{m}^3} \times V_{tank} = 980\frac{\text{kg}}{\text{m}^3} \times \frac{\pi}{4} D^2 H = 980\frac{\text{kg}}{\text{m}^3} \times \frac{\pi}{4} 0.5^2 \times 1\ \text{m}^3 = 192.42\ \text{kg}$$

Heat capacity of liquid water = 4184 J/(kg.K).

Applying energy balance around the tank:

$$\dot{E}_{in} - \dot{E}_{out} + \dot{Q} + \dot{W} = Accumulation \tag{10.18}$$

$$0 - 0 - UA(T - 298) = \frac{d(m_W C_W T)}{dt} \tag{10.19}$$

$$0 - 0 - UA(T - 298) = m_W C_W \frac{d(T)}{dt} \tag{10.20}$$

$$\frac{dT}{(T-298)} = -\frac{UA}{m_W C_W} dt \qquad \int_{273+80}^{30+273} \frac{dT}{(T-298)} = -\int_0^t \frac{UA}{m_W C_W} dt \tag{10.21}$$

$$ln\left(\frac{303-298}{353-298}\right) = -\frac{UA}{m_W C_W} t \tag{10.22}$$

$$t = \frac{-ln\left(\dfrac{303-298}{353-298}\right)}{\dfrac{UA}{m_W C_W}} = \frac{ln\left(\dfrac{353-298}{303-298}\right)}{\dfrac{UA}{m_W C_W}} = \frac{ln\left(\dfrac{55}{5}\right)}{\dfrac{120 \times 1.571}{192.4 \times 4184}} = \frac{2.3979}{2.34186 \times 10^{-4}}$$

$$= 10239.3\ \text{s} \times \frac{1\ \text{min}}{60\,\text{s}} = 170.65\ \text{min}$$

```
% This defines the function myode which has two input arguments which are:
% t and T and the output is the derivative of T, dT.
clear all;
U=120.0;
A=1.571;
mw=192.42;
cw=4184;
term1= '-(120*1.571*(T-298))./(192.42*4184)';
term2=inline(term1,'t','T');
dT = term2; % Defining dT/dt
options = odeset('RelTol',1e-5,'AbsTol',1e-5);
lower=input('Enter the lower limit for integration: ');
upper_min=input('Enter the higher limit for integration: ');
initialval=input('Enter the initial value for y: ');
upper=upper_min*60;
[t,T] = ode45(dT,[lower upper],[initialval],options);
Tanal=dsolve('DTanal = -(120*1.571*(Tanal-298))/(192.42*4184)','Tanal(0)=To');
To=initialval;
MyY=eval(Tanal);
plot(t/60,T,'k*',t/60,eval(Tanal),'k-','LineWidth',2);
ylabel('\bf \fontsize{14} T, K'),xlabel('\bf \fontsize{14} time, min');
hleg = legend('Tnumeric','Tanal','Location','Best');
display (['tfinal= ', num2str(t(end)/60),' min',' Tnum_final= ', ...
    num2str(T(end)),' K', ' Tanal_final= ', num2str(MyY(end)), ' K'])
```

Figure 10.6 **ST_ODE.m** file that defines the parameters and variables appearing in Eq. (10.20). MATLAB built-in **ode45** and **dsolve** methods are used.

Let us use MATLAB to solve this ODE. Figure 10.6 shows the MATLAB code (**ST_ODE.m**) for symbolically (analytically) and numerically solving the ODE [Eq. (10.20)] and Fig. 10.7 shows the output plot of T vs. t for both analytical and numerical solutions.

```
>> ST_ODE
Enter the lower limit for integration: 0
Enter the higher limit for integration: 170.65
Enter the initial value for y: 353
tfinal= 170.65 min Tnum_final= 303.0016 K  Tanal_final= 303.0016 K
```

Notice that there is no difference between the numerical and analytical solutions evaluated at the final value of time (i.e., 170.65 minutes).

Exercise 10.2 Re-run the code **ST_ODE.m** with different parameters in terms of U, A, and m_w and see the dynamic response of the tank to the cooling effect. Then, re-run the code with a tank fluid other than water. See how the final temperature, T_{final}, is affected by changing such parameters for the same t_{final}.

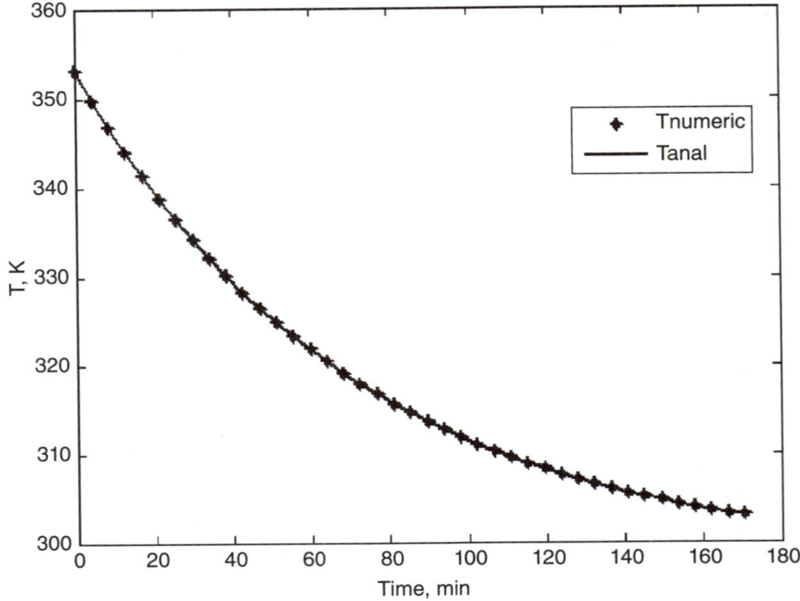

Figure 10.7 Plot of temperature vs. time for cooling a simple perfectly mixed tank.

10.4 Isothermal Batch Reactor

Let us consider a batch reactor in which the following reactions take place:

$$A \overset{k_1}{\Rightarrow} B \overset{k_2}{\Rightarrow} C$$

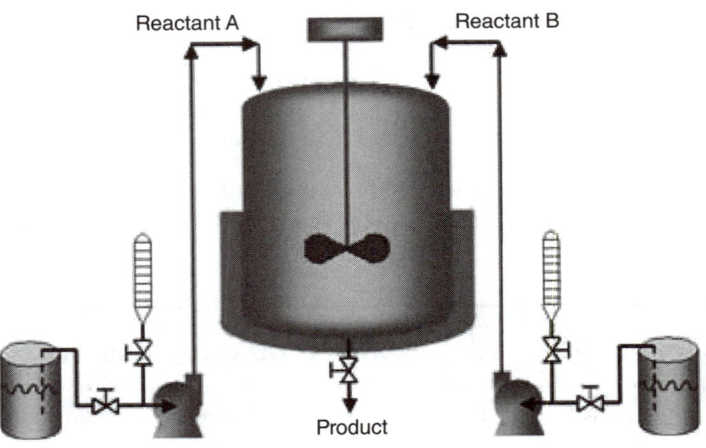

Figure 10.8 A homogeneous batch/semi-batch reactor.

Component molar balance:

$$\dot{n}_A\big|_{in} - \dot{n}_A\big|_{out} + \dot{n}_A\big|_{generated} - \dot{n}_A\big|_{consumed} = Accumulation \tag{10.23}$$

$$0 - 0 + 0 - k_1 C_A V = V\frac{dC_A}{dt} \tag{10.24}$$

Eq. (10.23) reduces to:

$$-k_1 C_A = \frac{dC_A}{dt} \tag{10.25}$$

Component molar balance:

$$\dot{n}_B\big|_{in} - \dot{n}_B\big|_{out} + \dot{n}_B\big|_{generated} - \dot{n}_B\big|_{consumed} = Accumulation \tag{10.26}$$

$$0 - 0 + k_1 C_A V - k_2 C_B V = V\frac{dC_B}{dt} \tag{10.27}$$

Eq. (10.27) reduces to:

$$+k_1 C_A - k_2 C_B = \frac{dC_B}{dt} \tag{10.28}$$

Component molar balance:

$$\dot{n}_C\big|_{in} - \dot{n}_C\big|_{out} + \dot{n}_C\big|_{generated} - \dot{n}_C\big|_{consumed} = Accumulation \tag{10.29}$$

$$0 - 0 + k_2 C_B V - 0 = V\frac{dC_C}{dt} \tag{10.30}$$

Eq. (10.30) reduces to:

$$k_2 C_B = \frac{dC_C}{dt} \tag{10.31}$$

Let us use MATLAB to analytically and numerically solve the set of ODEs given by Eqs. (10.25), (10.28), and (10.31).

To use the numerical technique, we need to put the equations in the following format:

$$\frac{dC_A}{dt} = -k_1 C_A \tag{10.32}$$

$$\frac{dC_B}{dt} = +k_1 C_A - k_2 C_B \tag{10.33}$$

$$\frac{dC_C}{dt} = +k_2 C_B \tag{10.34}$$

Figure 10.9 shows the code (**batch_reactor.m**) that defines parameters and calls the function file that contains the set of ODEs. Figure 10.10 shows the function file (**rxnode.m**) that contains the set of ODEs.

```
clear all;
% Batch reactor, multiple reaction and multiple species
% requires the odes be in function 'rxnode'
% define as global any variables used by any
% and all subroutines, functions
global k1 k2 % variables that we wish to share with the rxnode.m file
% Set parameters,
k1 = 2; k2 = 1;
ca0 = 2; % ca0 = initial concentration of species A
tfin = 2.0;
% c will be a matrix with three columns, one for each species.
% each row will be for another time point.
% initial conditions (c0) is a vector of three initial conds.
c0(1) = ca0 ; c0(2) = 0 ; c0(3) = 0;
set(gca,'NextPlot','replacechildren');
% Preallocate the struct array for the struct returned by getframe
F(50) = struct('cdata',[],'colormap',[]);
for loop = 1:50
    temp=0.0+(loop/50.0)*tfin;
options = odeset('RelTol',1e-4,'AbsTol',[1e-4 1e-4 1e-5]);
[t,c] = ode45(@rxnode,[0 temp],[2 0 0],options); %Numerical ODE solver
ca = c(:,1);
cb = c(:,2);
cc = c(:,3);
Canal=dsolve('DCA=-k1*CA','DCB=k1*CA-k2*CB','DCC=k2*CB','CA(0)=CAo', ...
    'CB(0)=CBo','CC(0)=CCo'); %Symbolic ODE solver
CAo=c0(1);CBo=c0(2);CCo=c0(3);
CA=eval(Canal.CA); %Function evaluation for the symbolic solution
CB=eval(Canal.CB);
CC=eval(Canal.CC);
subplot(311) % splits into three plots, selects plot 1 to draw in
plot (t,c(:,1),'k*',t,CA,'k-','LineWidth',2) % plots all rows column 1 of matrix c
% title('C_A'),
xlabel('\bf time'), ylabel('\bf C_A');
subplot (312) % advance to plot 2 and repeat
plot (t,c(:,2),'k*',t,CB,'k-','LineWidth',2) % plots all rows column 2 of matrix c
% title('C_B'),
xlabel('\bf time'), ylabel('\bf C_B');
subplot(313)
plot (t,c(:,3),'k*',t,CC,'k-','LineWidth',2) % plots all rows column 3 of matrix c
% title('C_C'),
xlabel('\bf time'), ylabel('\bf C_C')
%Since F is a struct array of 50 entries, getframe
% will get current figure (i.e., gcf) 50 times
F(loop) = getframe(gcf);
% to record the frames and play them back in the same or new figure with the
% form
end
 implay(F);
```

Figure 10.9 **batch_reactor.m** file that defines parameters appearing in the set of ODEs, calls MATLAB numerical and analytical ODE solver, and outputs the results.

```
function dC_dt = rxnode(t,C)
global k1 k2 % variables that we wish to share with the main script
dC_dt = zeros(3,1);
dC_dt(1) = -k1*C(1);
dC_dt(2) = k1*C(1)-k2*C(2);
dC_dt(3) = k2*C(2);
```

Figure 10.10 **rxnode.m** file that defines the set of ODEs that will be called upon by the MATLAB ode numerical integrator (**ode45**).

Figure 10.11 shows the concentration profile for the three species C_A, C_B, and C_C as a function of time. Notice that if B is the desired product, then the reaction has to be stopped (via quenching) when C_B reaches its maximum value. For example, $C_{Bmax} = 1.000$ unit of concentration occurs at $t = 0.6899$ unit of time, depending on the unit of k_1 (time^{-1}). On the other hand, if C is the desired product, then the reaction should go to completion.

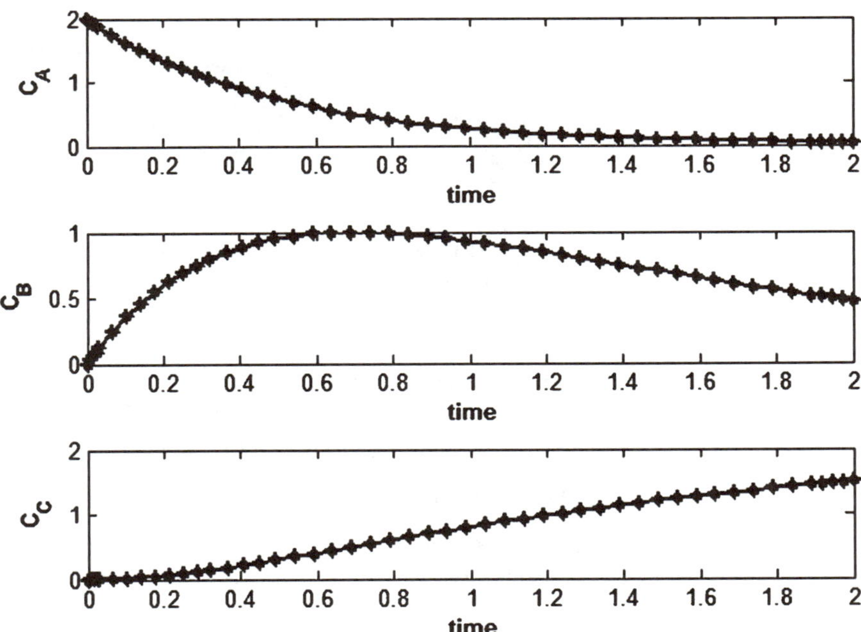

Figure 10.11 Concentration profile as a function of time for the three species C_A, C_B, and C_C, evaluated numerically (*) and analytically (—).

Figure 10.12 shows the movie player for the concentration profile that is created by the MATLAB command:

```
F(loop) = getframe(gcf);
```

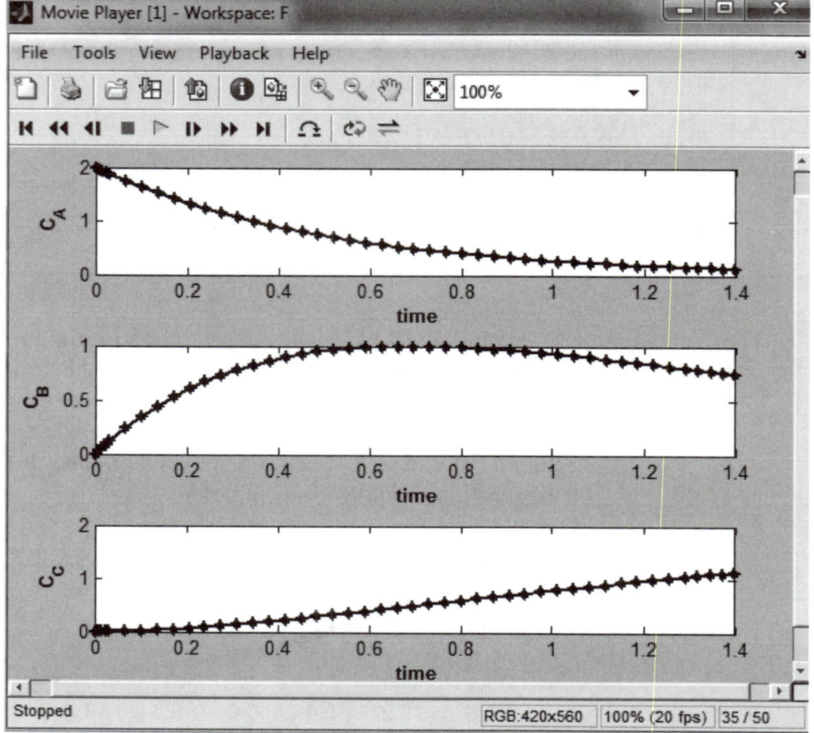

Figure 10.12 MATLAB creates a frame struct called **F** that can be played via **implay(F)**. Here, the frame #35 of 50 is shown in the picture as the movie is paused at that position.

This creates a frame (snapshot for the current figure) for every loop and stores it in the frame struct **F** that contains RGB information about each frame. The frame struct can be played by the command:

```
>>implay(F);
```

The frame struct **F**, once played, shows the instantaneous progress of the species concentration with time.

Exercise 10.3 Try different combinations of the rate constants: k_1 and k_2. Take, for example, when $k_1 > k_2$; $k_1 = k_2$; $k_1 < k_2$, and monitor the shift in the peak of C_B with time.

10.5 A Non-Isothermal CSTR with Water Cooling Jacket

Figure 10.13 A schematic of a continuous stirred-tank reactor (CSTR) equipped with a cooling jacket.

Let us consider a batch reactor in which the following liquid-phase reaction takes place:

$$A \xrightarrow{k_1} B$$

Component molar balance:

$$\dot{n}_A\big|_{in} - \dot{n}_A\big|_{out} + \dot{n}_A\big|_{generated} - \dot{n}_A\big|_{consumed} = Accumulation \quad (10.35)$$

$$FC_{Ao} - FC_A + 0\big|_{generated} - k_1 C_A V = \frac{d(VC_A)}{dt} = V\frac{dC_A}{dt} \quad (10.36)$$

In Eq. (10.36), the liquid hold-up in the tank is held constant (i.e., $dV/dt = Adh/dt = 0$). This simply means that $\rho_o F_o = \rho F$. In other words, the volume change as a result of a chemical reaction is negligible. However, both C_A and T are allowed to change as a function of time.

Energy balance around the reactor:

$$\dot{E}_{in} - \dot{E}_{out} + \dot{Q}_{Exchange} + \dot{Q}_{Rxn} + \dot{W}_{Shaft} = Accumulation \quad (10.37)$$

$$\rho F C_P T_o - \rho F C_P T - UA_H(T - T_J) - k_1 C_A V(\Delta H_r^0) = \frac{d(\rho V C_P T)}{dt} \quad (10.38)$$

$$k_1 = k_1^* e^{\frac{E_A}{R}\left(\frac{1}{T^*} - \frac{1}{T}\right)} \quad (10.39)$$

Because this is a liquid-phase reaction, ρ can be considered constant (i.e., constant density system). C_P is also assumed to be constant.

$$\rho F C_P T_o - \rho F C_P T - UA_H(T - T_J) - k_1 C_A V(\Delta H_r^0) = \frac{d(\rho V C_P T)}{dt} = \rho V C_P \frac{dT}{dt} \quad (10.40)$$

Energy balance around the jacket:

$$\dot{E}_{in} - \dot{E}_{out} + \dot{Q}_{Exchange} = Accumulation \quad (10.41)$$

$$\rho_J F_J C_J T_{Jo} - \rho_J F_J C_J T_J + UA_H(T - T_J) = \frac{d(\rho_J V_J C_J T_J)}{dt} = \rho_J V_J C_J \frac{d(T_J)}{dt} \quad (10.42)$$

Again in Eq. (10.42), ρ_J is factored out as the cooling medium is water; i.e., ρ_J is assumed to be constant. Moreover, V_J is factored out because there is a constant hold-up of water inside the jacket; i.e., the inlet and outlet flow rates of water are equal. C_J is also assumed to be constant. So we have three ordinary differential equations [(10.36), (10.40), and (10.42)] describing the time rate of change of the variables: C_A, T, and T_J, respectively. In addition, we have one algebraic equation [Eq. (10.39)] that describes how k_1 changes with T. All other terms appearing in such equations will be treated as parameters, in addition to the three initial values: $C_A(0)$, $T(0)$, and $T_J(0)$.

Moreover, a control system will be implemented such that as long as the temperature of the reactor is below 373 K (100°C), steam will be introduced to the jacket side for heating up the reaction medium as the rate constant is a function of T. If T exceeds 100°C, then cooling will take into effect. During the heating phase, the inlet valve for steam will be opened and the inlet valve for cooling water will be closed. The situation will be reversed if T exceeds 100°C. This can be implemented via having a temperature sensor for the reactor in which the temperature is converted into either a current or voltage value which will be transmitted to a control box (or unit) and be compared with the set point (i.e., 100°C). The switching from heating to cooling is, therefore, done via the temperature controller. During the heat up, it will keep the steam valve open and the cooling water valve shut. This is accomplished by using split-ranged valves. Also during the heat-up, the cooling-water outlet valve is kept closed and the condensate valve is kept open. When cooling is required, the temperature controller shuts the steam valve and opens the cooling-water valve just enough to make the reactor temperature follow the set point. The cooling-water outlet valve is now kept open and the condensate outlet valve is kept closed.

The overall heat transfer coefficient, U_S, in the presence of steam on the jacket side is higher than that when cooling water is present. Notice that during heating, there will be no need to worry about the dynamics of the jacket and the jacket will assume a constant temperature equal to the steam saturation (i.e., condensation) temperature; hence, $\frac{d(T_J)}{dt} = 0$. Here F_J is assumed to be zero as there is no entering cooling water during the heating phase.

Let us solve the system made of the three odes using MATLAB. The following values, shown in Table 10.1, are used for the given set of parameters:

Table 10.1 Typical values are shown for parameters appearing in the previous set of equations, which describes the non-isothermal CSTR.

Parameter	Value	Parameter	Value
k_1^*	$1.111 \times 10^{-4}\ s^{-1} = 0.4\ h^{-1}$	A_H	12 m^2
R	8.314 J/(mol.K)	ΔH_r^0	–30,000 J/mol
E_A	167,360 J/mol	V	4.0 m^3
U	38 W/(m^2.K) = 136,800 J/(h.m^2.K)	V_J	0.4 m^3
C_P	3500 J/(kg.K)	C_J	4184 J/(kg.K)
ρ	800 kg/m^3	ρ_J	1000 kg/m^3
F	$2.778 \times 10^{-4}\ m^3/s = 1.00\ m^3/h$	F_J	$8.333 \times 10^{-4}\ m^3/s = 3.0\ m^3/h$
T_{J_o}	298 K	$T_J(0) = T(0)$	298 K
$C_A(0)$	20,000 mol/m^3	T^*	350 K

Figure 10.14 shows the MATLAB code (**non_isotherm_reactor.m**) that specifies values for all parameters, except for F_J and U_S, appearing in the previous set of algebraic/differential equations, the initial and final time value in hour,

```
clear all;
global k1o EA R U AH CAo row V Cp F To delHR Tjo Vj Cj rowj
% The list of the above variables are shared with next CSTR.m file
k1o=0.4;
EA=167360;
R=8.314;
U=136800;
AH=12.0;
CAo=20.0e+3;
row= 800.0;
V=4.0;
Cp=3500.0;
F=1.0;
To=298.0;
delHR=-30000;
Tjo=298.0;
Vj=0.4;
Cj=4184;
rowj=1000.0;
tf=2.0;
options = odeset('RelTol',1e-6,'AbsTol',[1e-6 1e-6 1e-6]);
% [t,var] = ode45(@CSTR,[0 tf],[20000 298 298],options);
[t,var] = ode15s(@CSTR,[0 tf],[20000 298 298],options);
CA = var(:,1);
T = var(:,2);
Tj = var(:,3);
subplot(311) % splits into three plots, selects plot 1 to draw in
plot (t,CA,'k.','LineWidth',3) % plots all rows column 1 of matrix var
get(gcf,'CurrentAxes');
h=gca;
set(h,'YGrid','on');
xlabel('\bf time, h'), ylabel(['\bf C_A',' mol/m^3']);
subplot (312) % advance to plot 2 and repeat
plot (t,T,'kx','LineWidth',2) % plots all rows column 2 of matrix var
get(gcf,'CurrentAxes');
h=gca;
set(h,'YGrid','on');
xlabel('\bf time, h'), ylabel('\bf T, K');
subplot(313)
plot (t,Tj,'k-','LineWidth',3) % plots all rows column 3 of matrix var
get(gcf,'CurrentAxes');
h=gca;
set(h,'YGrid','on');
xlabel('\bf time, h'), ylabel('\bf Tj, K');
display (['tfinal= ', num2str(t(end)),' hr',' CA_final=', ...
    num2str(CA(end)),' mol/m^3'])
display ([' Treactor_final= ', num2str(T(end)),' K', ...
    ' Tjacket_final= ', num2str(Tj(end)), ' K '])
```

Figure 10.14 The MATLAB code (**non_isotherm_reactor.m**) specifying values of parameters, except for F_J and U_S, the initial and final time value in hours, the initial values for C_A, T, and T_J, respectively, calls the MATLAB ode solver, and finally outputs the results in the form of three subplots for C_A, T, and T_J, vs. time.

```
function [deriv] = CSTR(t,var)
global k1o EA R U AH CAo row V Cp F To delHR Tjo Vj Cj rowj
% The list of the above variables are shared with non_isotherm_reactor.m
% This defines the function CSTR which has two input arguments which are:
% t and var and the output is the derivative of var, deriv.
% Since we have three ode's, we initialize dy as 3 by 1 matrix (i.e.,
CA=var(1);
T=var(2);
Tj=var(3);
deriv = zeros(3,1);
k1=k1o*exp((EA/R)*(1/350-1/T));
deriv(1) = (F/V)*(CAo-CA)-k1*CA;
if T<=373
    Fj=0.0;
    US=1.0e6;
    deriv(2) =(F/V)*(To-T) +(US*AH*(373.0-T))/(row*V*Cp)-(k1*CA*delHR)/(row*Cp);
    deriv(3)=0.0;
elseif T>373
    Fj=3.0;
    deriv(2) =(F/V)*(To-T) -(U*AH*(T-Tj))/(row*V*Cp)-(k1*CA*delHR)/(row*Cp);
    deriv(3)=(Fj/Vj)*(Tjo-Tj)+(U*AH*(T-Tj))/(rowj*Vj*Cj);
end
```

Figure 10.15 The code for the CSTR function (**CSTR.m**) that defines the set of three derivatives [Eqs. (10.36, 10.40, and 10.42)] and the algebraic equation for k dependence on T [Eq. (10.39)]. The temperature-controlled switch on/off valve is presented here in the form of an if conditional statement.

the initial values for C_A, T, and T_J, calls the MATLAB ode solver, and finally outputs the results in the form of C_A, T, and T_J vs. time.

Figure 10.15 shows the code for the CSTR function (**CSTR.m**) that defines the set of three derivatives [Eqs. (10.36, 10.40, and 10.42)] and the algebraic equation for k dependence on T [Eq. (10.39)]. The temperature-controlled switch on/off valve is presented in the form of an "if conditional" statement.

Figure 10.16 shows the three sub-plots for C_A, T, and T_J, vs. time. One can notice that the steady-state condition for the three variables is attained after 1 hour of operation. Moreover, the code outputs the final values of the three variables C_A, T, and T_J. C_A is essentially zero.

```
>> non_isotherm_reactor
tfinal= 2 hr  CA_final=7.8222e-005 mol/m^3
 Treactor_final= 521.1846 K  Tjacket_final= 324.2455 K
```

NOTE There is more than one possible steady-state condition that may be attained by the non-isothermal CSTR. In other words, C_A^{SS}, T^{SS}, and T_J^{SS} may each assume a value that is different but dependent on the other two. Moreover, the ode solver: **ode15s** (for a stiff system of ODEs) was used instead of ode45, for the latter went in vain. Consequently, if the user is stuck with the "first-to-try method" then he or she ought to use other MATLAB ode solvers. Please refer to Table 7.3 in Sec. 7.3 to see when to use one method over others.

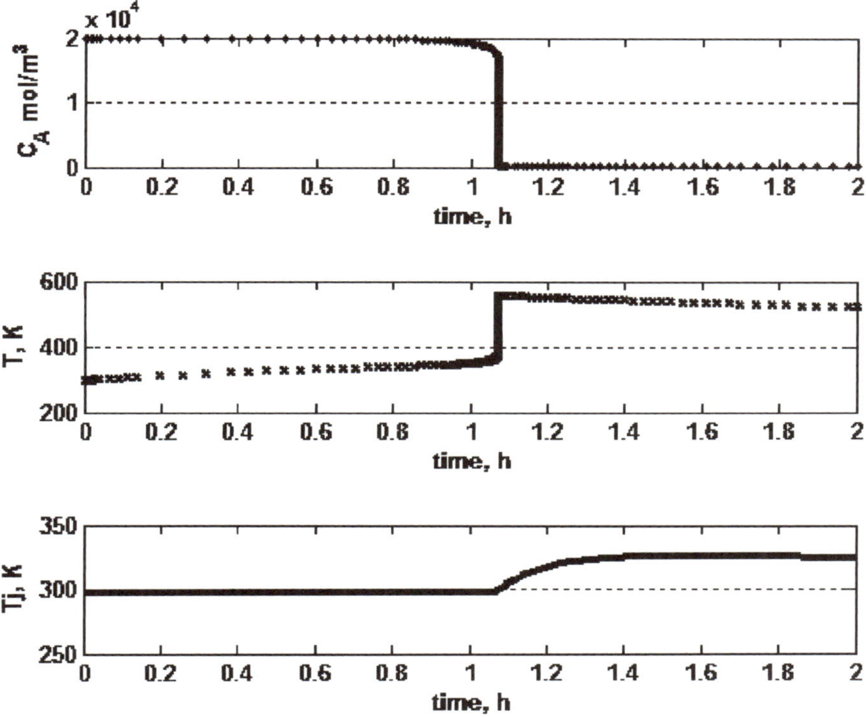

Figure 10.16 Plots of C_A, T, and T_J, vs. time. The steady-state condition is attained after 1 hour of operation.

Exercise 10.4 There are different parameters associated with the set of algebraic/differential equations describing the non-isothermal CSTR and they are shown in Table 10.1. To have a clearer vision and see how each such parameter affects the performance of the process, you may attempt to increase or decrease its value and see how soon the non-isothermal reactor evolves to steady-state conditions in terms of C_A, T, and T_J. Alternatively, you may monitor C_B as a function of time. For example, if the molecular weight of B is 40 g/mol, then the following set of commands can be added at the end of **non_isotherm_reactor.m** to generate C_B as a function of time.

```
figure;
CB=(row*1000/40)-CA;% row=CA*MWA+CB*MWB; MWA=MWB for A→B type
% of reaction
plot (t,CB,'k-','LineWidth',2);
title('\bf C_B vs. time');
xlabel('\bf time, h'), ylabel(['\bf C_B',' mol/m^3']);
```

10.6 A Pressurized, Isothermal Gas-Phase CSTR

Consider a pressurized gas-phase CSTR where the following reaction takes place:

$$A \underset{k_2}{\overset{k_1}{\rightleftarrows}} B \tag{10.43}$$

The reaction is first order for both directions. The reactor contains a mixture of reacting species and is perfectly mixed. Because the mixture of gases is made only of A and B, then the mole fraction of A will be denoted by y, with the understanding that the mole fraction of B will be $(1 - y)$. The pressure inside the vessel (GAS-CSTR in Fig. 10.17) is P. Hence, both P and y may change with time. On the other hand, because gases will occupy the entire space of the reactor and the reactor has rigid walls, the volume V will be constant. For simplicity, the temperature will be assumed to be constant.

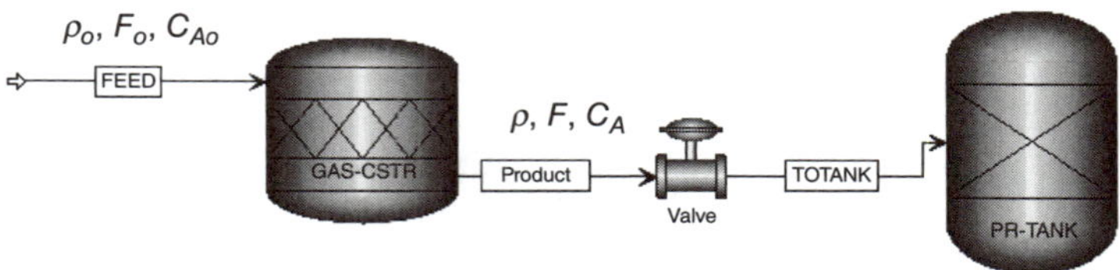

Figure 10.17 A schematic for a pressurized, isothermal gas-phase CSTR.

The flow out of the reactor passes through a restriction (control valve) into another vessel (PR-tank in Fig. 10.17), which is held at constant P_D. The outflow will vary with the pressure and the composition of the reactor. The flow through a valve can be given as:

$$F = C_v \sqrt{\frac{P - P_D}{\rho}} \tag{10.44}$$

C_V is the valve-sizing (discharge) coefficient. Density varies with pressure, composition, and temperature:

$$\begin{aligned} \rho = \frac{P \times MW}{R \times T} &= \frac{P \times (yMW_A + (1-y)MW_B)}{RT} \\ &= \frac{P \times (MW_B)}{RT} = \frac{P \times (MW_A)}{RT} \end{aligned} \tag{10.45}$$

where MW is the average molecular weight, MW_A is the molecular weight of A, and MW_B is the molecular weight of B.

NOTE For **A** reversibly or irreversibly giving **B** as the total mass before and after the chemical reaction is always conserved, and in this particular type of reaction in which 1 mole of **A** is consumed and in return 1 mole of **B** is generated, the total number of moles is also conserved and molecular weights are thus identical.

Assuming an ideal gas mixture, the concentration of reactant of A can be given by:

$$C_A = \frac{Py}{RT} \tag{10.46}$$

Total material balance:

$$\rho_o F_o - \rho F = \frac{d(\rho V)}{dt} = V\frac{d(\rho)}{dt} \tag{10.47}$$

Component material balance:

$$F_o C_{Ao} - FC_A - k_1 C_A V + k_2 C_B V = V\frac{d(C_A)}{dt} \tag{10.48}$$

C_B can be expressed as:

$$\rho = C_A MW_A + C_B MW_B \Rightarrow C_B = \frac{(\rho - C_A MW_A)}{MW_B} \tag{10.49}$$

Alternatively,

$$C_B = C_{tot} - C_A = \frac{P}{RT} - \frac{Py}{RT} = \frac{P}{RT}(1-y) \tag{10.50}$$

Variables are F, P, ρ, C_A, C_B and y. Parameters are C_v, P_D, MW_A, MW_B, R, T, $\rho_o, F_o, C_{Ao}, k_1, k_2$, and V. Initial values are $\rho(0)$ and $C_A(0)$.

Figure 10.18 shows the MATLAB code (**isotherm_gas_reactor.m**) that specifies all parameters: the initial and final time value and the initial values for ρ and C_A; calls the MATLAB default ode solver; and finally outputs the results in the form of four sub-plots for C_A, X_A, y, and C_B, vs. time. Notice that the conversion for A, X_A is defined as: $X_A = (C_{Ao} - C_A)/C_{Ao}$.

Figure 10.19 shows the code for the CSTR_Gas function (**CSTR_Gas.m**) that defines the set of the two derivatives [Eqs. (10.47 and 10.48)] for ρ and C_A and the algebraic equations for the variables P, F, y, and C_B.

Figure 10.20 shows the four sub-plots for C_A, X_A, y, and C_B, vs. time.

```
clear all;
global Cv P_D MWA R T rowo Fo CAo k1 k2 V P  F y CB
Cv=15.0;
P_D=10.0;
MWA=50.0;
MWB=50.0;
R=0.082057;
T=350.0;
Po=30.0;
rowo=(Po*MWA)/(R*T);
Fo=10.0;
CAo=rowo/MWA;
k1=0.5;
k2=0.05;
V=1000.0;
tf=30.0;
options = odeset('RelTol',1e-6,'AbsTol',[1e-6 1e-6]);
[t,var] = ode45(@CSTR_Gas,[0 tf],[rowo CAo],options);
row = var(:,1);
CA = var(:,2);
P=row*R*T/MWA;
F=Cv*sqrt((P-P_D)./row);
y=CA*R*T./P;
XA=(CAo-CA)/CAo;
CB=(P/(R*T)).*(1.0-y);
% CB=(row-CA*MWA)/MWB; % You may also use this to define CB.
subplot(411) % splits into four plots; selects plot #1 to draw in
plot (t,CA,'k.','LineWidth',3) % plot #1
ylabel(['\bf C_A',' mol/L']);
subplot (412) % advance to plot #2
plot (t,XA,'kx','LineWidth',2) % plot #2
ylabel('\bf Conversion, X_A');
subplot(413) % advance to plot #3
plot (t,y,'k-','LineWidth',3)  % plot #3
ylabel('\bf y');
subplot(414) % advance to plot #4
plot (t,CB,'k*','LineWidth',1)  %plot #4
xlabel('\bf time unit'), ylabel('\bf C_B, mol/L');
```

Figure 10.18 The MATLAB code (**isotherm_gas_reactor.m**) specifies all parameters: the initial and final time value and the initial values for ρ and C_A; calls MATLAB's default ode solver; and finally outputs the results in the form of four sub-plots for C_A, X_A, y, and C_B, vs. time.

```
function [deriv] = CSTR_Gas(t,var)
global Cv P_D MWA R T rowo Fo CAo k1 k2 V P F y CB
% This defines the function CSTR_Gas which has two input arguments which are:
% t and var and the output is the derivative of var, deriv.
% Since we have two ode's, we initialize dy as 2 by 1 matrix.
row=var(1);
CA=var(2);
deriv = zeros(2,1);
P=row*R*T/MWA; %row=P*MW/(RT);
F=Cv*sqrt((P-P_D)/row); % flow through a valve
y=CA*R*T/P; %yA=PA/P; yA=CA*RT/P;
CB=(P/(R*T))*(1-y); %CB=P*(1-yA)/(RT);
deriv(1) = (rowo*Fo/V)-(row*F/V);
deriv(2) =(Fo*CAo/V)-(F*CA/V)-k1*CA+k2*CB;
```

Figure 10.19 The code for CSTR_Gas function (**CSTR_Gas.m**) that defines the set of two derivatives [Eqs. (10.47 and 10.48)] for ρ and C_A and the algebraic equations for the variables P, F, y, and C_B.

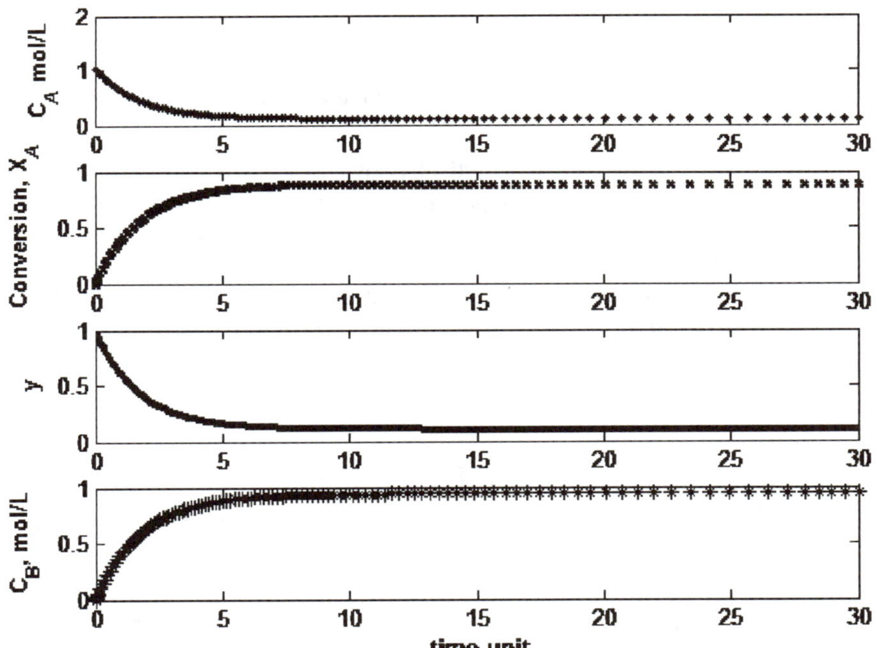

Figure 10.20 The plots for C_A, X, y, and C_B, vs. time.

10.7 One-Dimensional Heat Flow

A solid slab (or, cylindrical rod) is extending from $x = 0$ to $x = L$, and is initially at a temperature T_1. At time $t = 0$, the surface at $x = 0$ is suddenly raised to temperature T_0 and maintained at that temperature for $t > 0$. The slab (or, cylindrical rod) is *insulated* both at lateral sides and at the end, as shown in Fig. 10.21.

$$(Rate\ of\ energy\ in) - (Rate\ of\ energy\ out) = (Rate\ of\ accumulation)$$

$$-kA\left.\frac{\partial T}{\partial x}\right|_x - \left(-kA\left.\frac{\partial T}{\partial x}\right|_{x+\Delta x}\right) = \frac{\partial(\rho A \Delta x C_P T)}{\partial t} \tag{10.51}$$

Dividing by $A\Delta x$ and taking the limit as $\Delta x \to 0$, we obtain:

$$\frac{\partial T}{\partial t} = \alpha \frac{\partial^2 T}{\partial x^2} \tag{10.52}$$

where $\alpha = \dfrac{k}{\rho C_P}$ is called the heat (thermal) diffusivity.

Let $C^2 = \dfrac{k}{\rho C_P}$, then Eq. (10.52) becomes:

$$\frac{\partial T}{\partial t} = C^2 \frac{\partial^2 T}{\partial x^2} \quad \text{One-dimensional heat equation} \tag{10.53}$$

Eq. (10.53) can be analytically solved for given boundary and initial conditions using the method of separation of variables. Obviously, $T = f(x,t) = T(x,t)$. The open-form (i.e., power series) solution will be shown here for the sake of enriching chemical engineering students' math-based skills. The faculty or student may jump, however, to Eq. (10.70) and go from there for the sake of using MATLAB as a tool for expressing such an open-form solution.

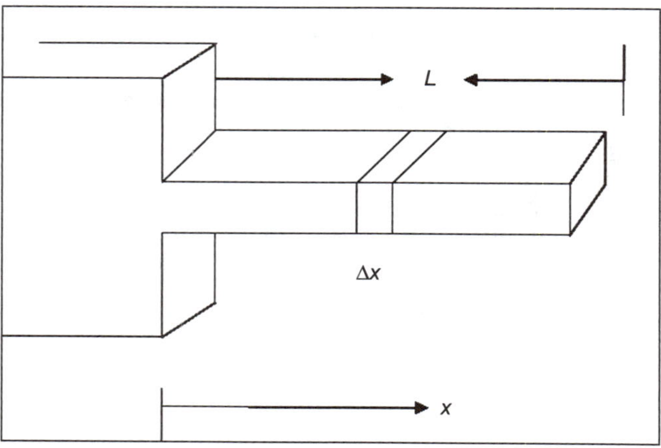

Figure 10.21 Heat transfer in a finite slab.

Let us consider Eq. (10.53) with the following initial and boundary conditions:

$$I.C.: @t = 0, T(x,0) = T_1 \ for \ all \ x$$

$$B.C.\#1: @x = 0, T(0,t) = T_0 \ for \ all \ t > 0$$

$$B.C.\#2: @x = L, \left.\frac{\partial T}{\partial x}\right|_{x=L} = 0 \ for \ all \ t > 0$$

Let us define a dimensionless temperature, θ, as $\theta = \dfrac{T - T_o}{T_1 - T_o}$, then Eq. (10.53) becomes:

$$\frac{\partial \theta}{\partial t} = C^2 \frac{\partial^2 \theta}{\partial x^2} \tag{10.54}$$

$$\text{t} \, B.C.\#1: @x = 0, \ T(0,t) = T_o \ \Rightarrow \theta(0,t) = 0 \ for \ all \ t > 0$$

$$B.C.\#2: @x = L, \left.\frac{\partial T}{\partial x}\right|_{x=L} = 0 \Rightarrow \left.\frac{\partial \theta}{\partial x}\right|_{x=L} = 0 \ for \ all \ t > 0$$

The First Step:

Assume that $\theta(x,t) = F(x) \times G(t)$

Carrying out partial differentiation *once* with respect to *t* and *twice* with respect to *x*:

$$\frac{\partial \theta}{\partial t} = FG' \qquad \frac{\partial^2 \theta}{\partial x^2} = GF''$$

Substitute into Eq. (10.54) to get:

$$FG' = C^2 GF'' \tag{10.55}$$

Rearrangement of Eq. (10.55) results in:

$$\frac{G'}{C^2 G} = \frac{F''}{F} = \phi = constant \tag{10.56}$$

In general, ϕ may assume zero, negative, positive, or even a complex value. The physical problem at hand will determine the potential value of ϕ.

Case 1:

$$\phi = 0$$

The trivial case.

Case 2:

$$\phi > 0$$

Solving for

$$\frac{G'}{C^2G} = \phi \Rightarrow G = C_I e^{(\phi C^2 t)}$$

will make

$$\theta(x,t) \rightarrow \infty \quad \text{as} \quad t \rightarrow \infty.$$

This will violate the physical condition that $\theta(x,t)$ is *finite* for all values of *t*.

Case 3:

$$\phi < 0$$

$$\frac{G'}{C^2G} = \frac{F''}{F} = \phi = -P^2 \tag{10.57}$$

The Second Step:

Determine solutions of the ordinary differential equation (ODE), which will satisfy the BCs:

$$\frac{F''}{F} = -P^2 \qquad \Rightarrow -P^2F = F'' \qquad \Rightarrow F'' + P^2F = 0 \tag{10.58}$$

Eq. (10.58) is considered a constant coefficient second-order ODE with a characteristic equation:

$$(D^2 + P^2)F = 0 \qquad \Rightarrow \qquad D = \pm\sqrt{-P^2} = \pm Pi$$

The solution for Eq. (10.58) will be of the form:

$$F(x) = Acos(Px) + Bsin(Px) \tag{10.59}$$

Applying $BC\#1: @x = 0,\ T(0,t) = T_0 \quad \Rightarrow \theta(0,t) = 0$ for all $t > 0$, one gets:

$$\theta(0,t) = 0 = F(0) \times G(t) \quad \Rightarrow F(0) = 0 \Rightarrow Acos(0) + Bsin(0) = 0 \quad \Rightarrow A = 0$$

$$F(x) = Bsin(Px) \tag{10.60}$$

Applying

$$BC\#2: @x = L,\ \left.\frac{\partial\theta}{\partial x}\right|_{x=L} = 0 \text{ for all } t > 0 \quad \Rightarrow GF'|_{x=L} = 0 \quad \Rightarrow F'|_{x=L} = 0$$

Keep in mind that $G(t)$ for all $t > 0$ cannot be zero; otherwise the solution will become trivial.

$$F(x) = Bsin(Px) \qquad \Rightarrow F'(x) = BPcos(Px) \qquad \Rightarrow BPcos(Px)|_{x=L} = 0$$

This implies that

$$\Rightarrow BPcos(PL) = 0$$

Again, notice that neither P nor B can be set to zero; otherwise a trivial solution is obtained.

Thus,

$$cos(PL) = 0 \qquad \Rightarrow PL = (2n-1)\frac{\pi}{2} \tag{10.61}$$

Notice that Eq. (10.61) gives us a formula for defining the parameter P.

$$P = \frac{(2n-1)}{2}\frac{\pi}{L} \tag{10.62}$$

Substituting P in Eq. (10.60), we get:

$$F(x) = Bsin\left(\frac{(2n-1)}{2}\frac{\pi}{L}x\right) \Rightarrow F(x) = B_n sin\left(\frac{(2n-1)}{2}\frac{\pi}{L}x\right) \quad n = 1, 2, 3, \ldots \tag{10.63}$$

$$\frac{G'}{C^2G} = -P^2 \qquad \Rightarrow \qquad G' = -C^2P^2G \qquad \Rightarrow G' + C^2P^2G = 0$$

Let us define $\beta^2 = C^2P^2$ which implies that: $G' + \beta^2 G = 0$ and the solution will be:

$$G(t) = C_I e^{(-\beta^2 t)} = C_I e^{-C^2\left(\frac{(2n-1)}{2}\frac{\pi}{L}\right)^2 t} \qquad \Rightarrow G(t) = C_n e^{-C^2\left(\frac{(2n-1)}{2}\frac{\pi}{L}\right)^2 t} \tag{10.64}$$

Since in the first step, it was assumed that $\theta(x,t) = F(x) \times G(t)$, then we have:

$$\theta(x,t) = F(x) \times G(t) = B_n sin\left(\frac{(2n-1)}{2}\frac{\pi}{L}x\right) \times C_n e^{-C^2\left(\frac{(2n-1)}{2}\frac{\pi}{L}\right)^2 t}$$

Combining the two constants as one constant such that $b_n = B_n \times C_n$, then we have:

$$\theta(x,t) = b_n sin\left(\frac{(2n-1)}{2}\frac{\pi}{L}x\right) \times e^{-C^2\left(\frac{(2n-1)}{2}\frac{\pi}{L}\right)^2 t} \qquad n = 1, 2, 3, \ldots \tag{10.65}$$

The Third Step:

Based on the fundamental theorem, the linear combination of such n solutions will also be a solution for the original PDE. Thus,

$$\theta(x,t) = \sum_{n=1}^{\infty} b_n sin\left(\frac{(2n-1)}{2}\frac{\pi}{L}x\right) \times e^{-C^2\left(\frac{(2n-1)}{2}\frac{\pi}{L}\right)^2 t} \qquad 0 \le x \le L \tag{10.66}$$

b_n must be chosen such that $\theta(x, t)$ becomes equivalent to $f(x)$ present in the half-range Fourier series, namely, the *Fourier sine series*. Exploiting *I.C.*: @ $t = 0$, $\theta(x, 0) = 1$ *for all x*, Eq. (10.66) becomes:

$$\theta(x,0) = 1 = \sum_{n=1}^{\infty} b_n sin\left(\left(\frac{2n-1}{2}\right)\frac{\pi}{L}x\right)e^{\left[-C\left(\left(\frac{2n-1}{2}\right)\frac{\pi}{L}\right)\right]^2(0)}$$

$$1 = \sum_{n=1}^{\infty} b_n sin\left(\frac{(2n-1)}{2}\frac{\pi}{L}x\right) \tag{10.67}$$

$$b_n = \frac{2}{L}\int_0^L 1 \times sin\left(\frac{(2n-1)}{2}\frac{\pi}{L}x\right)dx = \frac{2}{L}\times\frac{2L}{(2n-1)\pi}cos\left(\left(\frac{2n-1}{2}\right)\frac{\pi}{L}x\right)\Bigg|_L^0 = \frac{4}{(2n-1)\pi} \tag{10.68}$$

Eq. (10.66) becomes:

$$\theta(x,t) = \frac{4}{\pi}\sum_{n=1}^{\infty}\frac{1}{(2n-1)}sin\left(\frac{(2n-1)}{2}\frac{\pi}{L}x\right)\times e^{-C^2\left(\frac{(2n-1)}{2}\frac{\pi}{L}\right)^2 t} \tag{10.69}$$

The final solution is:

$$\theta(x,t) = \frac{T - T_o}{T_1 - T_o} = \frac{4}{\pi}\sum_{n=1}^{\infty}\frac{1}{(2n-1)}sin\left(\frac{(2n-1)}{2}\frac{\pi}{L}x\right)\times e^{-\frac{k}{\rho C_P}\left(\frac{(2n-1)}{2}\frac{\pi}{L}\right)^2 t} \tag{10.70}$$

Before we graphically present the solution, that is, Eq. (10.70), which represents the temperature profile as a function of space (x) and time (t), we would like to make the following points:

1. The finite slab was initially found at $\theta = 1$ (i.e., $T = T_1$) and at time $t \geq 0$, the left boundary is in a direct contact with a semi-infinite reservoir that is maintained at a constant temperature $T = T_o$ (i.e., $\theta = 0$).
2. The fact that the slab is finite and insulated from all sides except the side where it is in contact with the semi-infinite reservoir means that the system will eventually evolve toward thermal equilibrium with the semi-infinite reservoir. In other words, θ will change both with space and time such that it drops from the initial value, where $\theta = 1$ to the final equilibrium value where θ *is zero* everywhere, but of course, after a sufficiently long contact time with the semi-infinite reservoir.
3. There should be some sort of a discrepancy between the open-form analytical and any approximate (i.e., numerical) solution in terms of how fast or how soon the finite slab will eventually attain the thermal equilibrium with its neighbor, the semi-infinite reservoir. Generally speaking, the qualitative behavior will be the same.

4. If $T_o < T_1$, then there will be cooling of the finite slab by the semi-infinite cold reservoir. On the contrary, if $T_o > T_1$, then heating of the finite slab by the semi-infinite hot reservoir will take place. Always keep in mind that the heat transfer direction will be from the hot to the cold region.

Figure 10.22 shows the MATLAB code (**pde_10_7_Anal.m**) that defines the physical properties of aluminum, the time and space span, and calls upon the composite function **myseries(x,t),** which itself is a summation over 500 times for the product: $sin\left(\frac{(2n-1)}{2}\frac{\pi}{L}x\right)$ and $e^{-\frac{k}{\rho C_P}\left(\frac{(2n-1)}{2}\frac{\pi}{L}\right)^2 t}$ (i.e., the summation appearing in Eq. (10.70), is carried out up to $n = 500$). The code for the **myseries(x,t)** function is given in Fig. 10.23.

```
clear all;
global row Cp k
% The material is aluminum (Al).
row=2712;
Cp= 0.91e3;
k=118/1.731;
C2=k/(row*Cp);
myi=0;
for x = 0:0.04:2;
    myi=myi+1;
    myj=0;
    xscaled=x/2;
for t = 0:300:15000;
    myj=myj+1;
    mysol(myj,myi)=myseries(xscaled,t);
end
end
x = 0:0.04:2;
t = 0:300:15000;
surf(x,t,mysol)
title('\bf \fontsize{14} Analytical Solution');
ylabel('\bf \fontsize{14} Time t, s');
xlabel('\bf \fontsize{14} Distance x, m');
zlabel('\bf \fontsize{14} \theta(x,t)');
figure;
plot(x,mysol(1,:),'k-','LineWidth',2);
hold on;
plot(x,mysol(16,:),'ko','LineWidth',2);
plot(x,mysol(26,:),'k*','LineWidth',2);
plot(x,mysol(end,:),'k+','LineWidth',2);
title('\bf \fontsize{14} \theta(x) for Successive t Values');
xlabel('\bf \fontsize{14} Distance x, m');
ylabel('\bf \fontsize{14} \theta(x)');
hleg = legend('t=0 s','t=4500 s','t=7500 s','t=15000 s','Location','Best');
```

Figure 10.22 The MATLAB code (**pde_10_7_Anal.m**) that defines the physical properties of aluminum, the time and space span, and calls upon the composite function **myseries(x,t)**.

```
function [out1] = myseries(x,t)
global row Cp k
C2=k/(row*Cp);
sum=0.0;
for j=1:500
sinterm = sin(((2*j-1)/2)*pi*x);
expterm=exp(-(C2*(((2*j-1)/2)*pi)^2)*t);
multipTerm=sinterm.*expterm;
sum=sum+((4.0/pi)/(2*j-1))*multipTerm;
end
out1=sum;
```

Figure 10.23 The MATLAB code (**myseries.m**) that defines the **myseries** function, which itself is a summation over 500 times for the product: $sin\left(\frac{(2n-1)}{2}\frac{\pi}{L}x\right)$ and $e^{-\frac{k}{\rho C_P}\left(\frac{(2n-1)}{2}\frac{\pi}{L}\right)^2 t}$.

Figure 10.24 shows the 3D surface plot for θ as a function of distance x, measured in meters, and time t, in seconds. Notice that the variable x extends from 0 to L = 2; hence, it is scaled down to be from 0 to 1 by dividing x by L in the calculations, and finally rescaled back to be from 0 to $L = 2$.

Figure 10.25 shows different snapshots, taken at different t values, for θ as a function of distance x, measured in meters.

Notice how the curve (i.e., the cold front wave) is moving downward and toward the end of the slab as a result of the slab being in direct contact with the boundary of a cold semi-infinite reservoir at $x = 0$. Obviously, as time → ∞, then θ will eventually drop to zero everywhere.

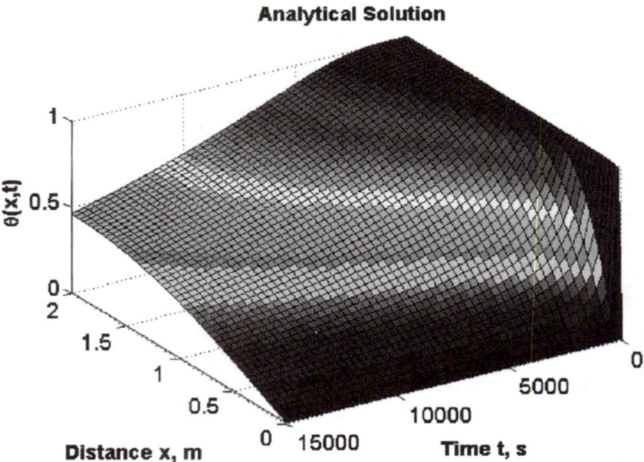

Figure 10.24 The 3D surface plot for θ as a function of distance x, measured in meters, and time t, in seconds.

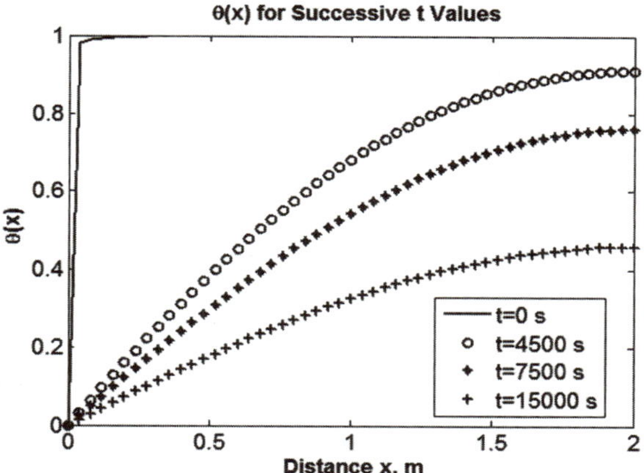

Figure 10.25 Different snapshots, taken at different t values, for θ as a function of distance x, measured in meters.

10.7.1 Solving 1D Heat Equation Using the MATLAB PDE Solver

The general formula for the partial differential equation is given as:

$$c\left(x,t,u,\frac{\partial u}{\partial x}\right)\frac{\partial u}{\partial t}=x^{-m}\frac{\partial}{\partial x}\left(x^{m}f\left(x,t,u,\frac{\partial u}{\partial x}\right)\right)+s\left(x,t,u,\frac{\partial u}{\partial x}\right) \tag{10.71}$$

The PDEs hold for $t_0 \le t \le t_f$ and $a \le x \le b$. The interval $[a, b]$ must be finite. *m can be* 0, 1, *or* 2, *corresponding to slab, cylindrical, or spherical symmetry, respectively*. If $m > 0$, then $a \ge 0$ must also hold. In Eq. (10.71), $f\left(x,t,u,\frac{\partial u}{\partial x}\right)$ is a flux term and $s\left(x,t,u,\frac{\partial u}{\partial x}\right)$ is a source term. The flux term must depend on $\frac{\partial u}{\partial x}$. The coupling of the partial derivatives with respect to time is restricted to multiplication by a diagonal matrix $c\left(x,t,u,\frac{\partial u}{\partial x}\right)$. The diagonal elements of this matrix are either identically zero or positive. *An element that is identically zero corresponds to an elliptic equation and otherwise to a parabolic equation.* There must be at least one parabolic equation. An element of c that corresponds to a parabolic equation can vanish at isolated values of x if they are mesh points.

At the initial time $t = t_0$, for all x, the solution components satisfy initial conditions of the form:

$$u(x,t_o)=u_o(x) \tag{10.72}$$

At the boundary $x = a$, or $x = b$, for all t, the solution components satisfy a boundary condition of the form:

$$p(x,t,u)+q(x,t)f\left(x,t,u,\frac{\partial u}{\partial x}\right)=0 \tag{10.73}$$

$q(x, t)$ is a diagonal matrix with elements that are either identically zero or never zero. Note that the boundary conditions are expressed in terms of the f rather than partial derivative of u with respect to x, $\frac{\partial u}{\partial x}$. Also, of the two coefficients, only p can depend on u.

After this brief summary, let us match Eq. (10.54) with Eq. (10.71) and define each term appearing in the latter equation. Notice that u is essentially θ.

$$\frac{1}{C^2}\frac{\partial u}{\partial t} = x^0 \frac{\partial}{\partial x}\left(x^0 \frac{\partial u}{\partial x}\right) + 0 \tag{10.74}$$

The following parameter and terms are drawn out:

$$m = 0 \qquad \text{(slab coordinates)} \tag{10.75}$$

$$c\left(x, t, u, \frac{\partial u}{\partial x}\right) = \frac{1}{C^2} = \frac{\rho C_P}{k} \tag{10.76}$$

$$f\left(x, t, u, \frac{\partial u}{\partial x}\right) = \frac{\partial u}{\partial x} \tag{10.77}$$

$$s\left(x, t, u, \frac{\partial u}{\partial x}\right) = 0 \tag{10.78}$$

Notice that m [given by Eq. (10.75)] is defined in the main program (**pde_10_7.m**) that is shown in Fig. 10.26. This main program calls upon *three separate*

```
clear all;
global row Cp k
% Ref: http://www.engineeringtoolbox.com/metal-alloys-densities-d_50.html
% For Aluminum
% row=2712 kg/m3
% For Cast Iron
% row=6800-7800 kg/m3
% For Copper
% row=8930 kg/m3
% For Iron
% row= 7850 kg/m3
% For Lead
% row= 11340 kg/m3
% For Silver
% row=10490 kg/m3

% Ref: http://www.engineeringtoolbox.com/specific-heat-metals-d_152.html
% For Aluminum
% Cp= 0.91e3 (J/kg.K)
% For Cast Iron
```

Figure 10.26 The main program (**pde_10_7.m**) that calls upon *three user-defined functions* pertaining to the problem of interest. In addition, it outputs the results in the form of 3D and 2D plots for the temperature profile as a function of x and t.

```
% Cp=0.46e3 J/(kg.K)
% For Copper
% Cp=0.39e3 J/(kg.K)
% For Iron
% Cp=0.46e3 J/(kg.K)
% For Lead
% Cp=0.13e3 J/(kg.K)
% For Silver
% Cp=0.23e3 J/(kg.K)

%Ref: http://www.engineeringtoolbox.com/thermal-conductivity-metals-d_858.html
%1 Btu/(hr oF ft2/ft) = 1 Btu/(hr oF ft) = 1.731 W/(m K) = 1.488 kcal/(h m oC)
% For Aluminum at 68oF
% k=118/1.731 W/(m.K)
% For Cast Iron, gray at 70oF
% k=27/1.731 - 46/1.731 W/(m.K)
% For Copper at 68oF
% k=223/1.731 W/(m.K)
% For Iron at 68oF
% k= 42/1.731 W/(m.K)
% For Lead at 68oF
% k=20/1.731 W/(m.K)
% For Silver at 68oF
% k= 235/1.731 W/(m.K)
%%%%%%%%%%%%%%%%%%%%%%%%%%%%%%%%%%%%%%
% For Aluminum
row=2712;
Cp= 0.91e3;
k=118/1.731;
x = 0:0.01:2;
t = 0:75:15000;
m = 0;
sol = pdepe(m,@pdex1pde,@pdex1ic,@pdex1bc,x,t);
u=sol(:,:,1);
surf(x,t,u);
title('\bf \fontsize{14} Numerical solution computed with 200 mesh points');
xlabel('\bf \fontsize{14} Distance x, m');
ylabel('\bf \fontsize{14} Time t, s');
zlabel('\bf \fontsize{14} \theta(x,t)');
figure;
plot(x,u(1,:),'k-','LineWidth',2);
hold on;
plot(x,u(61,:),'ko','LineWidth',2);
plot(x,u(101,:),'k*','LineWidth',2);
plot(x,u(end,:),'k+','LineWidth',2);
title('\bf \fontsize{14} \theta(x) for Successive t Values');
xlabel('\bf \fontsize{14} Distance x, m');
ylabel('\bf \fontsize{14} \theta(x)');
hleg = legend('t=0 s','t=4500 s','t=7500 s','t=15000 s','Location','Best');
```

Figure 10.26 *(Continued)*

user-defined functions that pertain to the partial differential Eq. (10.54) that we attempt to solve. I have chosen *aluminum* (Al) as a representative material; the user is free to try other materials as well. Their physical properties are commented at the beginning of the main program.

The first user-defined function, called the **pdex1pde** function, is shown in Fig. 10.27.

```
function [c,f,s] = pdex1pde(x,t,u,DuDx)
global row Cp k
c = row*Cp/k;
f = DuDx;
s = 0;
```

Figure 10.27 **pdex1pde.m**, which defines $c\left(x,t,u,\frac{\partial u}{\partial x}\right)$, $f\left(x,t,u,\frac{\partial u}{\partial x}\right)$, and $s\left(x,t,u,\frac{\partial u}{\partial x}\right)$.

This function defines the three terms ($c, f,$ and s) that are defined by Eq. (10.76) through (10.78).

The *second user-defined function* is **pdex1ic** that defines the initial condition. The initial condition for the PDE [Eq. (10.54)] is as follows:

$$IC: @t=0,\ T(x,0)=T_1\ \Rightarrow \theta(x,0) \Rightarrow uo=1.0\ \ for\ \ all\ \ x>0$$

The MATLAB code for this initial condition is shown in Fig. 10.28.

```
function u0 = pdex1ic(x)
u0 = 1.0;
```

Figure 10.28 **pdex1ic.m** file, which defines the initial condition for u (or, θ), that is $u(x,0)=u_o$.

The *third and last user-defined function* is **pdex1bc** that defines the boundary conditions at the left (lower) and right (upper) limit of x, as prescribed in Eq. (10.73).

Here are the two boundary conditions for x:

$$BC\#1: @x=0,\ T(0,t)=T_o\ \Rightarrow \theta(0,t)=u(0,t)=0 \text{ for all } t>0$$

$$BC\#2: @x=L,\ \left.\frac{\partial T}{\partial x}\right|_{x=L}=0 \Rightarrow \left.\frac{\partial \theta}{\partial x}\right|_{x=L}=\left.\frac{\partial u}{\partial x}\right|_{x=L}=0 \text{ for all } t>0$$

The boundary conditions, written in the same format as Eq. (10.73), are:

$$u(0,t)+0\times\left.\frac{\partial u}{\partial x}\right|_{x=0}=0 \quad \Leftrightarrow \quad p_l+q_l\times\left.\frac{\partial u}{\partial x}\right|_{x=0}=0$$

The conclusion is simply that $p_l = u_l$ and $q_l = 0$

and

$$0 + 1 \times \left.\frac{\partial u}{\partial x}\right|_{x=L} = 0 \quad \Leftrightarrow \quad p_r + q_r \times \left.\frac{\partial u}{\partial x}\right|_{x=L} = 0$$

The conclusion is simply that $p_r = 0$ and $q_r = 1$

The MATLAB code is shown in Fig. 10.29.

```
function [pl,ql,pr,qr] = pdex1bc(xl,ul,xr,ur,t)
pl = ul;% NOTE: pl (pL) not p1 and ul (uL) not u1
ql = 0;
pr = 0;
qr = 1;
```

Figure 10.29 **pdex1bc.m** file, which defines the four terms: p_l, q_l, p_r, and q_r appearing in the standard equation [Eq. (10.73)]. They describe the boundary condition at the left and right limits of x.

Figure 10.30 shows a 3D plot for u (i.e., θ) as a function of x and t, using 200 mesh points for each dimension. Notice that the general behavior or topology of the surface is very similar to that of the analytical solution shown earlier in Fig. 10.24. Keep in mind that the solution given by the MATLAB PDE solver is approximate.

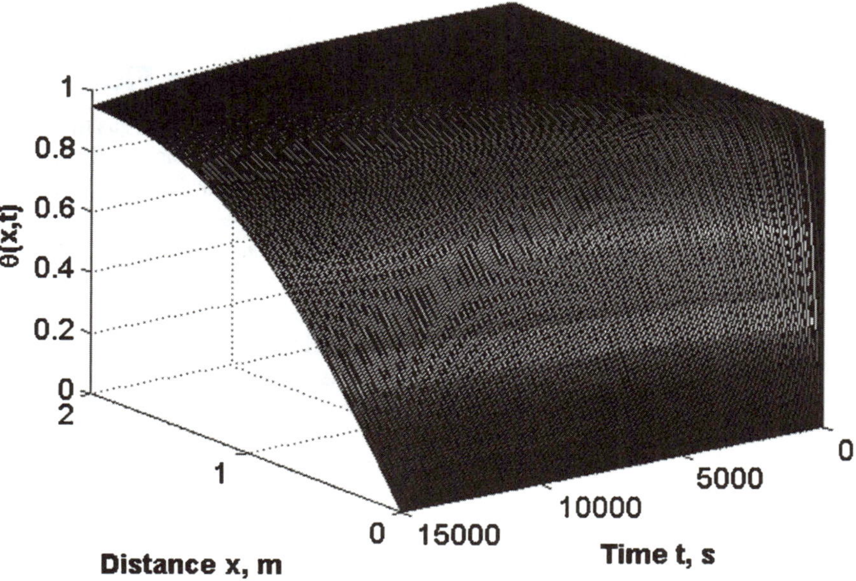

Figure 10.30 The 3D plot of u (i.e.,θ) as a function of x and t using 200 mesh points for each dimension.

Figure 10.31 shows different snapshots taken at different t values for θ as a function of distance x, measured in meters, which has similar behavior as that of the analytical solution shown in Fig. 10.25.

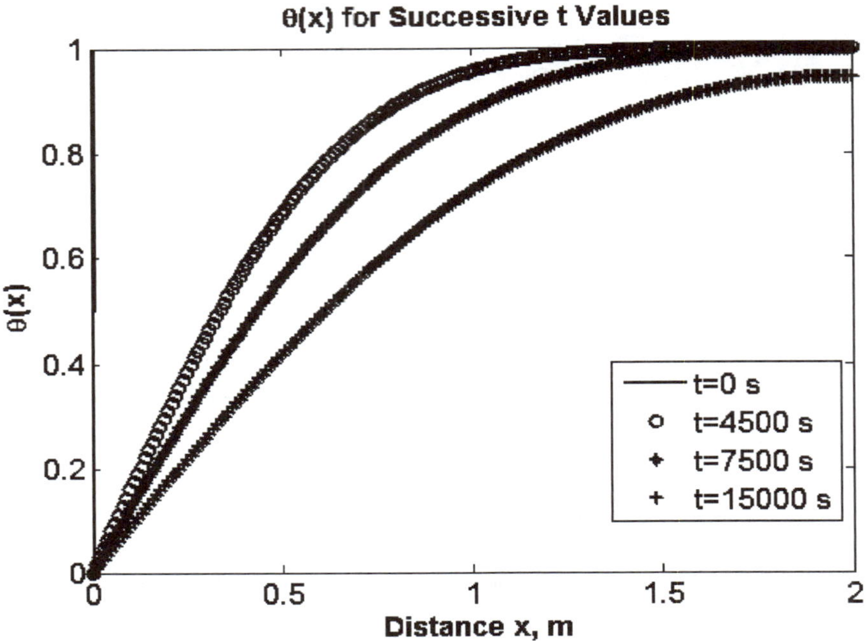

Figure 10.31 Different snapshots, taken at different t values, for θ as a function of distance x, measured in meters.

10.8 One-Dimensional Mass Flow

In an experimental work on properties of a gas mixture, the rate of mixing of gases in closed containers is important. Consider the case in which *propane* is introduced under a pressure into a standard compressed cylinder of length L. On top of propane, *air* is introduced in an equal molar amount. *Find an expression for the mole-fraction of air at any time and location.* It is assumed that a sharp, well-defined interface between air and propane occurs at the beginning of mixing. *No diffusion* takes place through both ends of the cylinder.

Let's take a differential volume element within the cylinder with an area equal to πR^2 and thickness Δz.

$$\text{In} - \text{Out} = \text{Accumulation}$$

$$N_{Az}A\big|_z - N_{Az}A\big|_{z+\Delta z} = \frac{\partial(VC_A)}{\partial t} = A\Delta z\frac{\partial(C_A)}{\partial t} \tag{10.79}$$

Divide by $A\Delta z$ and take the limit as $\Delta z \rightarrow 0$.

$$-\frac{\partial(N_{Az})}{\partial z} = \frac{\partial(C_A)}{\partial t} \quad \rightarrow\rightarrow\rightarrow \quad -\frac{\partial\left(-D_{AB}C\dfrac{\partial y_A}{\partial z}\right)}{\partial z} = \frac{\partial(Cy_A)}{\partial t} \tag{10.80}$$

For a constant temperature and pressure system, the total concentration of gas, C, will be constant, and if D_{AB} is assumed constant throughout the gas (i.e., D_{AB} does not change with location z), then Eq. (10.80) becomes:

$$D_{AB}\frac{\partial^2 y_A}{\partial z^2}=\frac{\partial y_A}{\partial t} \quad \rightarrow\rightarrow\rightarrow \quad D_{AB}\frac{\partial^2 y}{\partial z^2}=\frac{\partial y}{\partial t} \tag{10.81}$$

where D_{AB} is the diffusivity of air in propane, $y = y_A$ the mole-fraction of air.

$$IC:\ y(z,0)=\begin{cases}1 & 0\le z\le L/2\\ 0 & L/2\le z\le L\end{cases} \tag{10.82}$$

This means that air initially occupies the upper half of the cylinder

$$BC\#1:\ \left.\frac{\partial y}{\partial z}\right|_{z=0}=0 \tag{10.83a}$$

$$BC\#2:\ \left.\frac{\partial y}{\partial z}\right|_{z=L}=0 \tag{10.83b}$$

The top and bottom boundaries (BCs) are impermeable to diffusion or penetration by gases.

The detailed solution for Eq. (10.81) will not be shown here; only the final result will be presented. The details will be left as an exercise for the reader. Keep in mind that the procedure for solving such a type of PDE is similar to that shown in Sec. 10.7. However, in the final stage, the expansion for $f(x)$, *as the initial condition,* should become equivalent to the half-range Fourier series, namely, the *Fourier cosine series*.

The open-form analytical solution has the form of:

$$y(z,t)=\frac{1}{2}+\frac{2}{\pi}\sum_{n=1}^{\infty}\frac{(-1)^{n+1}}{(2n-1)}cos\left((2n-1)\pi\frac{z}{L}\right)\times e^{-D_{AB}\left((2n-1)\frac{\pi}{L}\right)^2 t} \tag{10.84}$$

Regarding the calculation of D_{AB}, we will use the Fuller et al. (Fuller EN, Schettler PD, Giddings J.C. *Ind Eng Chem* 58(5):18–27, 1966) equation:

$$D_{AB}=D_{BA}=\frac{0.00143T^{1.75}}{PM_{AB}^{1/2}\left[\left(\sum_V\right)_A^{1/3}+\left(\sum_V\right)_B^{1/3}\right]^2} \tag{10.85}$$

where D_{AB} is in cm²/s, P is in atm, and T is in K. A is air and B is propane (C_3H_8).

$$M_{AB}=\frac{2}{[(1/MW_A)+(1/MW_B)]}=\frac{2}{[(1/29)+(1/44)]}=34.96 \tag{10.86}$$

and $\sum V$= summation of atomic and structural diffusion volumes.

$$\left(\sum_V\right)_A^{1/3}=(19.7)_A^{1/3}=2.7008 \qquad \left(\sum_V\right)_B^{1/3}=(3\times15.9+8\times2.31)_B^{1/3}=4.0449$$

$T = 300\ K$ and $P = 1$ atm.

$$D_{AB} = D_{BA} = \frac{0.00143T^{1.75}}{PM_{AB}^{1/2}\left[\left(\sum_{V}\right)_{A}^{1/3} + \left(\sum_{V}\right)_{B}^{1/3}\right]^{2}} = \frac{30.9242}{269.0490} = 0.115\frac{\text{cm}^2}{\text{s}}$$

$$= 1.15\times 10^{-5}\,\frac{\text{m}^2}{\text{s}}$$

Figure 10.32 shows the MATLAB code (**pde_10_8_Anal.m**) that defines the diffusivity, D_{AB}, the time and space span, and calls upon a composite function **myseries_10_8(z,t)**, which itself is a summation over 500 times for the product of

```
clear all;
global DAB
% The mixture is air-propane
DAB=1.15e-5;
myi=0;
for z = 0:0.04:2;
    myi=myi+1;
    myj=0;
    zscaled=z/2;
for t = 0:300:15000;
    myj=myj+1;
    mysol(myj,myi)=myseries_10_8(zscaled,t);
end
end
z = 0:0.04:2;
t = 0:300:15000;
surf(z,t,mysol)
title('\bf \fontsize{14} Analytical Solution');
ylabel('\bf \fontsize{14} Time t, s');
xlabel('\bf \fontsize{14} Distance z, m');
zlabel('\bf \fontsize{14} y(z,t)');
figure
plot(z,mysol(1,:),'k-','LineWidth',2);
hold on;
plot(z,mysol(16,:),'ko','LineWidth',2);
plot(z,mysol(26,:),'k*','LineWidth',2);
plot(z,mysol(end,:),'k+','LineWidth',2);
title('\bf \fontsize{14} y(z) for Successive t Values');
xlabel('\bf \fontsize{14} Distance z, m');
ylabel('\bf \fontsize{14} y(z)');
hleg = legend('t=0 s','t=4500 s','t=7500 s','t=15000
s','Location','Best');
```

Figure 10.32 The MATLAB code (**pde_10_8_Anal.m**) that defines the diffusion, the time and space span, and calls upon a composite function (**myseries_10_8(z,t)**), which itself is a summation over 500 times for the product of the two functions $cos\left((2n-1)\frac{\pi}{L}z\right)$ and $e^{-D_{AB}\left((2n-1)\frac{\pi}{L}\right)^2 t}$.

the two functions: $cos\left((2n-1)\frac{\pi}{L}z\right)$ and $e^{-D_{AB}\left((2n-1)\frac{\pi}{L}\right)^2 t}$ (i.e., the summation appearing in Eq. [10.84], is carried out up to $n = 500$).

The code for the function: **myseries_10_8(z,t)** is defined in Fig. 10.33.

```
function [out1] = myseries_10_8(z,t)
global DAB
sum=0.5;
for j=1:500
preterm=((-1)^(j+1))/(2*j-1);
costerm = preterm*cos(((2*j-1))*pi*z);
expterm=exp(-(DAB*(((2*j-1))*pi)^2)*t);
multipterm=costerm*expterm;
sum=sum+(2/pi)*multipterm;
end
out1=sum;
```

Figure 10.33 The MATLAB code (**myseries_10_8.m**) that defines the function to which the main program in Fig. 10.32 refers.

Figure 10.34 shows the 3D surface plot for y as a function of distance, z measured in meters, and time t in seconds. Notice how the edge of the surface is sharp at low t and flattens out as t grows up, indicating the evolution toward equilibrium (i.e., uniform composition everywhere in the cylinder).

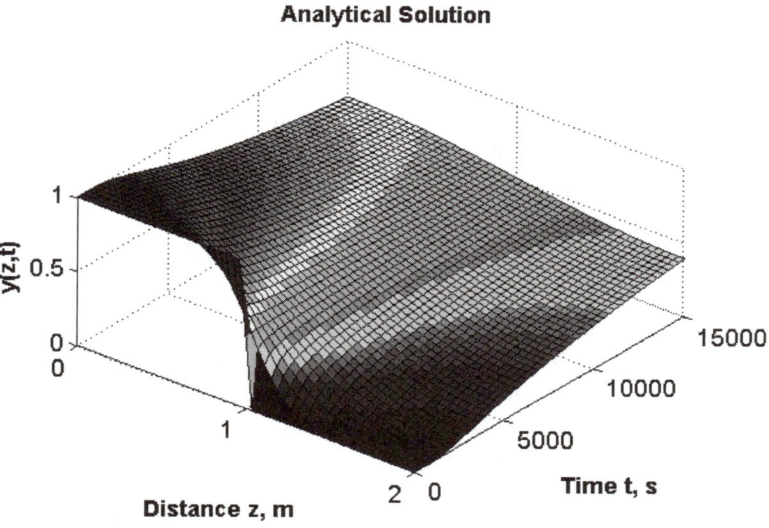

Figure 10.34 The 3D surface plot for y, the mole fraction of air, as a function of distance, z measured in meters and time t in seconds. Notice that the length of the cylinder L is 2.0 m.

Figure 10.35 shows different snapshots taken at different values of time, for y as a function of z. One can notice that the interface between air and propane is initially sharp (i.e., vertical interface that initially splits the cylinder into two separate zones of unmixed pure gases) and as time goes on, mixing by diffusion takes place and the system continues evolving toward equilibrium (i.e., the mole fraction of air will assume a uniform value of 0.5 everywhere throughout the cylinder.

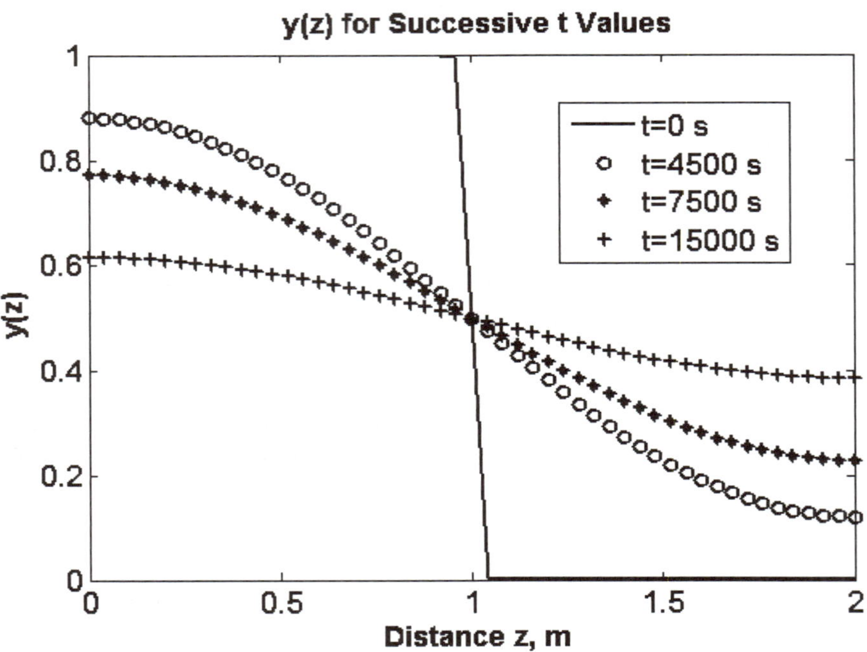

Figure 10.35 Different snapshots taken at different values of time, for y as a function of z. The sharp interface flattens out with time indicating gas mixing by diffusion.

10.8.1 *Solving 1D Mass Equation Using the MATLAB PDE Solver*

The code for the main calling program and the code for the three *user-defined functions* are shown here. The user is advised to refer to Sec. 10.7, if the matching process between the code and the physical problem in hand is not tractable or traceable.

Figure 10.36 shows the main program (**pde_10_8.m**).

The main program (Fig. 10.36) calls upon *three user-defined functions* that pertain to the partial differential Eq. (10.81) that we attempt to solve. The parameter m is already defined in the main program. The three terms (c, f, and s) are introduced in the *first user-defined* function called **`pdex1pde108 function,`** as shown in Fig. 10.37.

```
clear all;
global DAB
DAB=1.15e-5;
z = 0:0.01:2;
t = 0:75:15000;
m = 0;
sol = pdepe(m,@pdex1pde108,@pdex1ic108,@pdex1bc108,z,t);
u=sol(:,:,1);
surf(z,t,u)
title('\bf \fontsize{14} Numerical solution computed with 200 mesh points');
xlabel('\bf \fontsize{14} Distance z, m');
ylabel('\bf \fontsize{14} Time t, s');
zlabel('\bf \fontsize{14} y(z,t)');
figure;
plot(z,u(1,:),'k--','LineWidth',2);
hold on;
plot(z,u(61,:),'k.','LineWidth',1);
plot(z,u(101,:),'k-.','LineWidth',3);
plot(z,u(end,:),'k:','LineWidth',3);
title('\bf \fontsize{14} y(z) for Successive t Values');
xlabel('\bf \fontsize{14} Distance z, m');
ylabel('\bf \fontsize{14} y(z)');
hleg = legend('t=0 s','t=4500 s','t=7500 s','t=15000 s','Location','Best');
```

Figure 10.36 The main program (**pde_10_8.m**) that calls upon three user-defined functions pertaining to the problem of interest. In addition, it outputs the results in the form of 3D and 2D for the mole fraction profile of air as a function of z and t. u is essentially y.

```
function [c,f,s] = pdex1pde108(z,t,u,DuDx)
global DAB
c = 1/DAB;
f = DuDx;
s = 0;
```

Figure 10.37 **pdex1pde108.m** that defines $c\left(x,t,u,\frac{\partial u}{\partial x}\right)$, $f\left(x,t,u,\frac{\partial u}{\partial x}\right)$, and $s\left(x,t,u,\frac{\partial u}{\partial x}\right)$. See Eq. (10.71) for term-by-term comparison.

The *second user-defined function* is **pdex1ic108,** which defines the initial condition. The initial condition for the PDE [Eq. (10.81)] is as follows and its MATLAB code is shown in Fig. 10.38. Keep in mind that $L = 2$ m.

$$IC: y(z,0) = \begin{cases} 1 & 0 \le z \le L/2 \\ 0 & L/2 \le z \le L \end{cases} \tag{10.82}$$

```
function u0 = pdex1ic108(z)
if z<=1.0 % z<=L/2<=2/2=1.0
u0 = 1.0;
else
u0=0;
end
```

Figure 10.38 **pdex1ic108.m** file that defines the initial condition for u (or, y), that is $u(z,0)=u_o$. $L=2$m.

The *third and last user-defined function* is **pdex1bc108** that defines the boundary conditions at the lower and upper limits of z, as prescribed in Eq. (10.83).

Here are the two boundary conditions for z:

The boundary conditions, written in the same format as Eq. (10.73), are:

$$BC\ \#1\!: \left.\frac{\partial y}{\partial z}\right|_{z=0} = 0 \tag{10.83a}$$

$$0+1\times\left.\frac{\partial u}{\partial z}\right|_{z=0} = 0 \quad \Leftrightarrow \quad p_l + q_l \times \left.\frac{\partial u}{\partial z}\right|_{z=0} = 0$$

The conclusion is simply that $p_l = 0$ and $q_l = 1$.

$$BC\,\#2\!: \left.\frac{\partial y}{\partial z}\right|_{z=L} = 0 \tag{10.83b}$$

$$0+1\times\left.\frac{\partial u}{\partial z}\right|_{z=L} = 0 \quad \Leftrightarrow \quad p_r + q_r \times \left.\frac{\partial u}{\partial z}\right|_{z=L} = 0$$

The conclusion is simply that $p_r = 0$ and $q_r = 1$.

The MATLAB code is shown in Fig. 10.39.

```
function [pl,ql,pr,qr] = pdex1bc108(xl,ul,xr,ur,t)
pl = 0; %NOTE: pl (pL) not p1.
ql = 1; %NOTE: ql (qL) nor q1.
pr = 0;
qr = 1;
```

Figure 10.39 The **pdex1bc108.m** file that defines the four terms p_l, q_l, p_r *and* q_r appearing in the standard equation [Eq. (10.73)]. They describe the boundary condition at the lower and upper limits of z.

Figure 10.40 shows the 3D plot for y as a function of z and t, using 200 mesh points for each dimension. Notice that the general behavior or topology of the surface is very similar to that of the analytical solution shown earlier in Fig. 10.34. Again, keep in mind that the solution given by the MATLAB PDE solver is approximate.

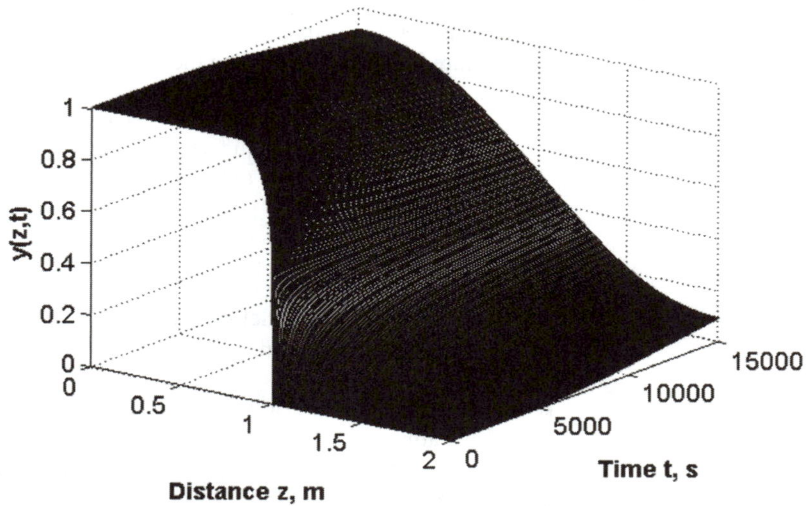

Figure 10.40 A 3D plot of y as a function of z and t using 200 mesh points for each dimension.

Figure 10.41 shows different snapshots taken at different t values for y as a function of z. The trends are essentially the same as those of the open-form, analytical solution, shown earlier in Fig. 10.35.

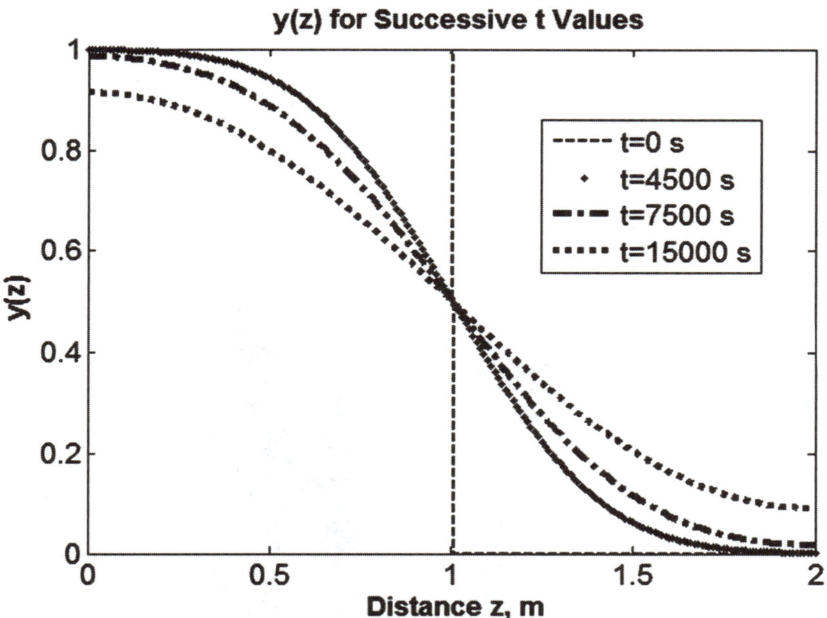

Figure 10.41 Different snapshots taken at different t values, for y as a function of the cylindrical axial distance z.

10.9 Unconstrained 1D Function Minimization

Let us consider a *two-stage compressor* in which the gas is cooled to the inlet gas temperature between stages. The theoretical work is given by:

$$W = \frac{kp_1V_1}{k-1}\left[\left(\frac{p_{middle}}{p_1}\right)^{\frac{k-1}{k}} - 2 + \left(\frac{p_3}{p_{middle}}\right)^{\frac{k-1}{k}}\right] \tag{10.87}$$

where

$k = \frac{C_P}{C_V}$ p_1 = inlet pressure p_{middle} = intermediate pressure
p_3 = outlet pressure V_1 = inlet volume

To minimize the amount of work, W, show that if $p_1 = 1$ atm and $p_3 = 4$ atm, then $p_{middle}^{optimum} = 2$ atm.

We show both solutions analytically and numerically via the MATLAB built-in **fminbnd** that finds the minimum of a *single-variable unconstrained function* on a fixed interval.

$$W(x) = \frac{kp_1V_1}{k-1}\left[\left(\frac{x}{p_1}\right)^{\frac{k-1}{k}} - 2 + \left(\frac{p_3}{x}\right)^{\frac{k-1}{k}}\right] \tag{10.88}$$

Take the first derivative of $W(x)$ with respect to x and equate the result to zero:

$$\frac{dW(x)}{dx} = \frac{k-1}{k} \times \frac{kp_1V_1}{k-1}\left[\frac{1}{p_1}\left(\frac{x}{p_1}\right)^{\frac{k-1}{k}-1} - \frac{p_3}{x^2}\left(\frac{p_3}{x}\right)^{\frac{k-1}{k}-1}\right] = 0 \tag{10.89}$$

Figure 10.42 A two-stage air compressor, engine power 13.0 HP. Free air flow at maximum pressure is 25.1 CFM (ft^3/min), maximum pressure 175 psi, and tank capacity is 30 gallons.

This implies that the bracketed term is zero:

$$\frac{1}{p_1}\left(\frac{x}{p_1}\right)^{\frac{k-1}{k}-1} - \frac{p_3}{x^2}\left(\frac{p_3}{x}\right)^{\frac{k-1}{k}-1} = 0 \quad \frac{1}{p_1}\left(\frac{x}{p_1}\right)^{\frac{k-1}{k}-1} = \frac{p_3}{x^2}\left(\frac{p_3}{x}\right)^{\frac{k-1}{k}-1} \tag{10.90}$$

The left-hand-side term can be put in the form of:

$$\frac{1}{x}\times\frac{x}{p_1}\left(\frac{x}{p_1}\right)^{\frac{k-1}{k}-1} = \frac{1}{x}\left(\frac{x}{p_1}\right)^{\frac{k-1}{k}} \tag{10.91}$$

The right-hand-side term can also be put in the form of:

$$\frac{p_3}{x^2}\left(\frac{p_3}{x}\right)^{\frac{k-1}{k}-1} = \frac{1}{x}\left(\frac{p_3}{x}\right)^{\frac{k-1}{k}} \tag{10.92}$$

Equating both sides, we get:

$$\frac{1}{x}\left(\frac{x}{p_1}\right)^{\frac{k-1}{k}} = \frac{1}{x}\left(\frac{p_3}{x}\right)^{\frac{k-1}{k}} \quad \rightarrow\rightarrow\rightarrow \quad \left(\frac{x}{p_1}\right)^{\frac{k-1}{k}} = \left(\frac{p_3}{x}\right)^{\frac{k-1}{k}} \tag{10.93}$$

Eq. (10.93) implies that:

$$\left(\frac{x}{p_1}\right) = \left(\frac{p_3}{x}\right) \quad \rightarrow\rightarrow\rightarrow \quad x^2 = p_1 \times p_3 = 1\text{ atm} \times 4\text{ atm} = 4\text{ atm}^2$$

$$x = 2\text{ atm} \tag{10.94}$$

Figure 10.43 shows **mywork.m** file that defines the two-stage compressor work as a function of the intermediate pressure, p_{middle}, which is denoted here as x. The output of the MATLAB built-in **fminbnd** gives x that minimizes the work. Note that 1 and 4 represent the lower and upper limit for the variable x, respectively.

```
function W = mywork(x)
k=1.4;
p1=1;
p3=4.0;
V1=100;
W=((k*p1*V1)/(k-1))*((x./p1).^((k-1)/k)-2+(p3/x).^((k-1)/k));
```

Figure 10.43 Defining the work of a two-stage compressor with inter-stage cooling.

```
>> p_middle = fminbnd(@mywork,1,4)

p_middle =

    2.0000
```

10.10 Unconstrained 2D Function Minimization

Let us consider a *three-stage compressor* in which the gas is cooled to the inlet gas temperature between stages. The theoretical work is given by:

$$W = \frac{kp_1V_1}{k-1}\left[\left(\frac{p_2}{p_1}\right)^{\frac{k-1}{k}} + \left(\frac{p_3}{p_2}\right)^{\frac{k-1}{k}} + \left(\frac{p_4}{p_3}\right)^{\frac{k-1}{k}} - 3\right] \tag{10.95}$$

where $k = \frac{C_P}{C_V}$, p_1 = inlet pressure, p_2 = inlet pressure to stage 2, p_3 = inlet pressure to stage 3, p_4 = outlet pressure, and V_1 = inlet volume.

To minimize the amount of work, W, show that if p_1 = 1 atm and p_4 = 10 atm, then the minimum occurs at a pressure ratio for each stage of $\sqrt[3]{10}$.

Figure 10.44 shows the **mywork3.m** file that defines the three-stage compressor work as a function of the intermediate pressures, p_2 and p_3, which are denoted here as $x(1)$ and $x(2)$. The output of MATLAB built-in **fminunc** gives x, which minimizes the work. The MATLAB built-in **fminunc** finds the minimum of a *multi-variable, unconstrained function* on a fixed interval.

```
function W = mywork3(x)
k=1.4;
p1=1;
p4=10.0;
V1=100;
W=((k*p1*V1)/(k-1))*((x(1)./p1).^((k-1)/k)-
2+(x(2)./x(1)).^((k-1)/k)+(p4/x(2)).^((k-1)/k));
```

Figure 10.44 Defining the work of a three-stage compressor with inter-stage cooling.

```
>> options = optimset('TolX',1e-8,'TolFun',1e-8)
```

The matrix [2, 4] in the following command represents the initial guess for $x(1)$ and $x(2)$.

```
>>x = fminunc(@mywork3,[2,4])
x =

   2.1544    4.6416
```

NOTE $x(1) = p_2$ = 2.1544 atm and $x(2) = p_3$ = 4.6416 atm. Moreover,

$$\frac{p_2}{p_1} = \frac{2.1544\text{ atm}}{1\text{ atm}} = \sqrt[3]{10} \qquad \frac{p_3}{p_2} = \frac{4.6416\text{ atm}}{2.1544\text{ atm}} = \sqrt[3]{10} \qquad \frac{p_4}{p_3} = \frac{10\text{ atm}}{4.6416\text{ atm}} = \sqrt[3]{10}$$

10.11 Constrained 2D Function Minimization

10.11.1 Inequality Constraint

In designing a can to hold a specified amount of food, the cost function for manufacturing one can is proportional to:

$$A(D, L) = f(D, L) = \left(\pi DL + 2\times\frac{\pi D^2}{4}\right) = \pi\left(DL + \frac{D^2}{2}\right) \tag{10.96}$$

and the constraints are:

$$V_{Cyl} = \frac{\pi D^2}{4}L > 400 \text{ mL} \tag{10.97}$$

$$3.0 \text{ cm} \le D \le 9 \text{ cm} \qquad 7 \text{ cm} \le L \le 18 \text{ cm} \tag{10.98a-b}$$

Here, the *total surface area* of a can is to be minimized without compromising the capacity. If Eq. (10.96) is multiplied by the thickness of the metallic shell of the can, then we will get the volume of the shell, and if the latter is multiplied by the density of the metal (i.e., the material of construction), then we will get the mass of an empty can. Basically, we try to find the minimum mass of a metal can that will satisfy both the capacity and geometry requirement.

Let us define the objective function; the inequality condition; and the bounds for both D and L. Figure 10.45 shows the MATLAB definition for the objective function given by Eq. (10.96).

```
function f = objecfun(x)
% x(1) stands for D and x(2) for L. A(D,L)=pix DL+2×{pi(D^2)/4}.
f = (pi*x(1)*x(2) + (pi/2)*x(1)^2);
```

Figure 10.45 **Objecfun.m** file declares the objective function in terms of vector **x** that has two components: $x(1)$ stands for D and $x(2)$ stands for L.

Figure 10.46 shows the nonlinear inequality constraint represented by Eq. (10.97).

```
function [c,ceq] = nonlconstr(x)% see Product Help for nonlconstr
% inequality constraint: (400-(pi/4)*(x(1)^2)*x(2))<=0
c =[400-(pi/4)*(x(1)^2)*x(2)];
ceq = []; % There is no equality constraint in this problem
```

Figure 10.46 **Nonlconstr.m** file for representing the nonlinear inequality constraint given by Eq. (10.97).

Go to: **MATLAB Start menu → Toolboxes → Optimization → Optimization Tool (optimtool)**. Alternatively, you may run in the MATLAB Command Window:

```
>> optimtool
```

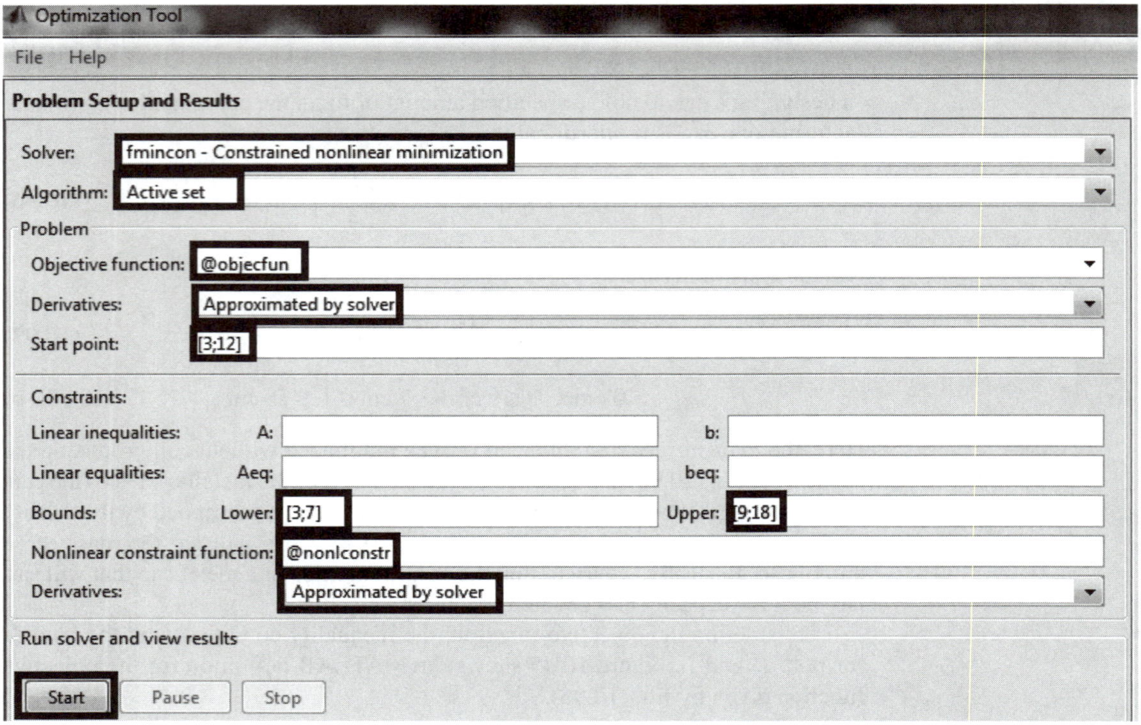

Figure 10.47 A portion of optimtool Window in which the user decides on the type of solver and algorithm to be used by MATLAB, the objective function file; the start point vector, the lower and upper bound vectors; the nonlinear constraint function; and finally the derivative evaluation method.

Figure 10.47 shows a portion of the pop-up **optimtool** Window, in which the user inputs the type of solver to be used by MATLAB; the solver (choose fmincon); the algorithm (choose Active set); the objective function file (choose a handle to **objecfun** function); the derivatives (choose Approximated by solver); the start point vector ([3;12]) initial guesses for D and L, respectively; the lower ([3;7]) and upper ([9;18]) bound vectors; the nonlinear inequality constraint (via a handle to **nonlconstr** function); and finally the derivatives (choose Approximated by solver).

Click the Start button as shown in Fig. 10.47 to see the results of the optimization (i.e., minimization in this case) for the objective function while fulfilling the nonlinear inequality constraint and also without violating the lower and upper bounds imposed on the independent variables D and L. Figure 10.48 shows the results of the optimization process carried out by the MATLAB solver. The local minimum for the objective function was found. This value of surface area, equal to 300.53 cm^2, is the minimum feasible surface area that can be obtained without really violating the restrictions imposed on the objective function in terms of the capacity requirement [Eq. (10.97)] and geometrical requirement [Eq. (10.98a–b)].

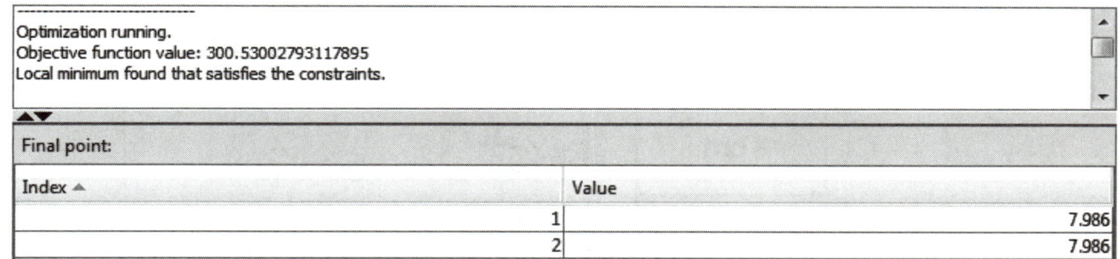

Figure 10.48 The minimum for the objective function evaluated at $D = 7.986$ cm and $L = 7.986$ cm. Notice here that $D = L$.

One last thing regards the level of displaying details for the MATLAB Command Window. It can be selected by the user. Figure 10.49 shows a portion of **optimtool** Window in which the user can pick up his or her choice under the **Options** frame or subwindow.

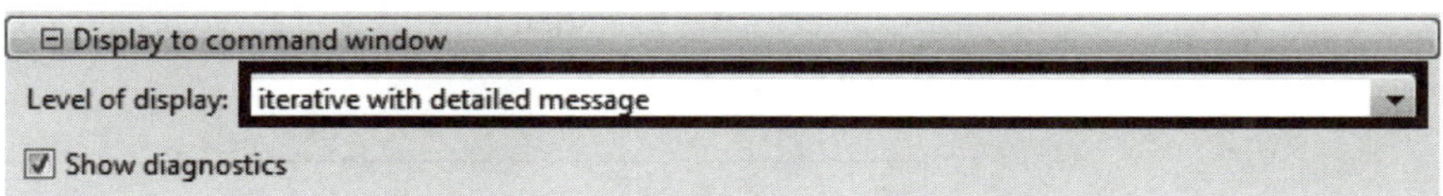

Figure 10.49 The level of displaying the results to the MATLAB Command Window.

The user will benefit from such pieces of information should the combination of selected solver, algorithm, and derivative evaluation method diverge.

10.11.2 *Equality Constraint*

A poster is to contain 400 cm^2 of printed matter with margins of 6 cm at the top and bottom and 8 cm at each side. Find the overall dimensions that will minimize the total area of the poster (Fig. 10.50).

The objective function to be minimized is the total area of the poster

$$A_{poster} = (x+16)\times(y+12) = xy + 16y + 12x + 192 \tag{10.99}$$

subject to the following equality constraint:

$$400 = x \times y \tag{10.100}$$

Figure 10.51 shows the MATLAB definition for the objective function given by Eq. (10.99).

Figure 10.52 shows the equality constraint represented by Eq. (10.100).

Figure 10.53 shows a portion of the pop-up **optimtool** Window, where the initial guess vector, the lower and upper bounds, the objective function, and the equality constraint are entered.

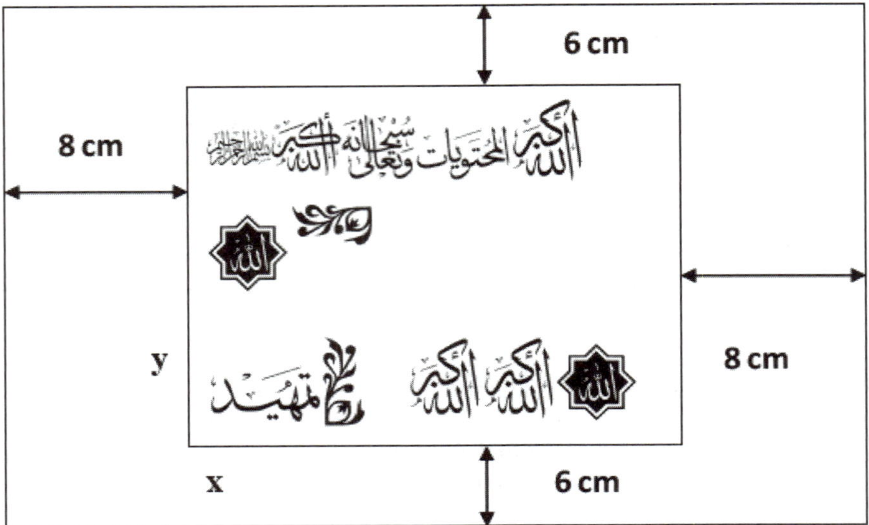

Figure 10.50 The printed matter area has to occupy an area of 400 cm^2 while bounded by margins from all sides.

```
function f = objecfun2(x)
% x(1) stands for x and x(2)for y in Figure 10.50
f = (x(1)+16).*(x(2)+12);
```

Figure 10.51 **objecfun2.m** file declares the objective function in terms of vector **x** that has two components: $x(1)$ stands for x and $x(2)$ stands for y, as shown in Figure 10.50.

```
function [c,ceq] = eqconstr(x)
c =[];
ceq = x(1)*x(2)-400;
```

Figure 10.52 **eqconstr.m** file for representing the equality constraint given by Eq. (10.100).

The solution is found to be $x = 23.094$ and $y = 17.321$ ($23.094 \times 17.321 = 400$ cm^2) with $A_{poster} = (x+16)\times(y+12) = (23.094+16)\times(17.321+12) = 1146.3\ \text{cm}^2$.

Exercise 10.5 In Eq. (10.99), replace y by x using Eq. (10.100), then differentiate A_{poster} with respect to x once and find the critical point(s) of the first derivative. Check the second derivative to see if you have minimum or maximum at the critical point(s). Compare your answer with that of the MATLAB **optimtool**.

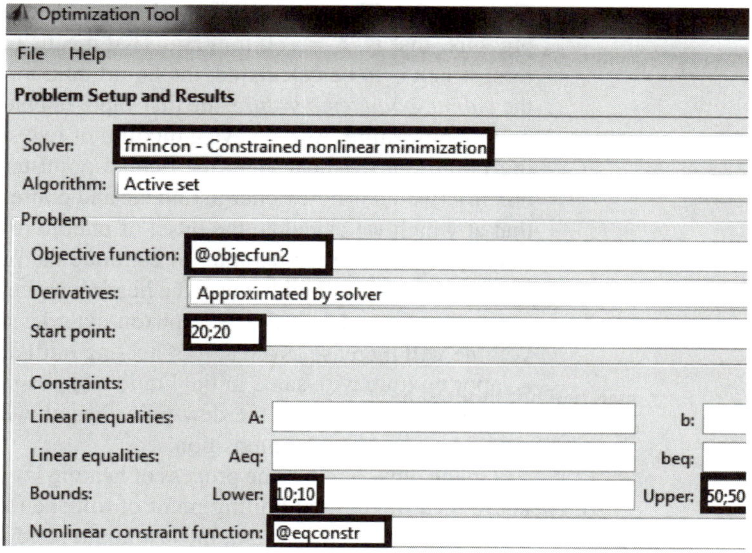

Optimization Tool

File Help

Problem Setup and Results

Solver: fmincon - Constrained nonlinear minimization

Algorithm: Active set

Problem

Objective function: @objecfun2

Derivatives: Approximated by solver

Start point: 20;20

Constraints:

Linear inequalities: A: b:

Linear equalities: Aeq: beq:

Bounds: Lower: 10;10 Upper: 50;50

Nonlinear constraint function: @eqconstr

Figure 10.53 **optimtool** Window in which the user enters the initial guess vector, the lower and upper bounds, the objective function, and the equality constraint.

10.12 Bubble-Point Calculation for Benzene-Toluene Mixture

Consider a liquid mixture of benzene and toluene at a mole fraction of 0.5 for both and a total pressure of 1 atmosphere (760 mm Hg). We would like to calculate the bubble-point temperature for such a liquid-mixture. For simplicity, we will assume an ideal liquid mixture. In other words, the activity coefficient for either component will be set to one $\gamma_B = \gamma_T = 1.0$.

Hence, the bubble point pressure as a function of temperature is given by:

$$P_{Bubble} = \sum_{i=1}^{N} X_i P_i^{sat}(T) = X_B P_B^{sat}(T) + X_T P_T^{sat}(T) = X_B P_B^{sat}(T) + (1 - X_B) P_T^{sat}(T) \tag{10.101}$$

The vapor pressure of either benzene or toluene is given by Antoine's equation as a function of temperature as in the following equations:

$$P_B^{sat}(T) = \exp\left(\frac{A1}{t\ [°C] + 273.15} + B1\right) = \exp\left(\frac{-3848.09}{t\ [°C] + 273.15} + 17.5318\right) \tag{10.102}$$

$$P_T^{sat}(T) = \exp\left(\frac{A2}{t\ [°C] + 273.15} + B2\right) = \exp\left(\frac{-4328.12}{t\ [°C] + 273.15} + 17.9130\right) \tag{10.103}$$

The mixture will start boiling once the bubble pressure hits the value of the applied pressure, that is, P = 760 mm Hg. That means below the bubble-point temperature, which is to be calculated, the liquid mixture is considered as sub-cooled. At the *bubble-point temperature*, the first vapor bubble forms; hence, the name bubble-point temperature. As we have a mixture of two substances having different boiling point temperatures, then the bubble-point temperature for the *rest of liquid mixture* (which is now richer in toluene and poorer in benzene) will be higher than that at which we calculate the onset of bubble formation. That simply means we have to elevate the temperature to continue the process of vaporization. Because we initially have a liquid phase, the heating process will be carried out in a closed system under constant pressure piston-cylinder arrangement, in which the total volume will increase as a result of heating until all liquid vaporizes and we reach a vapor mixture with same initial liquid composition but at its *dew-point temperature*. Keep in mind that the dew-point lies above the bubble-point for the same applied pressure and composition.

On the other hand, if the process of heating is continued in an open-atmosphere kettle then the normal boiling point of toluene (around 110.6°C) will be eventually reached. This means that the last liquid residue remaining in the kettle will be substantially toluene and almost null in benzene.

Because the calculated bubble-point pressure given by Eq. (10.101) is not linear in T, we will use MATLAB to calculate the root or zero of the following equation:

$$f(T) = P_{diff} = P_{Bubble} - P_{given} = X_B P_B^{sat}(T) + (1 - X_B) P_T^{sat}(T) - 760.0 \qquad (10.104)$$

Now, the left-hand side of Eq. (10.104) should essentially reduce to zero once we hit the right value of T. If the assumed T is either higher, or lower, then $f(T)$ will be either positive, or negative, respectively.

Figure 10.54 shows the MATLAB code (**BubbleT.m**) for defining $f(T)$ function. The root of $f(T)$ gives the true bubble-point temperature.

```
function Pdiff = BubbleT(T)
Pgiven=760.0; %Total Pressure is 760 mm Hg.
XB=0.5; % Mole-fraction of Benzene.
A1=-3848.09;% A1 and B1 are Antoine's constants for benzene
B1=17.5318;
A2=-4328.12;%% A2 and B2 are Antoine's constants for toluene
B2=17.913;
P1sat=exp(A1/(T+273.15)+B1); % Antoine's Eq. for benzene
P2sat=exp(A2/(T+273.15)+B2);% Antoine's Eq. for toluene
Pcalc=(XB*P1sat+(1-XB)*P2sat);
Pdiff=Pcalc-Pgiven;
```

Figure 10.54 The **BubbleT.m** file that defines $f(T)$, the root of which represents the bubble-point temperature for the given liquid mixture at the specified conditions of mole fraction and total pressure.

```
>> CalcT=fzero(@BubbleT,90.0)

CalcT =

   92.0481
```

Exercise 10.6 Try different values of X_B for a given P_{given}. Also, try different values of P_{given} for a given X_B. After running the code and seeing the results, try to see how the mole-fraction of benzene affects the value of the calculated bubble-point temperature. Does the bubble-point temperature increase or decrease with X_B? On the other hand, does it increase, or decrease with the applied pressure, P_{given}, on the top of the liquid?

10.13 Dew-Point Calculation for Benzene-Toluene Mixture

Consider now a vapor mixture of benzene and toluene at a mole fraction of 0.5 for both and a total pressure of 1 atmosphere (760 mmHg). We would like to calculate the dew-point temperature for such a vapor mixture. For simplicity, we will assume an ideal gas mixture. In other words, the fugacity coefficient for either component will be set to one $\hat{\phi}_B = \hat{\phi}_T = 1.0$.

Hence, the bubble-point pressure as a function of temperature is given by:

$$\frac{1}{P_{Dew}} = \sum_{i=1}^{N} \frac{y_i}{P_i^{sat}(T)} = \frac{y_B}{P_B^{sat}(T)} + \frac{y_T}{P_T^{sat}(T)} = \frac{y_B}{P_B^{sat}(T)} + \frac{1-y_B}{P_T^{sat}(T)} \qquad (10.105)$$

The vapor pressure of either benzene or toluene is given by Antoine's equation as a function of temperature as in the previous equations: Eq. (10.102) and (10.103).

The mixture will start condensing once the dew pressure hits the value of the applied pressure, that is, $P = 760$ mm Hg. That means above the dew-point temperature, which is to be found, the gas mixture is considered to be superheated. At the *dew-point temperature*, the first liquid droplet starts to form, hence the name dew-point temperature. As we have a mixture of two substances having different condensing point temperatures, then the dew-point temperature for the *rest of the vapor mixture* (which is now richer in benzene and poorer in toluene) will be lower than that at which we calculate the formation of the first droplet. That simply means we have to decrease the temperature to continue the process of condensation. Because we initially have a gas phase, the cooling process will be carried out in a closed system under constant pressure piston-cylinder arrangement in which the total volume will decrease as a result of cooling until all vapor condenses and we reach a liquid mixture with same initial vapor composition but at its *bubble-point temperature*.

Because, the calculated dew-point pressure given by Eq. (10.105) is not linear in T, we will use MATLAB to calculate the root or zero for the following equation:

$$f(T) = P_{diff} = P_{dew} - P_{given} = [y_B / P_B^{sat}(T) + (1-y_B)/P_T^{sat}(T)]^{-1} - 760.0 \qquad (10.106)$$

The left-hand side of Eq. (10.106) should essentially reduce to zero once we hit the right value of T. If the assumed T is either higher or lower, then $f(T)$ will be either positive or negative, respectively.

Figure 10.55 shows the MATLAB code (**DewT.m**) for defining the $f(T)$ function. The root of $f(T)$ gives the true dew-point temperature. Also shown is the calculation of X_B and X_T for the first formed droplet. Notice that the sum of X_i ($X_B + X_T =$ `0.2872 + 0.7128`) adds up to 1.00. Moreover, notice that the bubble-point temperature in the previous section was found to be `92.0481°C`, which is definitely lower than the dew-point temperature (`98.8634°C`) at the same mole fraction of benzene ($X_B = y_B = 0.5$) and a total applied pressure of 1 atmosphere.

```
function Pdiff = DewT(T)
Pgiven=760.0; %Total Pressure is 760 mm Hg.
yB=0.5; % Mole-fraction of Benzene.
A1=-3848.09;% A1 and B1 are Antoine's constants for benzene
B1=17.5318;
A2=-4328.12;%% A2 and B2 are Antoine's constants for toluene
B2=17.913;
P1sat=exp(A1/(T+273.15)+B1); % Antoine's Eq. for benzene
P2sat=exp(A2/(T+273.15)+B2);% Antoine's Eq. for toluene
Pinv=(yB/P1sat+(1-yB)/P2sat);
Pcalc=1/Pinv;
XB=(yB*Pgiven)/P1sat
XT=((1-yB)*Pgiven)/P2sat
Pdiff=Pcalc-Pgiven;
```

Figure 10.55 The **DewT.m** file that defines $f(T)$, the root of which represents the dew-point temperature for the given vapor mixture at the specified conditions of mole fraction and total applied pressure.

```
>> CalcT=fzero(@DewT,90.0)
XB =
     0.2872
XT =
     0.7128
CalcT =
   98.8634
```

Exercise 10.7 Try different values of y_B for a given P_{given}. Also, try different values of P_{given} for a given y_B. After running the code and seeing the results, try to see how the mole fraction of benzene affects the value of the calculated dew-point temperature. Does the dew-point temperature increase or decrease with y_B? On the other hand, does it increase or decrease with the applied pressure, P_{given}, of the vapor phase?

10.14 Problems

10.1 Dissolution and reaction of gas in liquids (described by a second-order ODE).

Oxygen dissolves into and reacts irreversibly with aqueous sodium sulfite solutions. If the gas solubility is denoted as C_A^* at the liquid-gas interface, derive the elementary differential equation that describes the *steady-state composition profile of oxygen in the liquid phase* when the rate of oxygen reaction is represented by $R_A = -kC_A^n$ and the local oxygen diffusion flux is described by $J_A = -D_A \dfrac{dC_A}{dz}$, where D_A is diffusivity and z is distance from the interface into the liquid.

Answer:

$$D_A \frac{d^2C_A}{dz^2} - kC_A^n = 0 \tag{10.107}$$

Notice that n, the reaction order, can be any real number, including zero. Solve Eq. (10.107) via transforming the second-order nonlinear ODE into a set of two first-order ODEs (see Sec.7.5).

Use:

$$y_1 = C_A \qquad y_2 = \frac{dC_A}{dz} \quad \rightarrow \quad \frac{dy_2}{dz} = \frac{d^2C_A}{dz^2}$$

You will have two ODEs with following boundary conditions:

$$y_1(0) = C_A(0) = C_A^* \tag{10.108}$$

$$y_2(0) = \left.\frac{dC_A}{dz}\right|_{z=0} = s \tag{10.109}$$

In terms of MATLAB notation, z will be replaced by t and C_A by y, or it can be used as is. For example, use C_A^* as 10 mol/L and s [in Eq. (10.109)] will represent the slope of C_A with respect to z. This can be zero; finite (small negative; say −1), or almost infinity (large negative; say −100), depending on the balance between mass transfer (replenishment, as it is an input to liquid medium from the gas side) and reaction kinetics (consumption). n can be 0, 0.5, 1, 1.5, or 2.

Depending on the reaction order, k will have units of

$$\frac{1}{s} \times \left(\frac{mol}{L}\right)^{1-n}. \text{ Use } k = 0.1\frac{1}{s} \times \left(\frac{mol}{L}\right)^{1-n}.$$

10.2 Steady-state catalytic chemical reactor (described by a single ODE).

Your task as a design engineer in a chemical company is to model a fixed bed reactor packed with the company proprietary catalyst of spherical shape. The catalyst is specific for the removal of a toxic gas at very low concentration in air, and the information provided from the catalytic division is that the reaction is *first-order with respect to the toxic gas concentration*. The reaction rate has units of moles of toxic gas removed per mass of catalyst per time.

Your first attempt, therefore, is to model the reactor in the simplest possible way so that you can develop some intuition about the system before any further modeling attempts are made to describe it exactly.

a) For simplicity, assume that there is no significant diffusion inside the catalyst and that diffusion along the axial direction is negligible. An isothermal condition is also assumed.

(This assumption is known to be invalid when the reaction is very fast and the heat of reaction is high). The coordinate z is measured from entrance of the packed bed. Perform a steady-state mass balance around a thin shell at the position z with the shell thickness of Δz and show that when the shell thickness approaches zero, the following relation is obtained:

$$u_o \frac{dC}{dz} = -(1-\epsilon)\rho_p(kC) \tag{10.110}$$

Where u_o is the superficial velocity, C is the toxic gas concentration, ϵ is the bed porosity, ρ_p is the catalyst density, and k is the rate constant for a heterogeneous reaction, which has units of $\frac{m^3}{(\text{kg catalyst} \times s)}$ for such a case (i.e., first-order reaction).

b) Show that Eq. (10.110) (i.e., a catalytic chemical reactor under a steady-state, isothermal condition) has the solution of the form:

$$C = C_o exp\left(-\frac{(1-\epsilon)\rho_p k}{u_o} z\right) \tag{10.111}$$

Where C_o is inlet concentration (mol/L) of toxic gas at $z = 0$.

c) Find the height of the fixed-bed reactor (expressed in m) such that the final concentration is equal to 1% of the initial value (i.e., $C = 0.01 \times C_o$). The fractional conversion is then given by:

$$X = \frac{C_o - C}{C_o} = 1 - \frac{C}{C_o} = 1 - exp\left(-\frac{(1-\epsilon)\rho_p k}{u_o} z\right) \tag{10.112}$$

Use the following numerical values for parameters appearing in Eq. (10.112) to find the height of the bed to achieve 99% fractional conversion:

u_o: 10 m/s; ϵ: 0.5; ρ_p: 3,000 kg/m^3; and k: 0.01 $\frac{m^3}{(\text{kg catalyst} \times s)}$

d) Use the MATLAB ode solver to solve Eq. (10.110). Notice that Eq. (10.110) can be solved using both the MATLAB symbolic and numerical ODE solver. Do not forget to put Eq. (10.110) in the standard form for solving an ODE (see Secs. 7.3 and 7.6). Use the same numerical values mentioned in part (c) plus you need to assume a value of C at $z = 0$ or (i.e., $y(0)$). You may assume that $y(0) = 1 \times 10^{-3}$ mol/L and terminate at t_f such that $y_f = 0.01 \times 1 \times 10^{-3} = 1 \times 10^{-5}$. Of course, t_f is nothing but the height of the fixed-bed reactor. Notice that for MATLAB ode solver, z is replaced by t, whereas C may be used as is, or replaced by y.

NOTE Do not forget to convert from mol/L to mol/m^3

10.3 Unsteady-state thermal model for thermocouple (described by a single ODE).

A cold thermocouple probe suddenly placed in a hot flowing water stream for the purpose of temperature measurement. The probe consists of two dissimilar metal wires joined by soldering at the tip, and the wires are then encased in a metal sheath and the tip is finally coated with a bead of plastic to protect it from corrosion. Take the mass of the soldered tip plus plastic bead to be m, with specific heat Cp. Denote the heat transfer coefficient as h.

a) If the effects of thermal conductivity can be ignored, show that the temperature response of the probe is described by:

$$mC_p \frac{dT}{dt} = hA(T_W - T) \tag{10.113}$$

Where T_W is water temperature and A is the surface area of the spherical probe, immersed in water, which is made of the soldered tip plus the plastic bead.

b) Given that the initial condition of T (i.e., $T(0)$) is T_0, show that the solution for Eq. (10.113) is:

$$\frac{T-T_W}{T_0-T_W}=\frac{T_W-T}{T_W-T_0}=e^{-\left(\frac{hA}{mC_p}\right)t} \tag{10.114}$$

The bracketed exponential term, appearing in Eq. (10.114), should have unit of time inverse so that it will cancel out the unit of t. Thus, the inverse of such a term will have units of time (i.e., $\left(i.e., \tau=\left(\frac{mC_p}{hA}\right)\right)$. τ is called the time constant of the thermocouple. For example, a miniature epoxy coated thermistor, from U.S. Sensor Corp., has a thermal time constant of 1 second maximum in a well-stirred oil bath and 10 seconds maximum in still air (see the figure that follows).

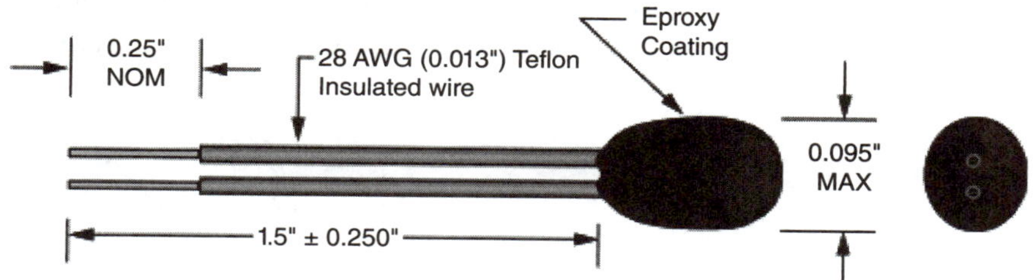

c) Use both the MATLAB symbolic and numerical ode solver to solve Eq. (10.113). Do not forget to put it in the standard form for solving an ODE (see Secs. 7.3 and 7.6). Use the time constant $\tau=\left(\frac{mC_p}{hA}\right)=1$ second in part (**b**) plus you need to assume a value of T at $t = 0$ or (i.e., $y(0)$). T_W is 90°C.

You may assume that $T(0) = 25°C$ and terminate at t_f such that $T_f = y_f = 89.997°C$. Of course, t_f is the time needed in second for the thermocouple (e.g., thermistor) to almost exactly measure or reach the temperature of water, where initially its temperature is 25°C. Notice that for MATLAB ode solver, T can be used as is, or replaced by y.

10.4 Steady-state dissolution of benzoic acid (described by a single ODE).

Water is initially passed through a tube constructed of solid benzoic acid. Because benzoic acid is slightly soluble in water with a maximum (i.e., saturation) solubility, C^{sat} moles acid/L solution at the given temperature. The inner walls of the tube will dissolve very slowly. By weighing the dried tube before and after exposure, it is possible to calculate the rate of mass transfer.

a) Take a quasi-steady state material balance for plug velocity profiles and show that the ODE obtained is:

$$u_o\frac{dC}{dz}-k_c\left(\frac{4}{D}\right)(C^{sat}-C)=0 \tag{10.115}$$

Where D denotes the inner tube diameter (assuming that it barely changes with dissolution process, hence constant), u_o is liquid velocity, and k_c is the mass transfer coefficient.

b) Show that the solution of Eq. (10.115) is given by:

$$\frac{C_L-C^{sat}}{C_0-C^{sat}}=\frac{C^{sat}-C_L}{C^{sat}-C_0}=e^{-\left(\frac{4k_c}{Du_o}\right)L} \tag{10.116}$$

Where C_0 is the inlet concentration of benzoic acid solution at $z = 0$ and C_L is the outlet concentration at $z = L$.

c) Use both the MATLAB symbolic and numerical ode solver to solve Eq. (10.115). Do not forget to put it in the standard form for solving an ODE (see Secs. 7.3 and 7.6). Use the following numerical values for parameters appearing in Eq. (10.115):

$$k_c = 0.0062\frac{\text{m}}{\text{s}}; \quad u_o = 1.5 \text{ m/s}; \quad D = 0.05 \text{ m}; \quad \& \quad C^{sat} = 0.0034 \text{ g/cm}^3 = 0.0278 \text{ mol/L}$$

Because fresh water is introduced at the inlet of the tube, the value of C at $z = 0$ will be zero (i.e., $C(0) = y(0) = 0$).

Terminate at t_f such that $C_L = y_f = 0.99 \times 0.0278$ mol/L (or, 0.99×0.0034 g/cm³). Of course, t_f is nothing but the length of the tube needed to achieve 99% of the saturated value. Notice that for the MATLAB ode solver, z is replaced by t, whereas C can be used as is, or replaced by y.

10.5 Steady-state reaction and diffusion in a porous solid phase (described by an ODE).

Let us consider a porous sphere where the substrate concentration at the surface is $C_{s,R}$. See the figure that follows:

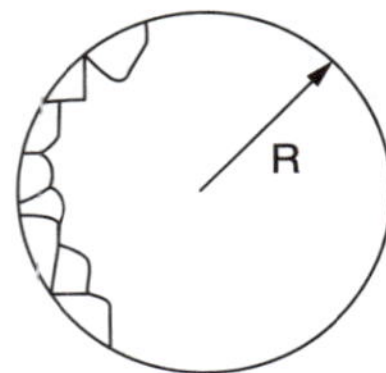

As we penetrate deeper inside the sphere, the substrate concentration will drop because of depletion. A reaction takes place within the porous solid sphere. Such a sphere may represent a biological cell that consumes certain substrates like glucose or oxygen and the surface will be the tissue of that cell. Drug/nutrient delivery through tissues is another example. For simplicity, we will consider the first-order reaction kinetics. The reaction rate is given by: $-r_S\Delta V = kC_s 4\pi r^2 \cdot \Delta r$ and diffusion by: $N_s = -D\frac{dC_s}{dr}$.

a) Carry out a steady-state shell balance using a thin shell with an area equal to $4\pi r^2$ and thickness Δr. Assume a pseudo-homogeneous medium and Fick's law describes diffusion. Divide through by $4\pi\Delta r$ and take the limit as $\Delta r \rightarrow 0$ to reach at:

$$\frac{d^2C_s}{dr^2} + \frac{2}{r}\frac{dC_s}{dr} - \frac{k}{D}C_s = 0 \tag{10.117}$$

Equation (10.117) is a second-order ODE, which requires two boundary conditions:

$$C_s\big|_{r=R} = C_{s,R} \tag{10.118}$$

C_s is finite everywhere, or

$$\left.\frac{dC_s}{dr}\right|_{r=0} = 0 \tag{10.119}$$

b) Define the following dimensionless variables:

$$\zeta = \frac{r}{R} \text{ and } \theta = \frac{C_s}{C_{s,R}}$$

Re-arrange Eq. (10.117) so that it becomes:

$$\frac{d^2\theta}{d\zeta^2}+\frac{2}{\zeta}\frac{d\theta}{d\zeta}-\phi^2\theta=0 \tag{10.120}$$

The transformed boundary conditions are:

$$\theta|_{\zeta=1}=1 \tag{10.121}$$

θ is finite everywhere, or

$$\left.\frac{d\theta}{d\zeta}\right|_{\zeta=0}=0 \tag{10.122}$$

where

$$\phi^2=\frac{R^2/D}{1/k}=\frac{\textit{characteristic diffusion time}}{\textit{characteristic reaction time}}=\frac{\textit{mass transfer resistance}}{\textit{reaction resistance}} \tag{10.123}$$

ϕ itself is the Thiele modulus that is used in expressing effectiveness factor of different catalyst shapes. By-product formation, in solid catalyzed reactions, depends upon the Thiele modulus. Activation energy analysis depends upon the Thiele modulus.
If $\phi^2 \ll 1 \rightarrow$ reaction limited regime
If $\phi^2 \gg 1 \rightarrow$ diffusion limited regime

c) Solve Eq. (10.120) via transforming the second-order ODE into a set of two first-order ODEs (see Sec. 7.5).

Use

$$y_1=\theta \qquad y_2=\frac{d\theta}{d\zeta} \qquad \rightarrow \frac{dy_2}{d\zeta}=\frac{d^2\theta}{d\zeta^2}$$

You will have two ODEs with the following boundary conditions:

$$y_1(0)=\theta|_{\zeta=0}=\theta_o \tag{10.124}$$

$$y_2(0)=\left.\frac{d\theta}{d\zeta}\right|_{\zeta=0}=0 \tag{10.125}$$

Depending on the value of ϕ^2, θ_o *will accordingly vary between zero as the lower limit, and one as the upper limit*. If the process is diffusion limited (e.g., $\phi^2=100$) then θ_o will essentially drop to zero. On the contrary, if the process is reaction limited (e.g., $\phi^2=0.01$), then the entire sphere will saturate (θ is unity everywhere).

So, assume a value of θ_o for a given value of ϕ^2. Your assumption will be verified via what is called the *shooting method* (see Sec. 7.7), which means after you assume a value of θ_o at $\zeta=0$, you carry out the solution of such two ODEs and verify that θ = 1 (or, $y_1=1$) at $\zeta=1$ [see Eq. (10.121)]. Adjust the value of θ_o until you satisfy the boundary condition given by Eq. (10.121).

10.6 Unsteady-state reaction and diffusion in a porous solid phase (described by a PDE).

Consider the physical case of reaction and diffusion in Problem 10.5, but this time under an *unsteady-state* condition.

a) Carry out an unsteady-state shell balance using a thin shell with an area equal to $4\pi r^2$ and thickness Δr. Assume pseudo-homogeneous medium and Fick's law describes diffusion. Divide through by $4\pi\Delta r$ and take the limit as $\Delta r \rightarrow 0$ to reach:

$$\frac{\partial^2 C_s}{\partial r^2} + \frac{2}{r}\frac{\partial C_s}{\partial r} - \frac{k}{D}C_s = r\frac{\partial C_s}{\partial t} \tag{10.126}$$

Equation (10.126) is a partial differential equation (PDE), which requires two boundary conditions and one initial condition:

C_s is finite everywhere, or

$$\left.\frac{dC_s}{dr}\right|_{r=0,\ for\ all\ t\geq 0} = 0 \tag{10.127}$$

$$C_s\big|_{r=R,\ for\ all\ t>0} = C_{s,R} \tag{10.128}$$

$$C_s\big|_{t=0,\ for\ all\ r\geq 0} = 0 \tag{10.129}$$

b) Refer to Eq. (10.130):

$$c\left(x,t,u,\frac{\partial u}{\partial x}\right)\frac{\partial u}{\partial t} = x^{-m}\frac{\partial}{\partial x}\left(x^m f\left(x,t,u,\frac{\partial u}{\partial x}\right)\right) + s\left(x,t,u,\frac{\partial u}{\partial x}\right) \tag{10.130}$$

Match term by term Eq. (10.126) with (10.130), after putting Eq. (10.126) in a form similar to that of Eq. (10.130). Find the identity of each of the following terms that appear in Eq. (10.130): c, m, f, and s. *Hint:* Since we have spherical coordinates, what is the correct value of m?

Remember that x is replaced by r and u is replaced by C_s.

c) At the initial time $t = t_0$, for all r the solution components satisfy initial conditions of the form:

$$u(r,t_o) = u_o(r) = C_s\big|_{t=0,\ for\ all\ r\geq 0} = 0 \tag{10.131}$$

d) At the boundary $r = 0$, or $r = R$, for all t the solution components satisfy a boundary condition of the form:

$$p(r,t,C_s)\big|_{r=0} + q(r,t)f\left(r,t,C_s,\frac{\partial C_s}{\partial r}\right)\bigg|_{r=0} = 0 \tag{10.132}$$

$$p(r,t,C_s)\big|_{r=R} + q(r,t)f\left(r,t,C_s,\frac{\partial C_s}{\partial r}\right)\bigg|_{r=R} = 0 \tag{10.133}$$

Match Eq. (10.132) with Eq. (10.127) to find p_l and q_l, which refer to the p and q terms appearing in Eq. (10.132). The subscript l refers to the left side of the variable r (radial direction).

On the other side, match Eq. (10.133) with Eq. (10.128) to find p_r and q_r, which refer to p and q terms appearing in Eq. (10.133). The subscript r refers to the right side of the variable r (radial direction).

e) Follow in a similar fashion the outlined procedure shown by the example given in Sec. 10.7 to solve Eq. (10.126), where the MATLAB M code is shown in Fig. 10.26 to 10.29. However, you need to assume some numerical values for $C_{s,R}$, R, D, and k.

Choose R, D, and k values such that $\phi^2 = \dfrac{R^2/D}{1/k}$ is 0.05, 1, and 100. $C_{s,R}$ may be assumed to be 100 mol/m^3. Of course, r and R values should be expressed in m. Moreover, t unit should be in s; k s^{-1}; and D in m^2/s.

You should be able to obtain the solution for Cs as a function of r and t for three different values of ϕ^2, as shown in the figures that follow.

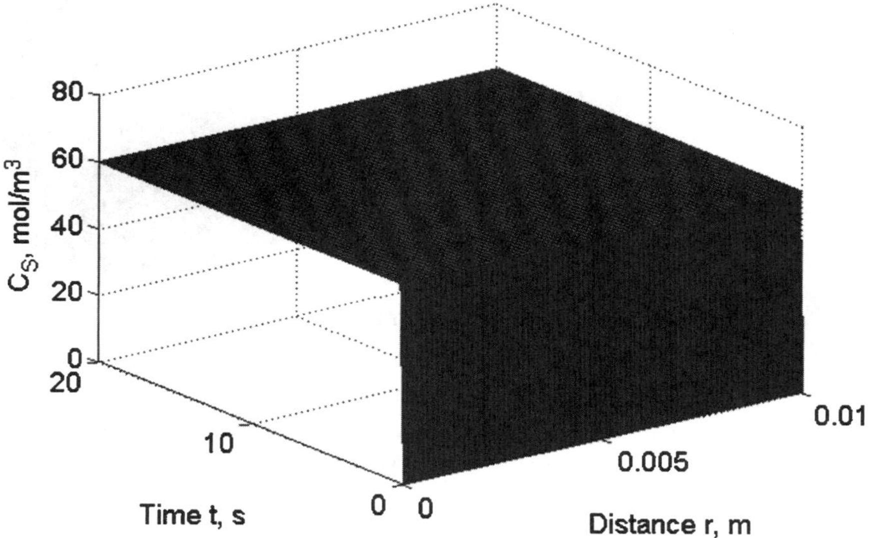

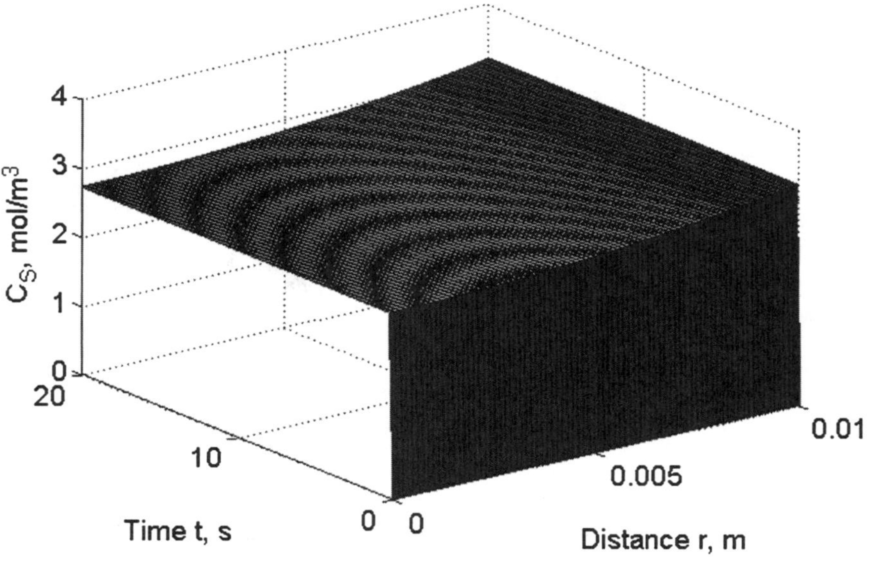

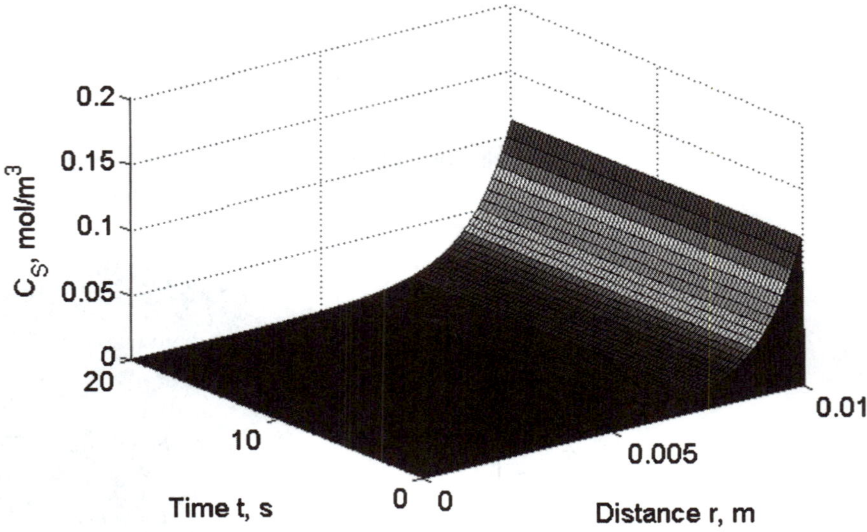

10.7 Complex batch reactor (described by a set of ODEs).

The complex chemical reaction shown in the following is carried out in an isothermal, constant-volume batch reactor. All the reactions follow simple first-order kinetic rate relationships, in which the rate of reaction is directly proportional to concentration, as shown in the following figure.

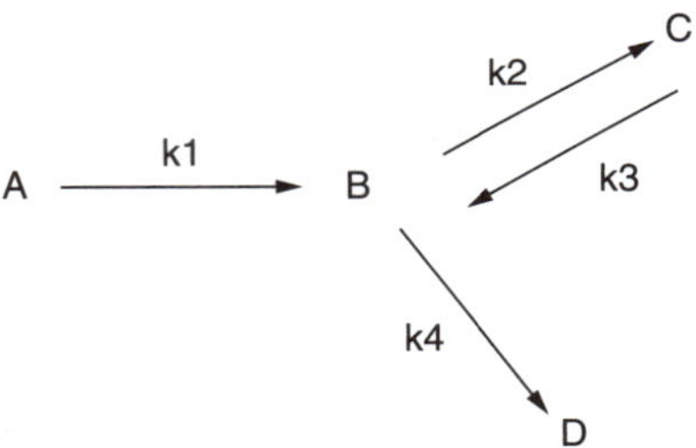

The model equations for each component of this system are:

$$\frac{dC_A}{dt} = -k_1 C_A \tag{10.134}$$

$$\frac{dC_B}{dt} = k_1 C_A - k_2 C_B + k_3 C_C - k_4 C_B \tag{10.135}$$

$$\frac{dC_C}{dt} = k_2 C_B - k_3 C_C \tag{10.136}$$

$$\frac{dC_D}{dt} = k_4 C_B \tag{10.137}$$

where C_A, C_B, C_C and C_D are the concentrations of components, A, B, C and D, respectively, and k_1, k_2, k_3 and k_4 are the respective kinetic rate coefficients.

a) Solve the set of ODEs, given by Eqs. (10.134), (10.135), (10.136), and (10.137), using both the MATLAB symbolic and numeric ode solver (see Secs. 7.4 and 7.6). Use the following rate constants and initial conditions for the variables:

$$k_l = 0.03\ \text{s}^{-1} \quad k_2 = 0.04\ \text{s}^{-1} \quad k_3 = 0.02\ \text{s}^{-1} \quad k_4 = 0.01\ \text{s}^{-1}$$

$$C_A(0) = 10\ \text{mol/L} \qquad C_B(0) = C_C(0) = C_D(0)$$

b) Show a combined plot for the variables (C_A, C_B, C_C, and C_D) as a function of time.

c) Show a plot for C_{tot} as a function of time, where $C_{tot} = C_A + C_B + C_C + C_D$. Does C_{tot} increase, remain constant, or decrease with time?

d) Calculate $\frac{dC_{tot}}{dt} = \frac{dC_A}{dt} + \frac{dC_B}{dt} + \frac{dC_C}{dt} + \frac{dC_D}{dt}$ by summing up the right-hand side of Eqs. (10.134), (10.135), (10.136), and (10.137). What do you conclude? Is your conclusion in harmony with that of (**c**)?

10.8 Benzene vaporizer (described by a set of algebraic/differential equations).

Consider benzene vaporizor. Assume that the volume of the vapor phase is small enough to make its dynamics negligible. If only a few moles of liquid have to be vaporized to change the pressure in the vapor phase, we can assume that this pressure is always equal to the vapor pressure of the liquid at any temperature ($P = P_v$, and $W_V = \rho_v F_v$). An energy equation for the liquid phase gives the temperature (as a function of time), and the vapor-pressure relationship gives the pressure in the vaporizer at that temperature. The following set of algebraic/differential equations describe the dynamics of liquid-phase while neglecting the gas-phase dynamics (i.e., gas-phase properties do not essentially change with time).

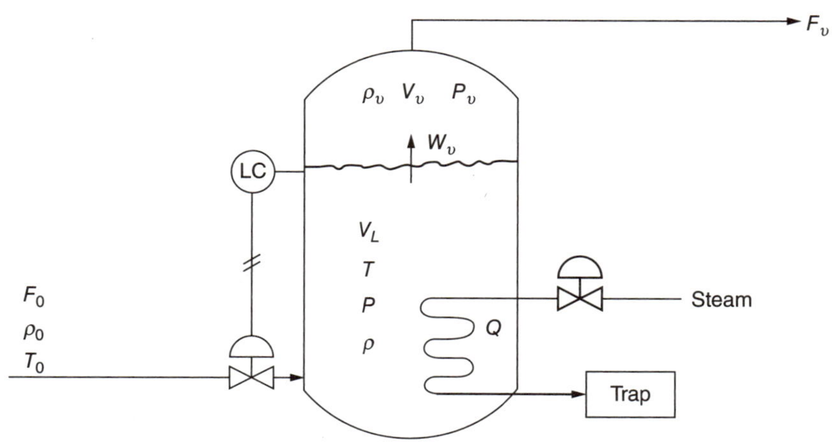

Total material balance:

$$\rho \frac{dV_L}{dt} = \rho_o F_o - \rho_v F_v \tag{10.138}$$

Energy balance:

$$C_p \rho \frac{d(V_L T)}{dt} = C_p \rho V_L \frac{d(T)}{dt} + C_p \rho T \frac{d(V_L)}{dt} = \rho_o C_{po} F_o T_o - \rho_v F_v (C_p T + \lambda_v) + Q \tag{10.139}$$

Equation of state:

$$\rho_v = \frac{P_v MW}{RT} \tag{10.140}$$

Equilibrium relationship:

$$P_v = \exp\left(\frac{A1}{t\ [°C] + 273.15} + B1\right) = \exp\left(\frac{-3848.09}{T\ [K]} + 17.5318\right) \tag{10.141}$$

Where P_v is in mmHg.

Energy balance on the coil side:

$$Q = \dot{m}_S \lambda_S \tag{10.142}$$

Let us say that the total volume of the tank is 5 m^3 and we want to keep the volume of liquid around 4 m^3. It should not exceed this limit. A flow control device is intalled such that if such a limit is reached the valve will be shut down. Here is the control device equation:

$$F_o = C_v(4 - V_L) \tag{10.143}$$

So we have six equations with six unknown quantites: V_L, F_o, T, ρ_v, P_v, and Q.

The parameters are: T_o, ρ_o, ρ, F_v, C_{pc}, C_p, λ_v, MW, R, $\dot{m}_S$, λ_S, C_v.

a) Put Eqs. (10.138) and (10.139) in their standard forms such that the time derivative is kept on the left-hand side of the equation. Equation (10.139) has two time derivatives: one for T and another for V_L. Replace the volume derivative using Eq. (10.138) ending up with only the temperature derivative. Theses two derivatives plus the other four algebraic equations will be plugged in the M-file that defines the derivatives (i.e., deriv(1) and derive(2)) (see Figs. 10.14 and 10.15 in Sec. 10.5).

b) Use the following numerical values for parameters appearing in Eqs. (10.138) through (10.143):

$$T_o = 298.15K \quad \rho_o = 873.8\frac{\text{kg}}{\text{m}^3} \quad \rho = 813\frac{\text{kg}}{\text{m}^3} \quad F_v = 8\times10^{-3}\frac{\text{m}^3}{\text{s}}$$

$$C_{p_o} = 1690\frac{J}{\text{kg}\cdot\text{K}} \quad C_p = 1908\frac{J}{\text{kg}\cdot\text{K}} \quad \lambda_v = 394{,}316\frac{J}{\text{kg}} \quad MW = 78.11\ \text{g/mol}$$

$$R = 0.08206\frac{\text{atm}\cdot\text{L}}{\text{mol}\cdot\text{K}} \quad \dot{m}_S = 1.05\times10^{-2}\frac{\text{kg}}{\text{s}} \quad \lambda_S = 2.26\times10^6\frac{\text{J}}{\text{kg}} \quad C_v = 5\times10^{-2}\frac{1}{\text{s}}$$

Initial conditions:

$$V_L(0) = 4.0\ \text{m}^3;\ T(0) = 298.15\ \text{K}$$

You should be able to obtain V_L, T, and P_v profiles similar to those shown in the following figure.

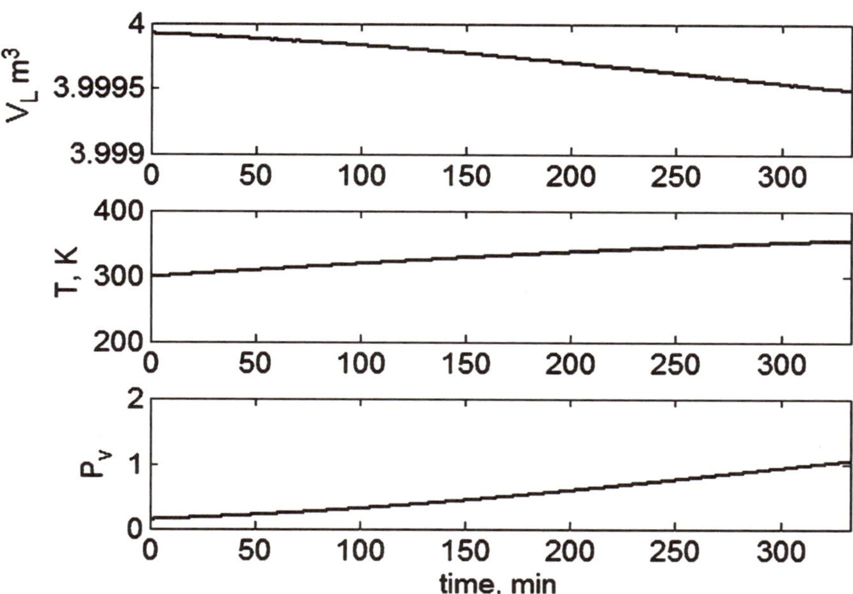

10.9 A box with a square base and open top is to hold 1000 cm^3. Find the dimensions that require the least material (assume uniform thickness of material) to construct the box.

Hint: Minimize the total surface area of the box with an equality constraint. Use the MATLAB **optimtool** to find the minimum of the objective function while satisfying the equality constraint.

10.10 Use the MATLAB **optimtool** to find the minimum of the objective function, $f(x)$, while satisfying the equality and inequality constraints.

$$f(x) = 8x_1 - 6x_2^2 - 16$$

$$36 - x_1^2 - x_2^2 = 0$$

$$(x_1 - 4)^2 + (x_2 - 2)^2 \geq 0$$

$$x_1 \geq 0$$

$$x_2 \geq 0$$

11 MATLAB Graphical User Interface Design Environment (GUIDE)

Chapter Outline

This chapter is intended for advanced users (i.e., developers) to demonstrate the advantage of using the MATLAB built-in Graphical User Interface Design Environment (GUIDE) as a tool to create a code working behind a friendly Graphic User Interface (GUI).

11.1 What Is the MATLAB Graphical User Interface Design Environment?

The Graphical User Interface Design Environment (GUIDE) refers to the platform that allows the creation of icons, buttons, and so on that are visually presented to a user as the "front end" of a software application. A software application that accepts only keyboard-entered commands is considered to be quite outdated and primitive. We prefer to point our mouse pointer to a graphical representation of some aspect of the application, click on it (invoking some event), and continue working with the application through interactive and successive prompts. We are also accustomed to windows, pull-down menus, slider controls, and check boxes.

There can be many reasons for creating a Graphical User Interface (GUI). For instance, you might want to automate a function that you use many times, or perhaps you want to share it with your colleagues who don't really need, want, or care about knowing MATLAB. The new versions of MATLAB are equipped with a set of structured event-driven components in the form of user interface controls (uicontrols) and menus (uimenus) that can easily be assembled and used to create GUIs. The fundamental power of GUIs is that they provide a means through which individuals can communicate with the computer without the hassle of programming commands. The graphical elements have become quite standardized and were developed into a user-friendly and intuitive set of tools. These tools can be used to increase the productivity of a user or provide a window to the sophistication and power of MATLAB application for people with little or no MATLAB programming experience.

The set of user interface components supplied with MATLAB allows the user to design GUIs that match those used in sophisticated software packages. Such components are graphics objects with handles and properties. They come in two classes: user interface controls (uicontrols) and user interface menus (uimenus). The uicontrols and uimenus can be combined with other graphics objects to create an informative, intuitive, and user-friendly interface. This chapter is designed to make you aware of MATLAB GUIDE capabilities and show you how to program fully functional GUIs that meet your needs or your customers' needs.

11.2 Invoking the MATLAB GUIDE

In the Command window, write at the prompt:

```
>>guide
```

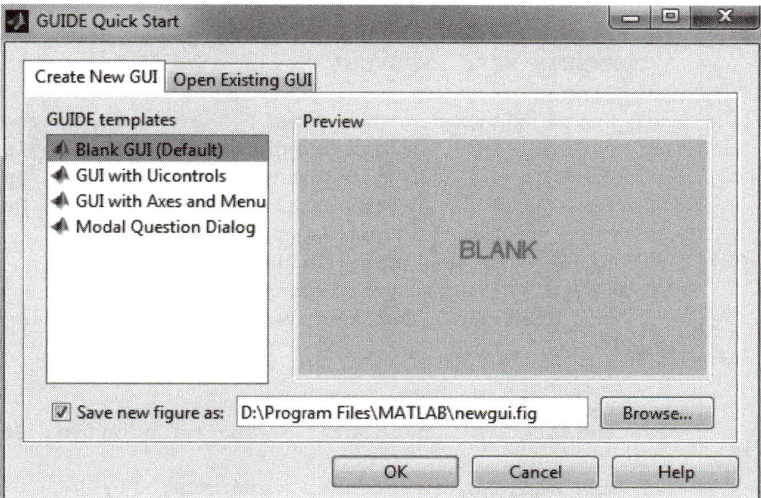

Figure 11.1 The window that pops up upon invoking **guide** command in the Command Window.

A new window will pop up as shown in Fig. 11.1. Here, you may choose either the Create New GUI tab or Open Existing GUI tab. Choose the Create New GUI tab and name the file newgui.fig. Click **OK**.

As a result, the **newgui.m** file will open up via the MATLAB M-file editor, in addition to the **newgui.fig** itself as shown in Figs. 11.2 and 11.3, respectively.

At this stage, you do not have to bother yourself a lot with the generated code, shown in Fig. 11.2. Soon, I will show you how to customize the interface in a way that suits your needs. Click on the run icon (the black-boxed arrow in Fig. 11.3), and Fig. 11.4 is the result.

11.3 Simple Tasks with the MATLAB GUIDE

Let us make GUI more functional. From the left pane, click on the Push Button icon (the top black-box OK in Fig. 11.3) and drag it to the development environment (i.e., gridded area) and then release the mouse. In addition, drag the Edit Text icon (the left black-box Edit in Fig. 11.3) to the gridded area and insert it below the **Push Button** graphical control. Figure 11.5 shows the GUIDE after adding the **Push-Button** and **Edit Text** controls. Notice that I changed the font size of both labels to make the picture clearer. To modify properties of a graphical control, double-click on the control itself and the **Inspector: uicontrol (name of the control)** window will pop up where you can modify many attributes of the selected control. Alternatively, while the mouse is positioned over the control, right-click the mouse and a shortcut menu will pop up from which you can select the **Property Inspector** item.

```
function varargout = newgui(varargin)
% COMMENTS WERE REMOVED!
gui_Singleton = 1;
gui_State = struct('gui_Name',       mfilename, ...
                   'gui_Singleton',  gui_Singleton, ...
                   'gui_OpeningFcn', @newgui_OpeningFcn, ...
                   'gui_OutputFcn',  @newgui_OutputFcn, ...
                   'gui_LayoutFcn',  [] , ...
                   'gui_Callback',   []);
if nargin && ischar(varargin{1})
    gui_State.gui_Callback = str2func(varargin{1});
end

if nargout
    [varargout{1:nargout}] = gui_mainfcn(gui_State, varargin{:});
else
    gui_mainfcn(gui_State, varargin{:});
end
% End initialization code - DO NOT EDIT


% --- Executes just before newgui is made visible.
function newgui_OpeningFcn(hObject, eventdata, handles, varargin)
%COMMENTS WERE REMOVED!
handles.output = hObject;

% Update handles structure
guidata(hObject, handles);

% UIWAIT makes newgui wait for user response (see UIRESUME)
% uiwait(handles.figure1);


% --- Outputs from this function are returned to the command line.
function varargout = newgui_OutputFcn(hObject, eventdata, handles)
% COMMENTS WERE REMOVED
varargout{1} = handles.output;
```

Figure 11.2 MATLAB generated code (**newgui.m**) associated with **newgui.fig**.

If you click run, MATLAB will present you with a window similar to that shown in Figure 11.6. Click the **Push-Button** and you will notice that it is clickable. Nevertheless, the clicking event produces nothing. This is simply because we have not yet instructed MATLAB to do any action upon clicking the **Push-Button** control.

Let us make more cosmetic changes using the **Property Inspector** for a given control. Double-click (or, right-click followed by the **Property Inspector** item selection) on the **Push-Button** control and the inspector window will pop up. We

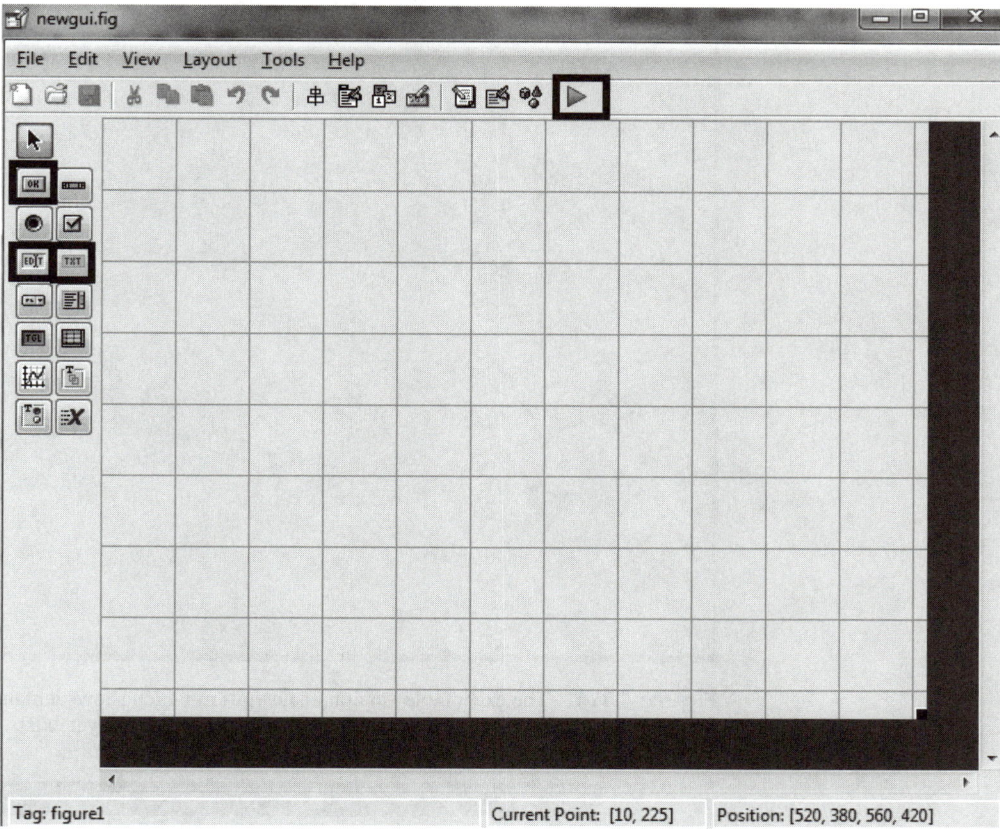

Figure 11.3 MATLAB guide equipped with the side toolbar for adding different icons where each symbolizes a different graphical object.

would like to change four things. First, the tag (i.e., name) of the control object or element, which can be referenced by the user upon writing the proper code needed to perform an action in response to an event triggered by the user. Second, the string that appears on the canvas (or the top) of the button itself. Third, the background color of the button to make it look different from the neighborhood where it resides. Fourth, the tooltip string that appears if the user hovers over or moves his or her mouse anywhere on the canvas of the **Push Button**.

For the first change, you will notice that the **BackgroundColor** item is the first entry appearing in the inspector window. Click on it and select the color you like from the color palette. For the second, third, and fourth changes, drop down to the required attribute and change the existing value to the new value you choose. For example, I chose **Click Me!** for the String attribute, **PushBut1** for the tag attribute, and **I challenge you to click me** for the tooltip text. Figure 11.7 shows a portion of the inspector window pertaining to the **Push Button** control.

After you are done with the changes you have already made to GUI, do not forget to save the changes. This is done via clicking on the save **blue diskette** icon

Figure 11.4 The executable version of **newgui.m** which shows a blank window. Such behavior is expected because no graphical object (control) yet exists.

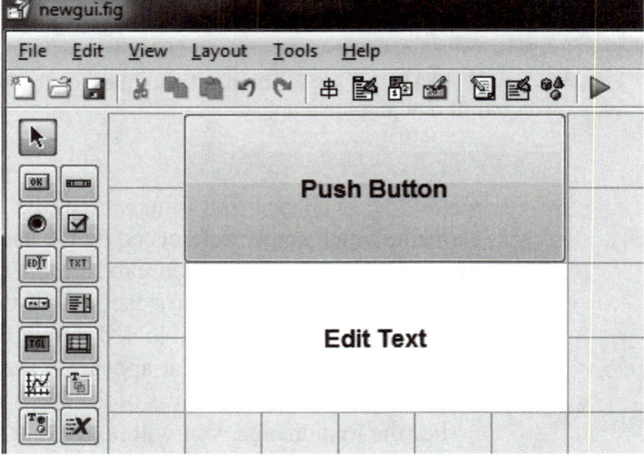

Figure 11.5 GUIDE is populated with two graphical controls: **Push Button** and **Edit Text**.

(shown at the top left corner, or use the CTRL+S key combination), then you may run the program by clicking on the run **green arrow** button. Figure 11.8 shows the new outlook of our **newgui** interface. This is how it is seen by the end-user, as opposed to the developer.

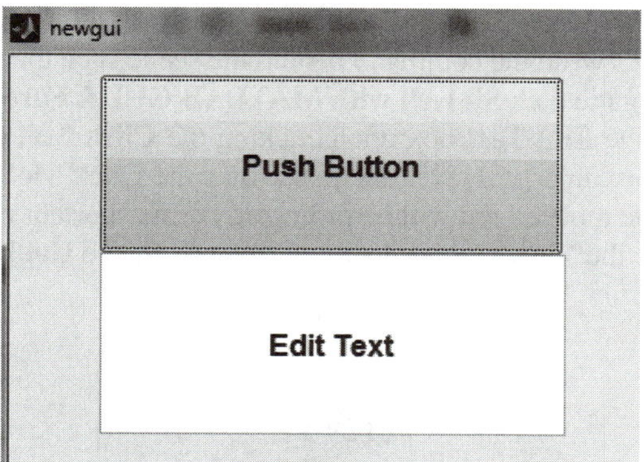

Figure 11.6 The executable GUI has two graphical objects but no action takes place upon clicking the **Push-Button** control.

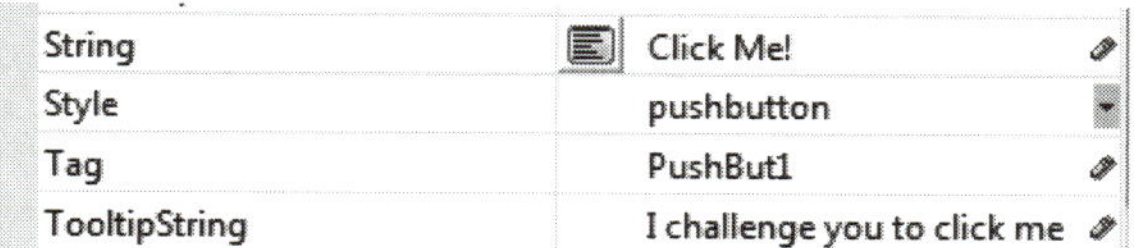

String	Click Me!
Style	pushbutton
Tag	PushBut1
TooltipString	I challenge you to click me

Figure 11.7 The modified attributes of the **Push-Button** control.

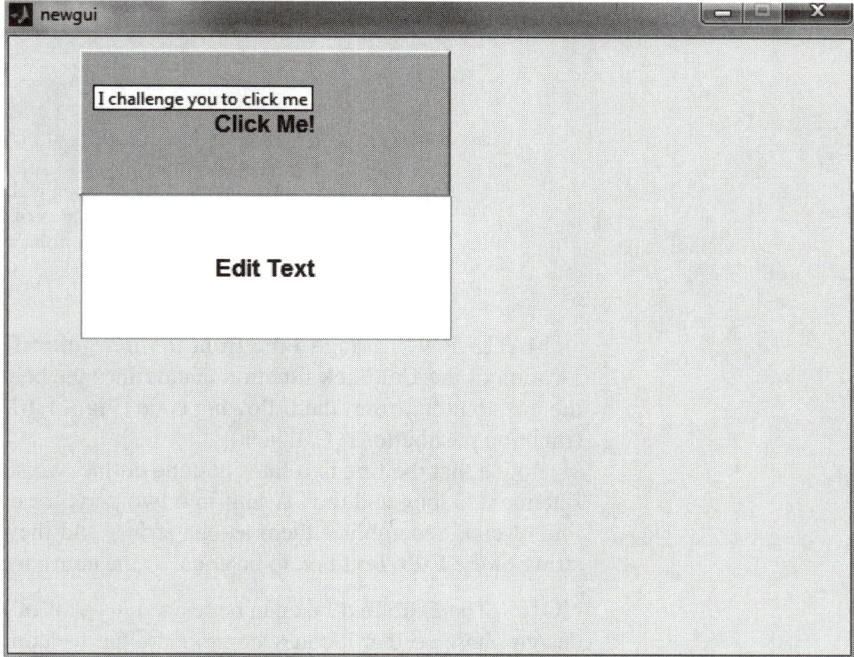

Figure 11.8 The new look for GUI where the background color for the **Push Button** was changed. Once the user moves the mouse over it, the pop-up tooltip text shows up.

Let us make GUI behave friendly, showing the minimum hostility, to the user. So, let us write the command behind the scene such that GUI will greet the user via showing the text: **Sit well with MATLAB GUI & Farewell to MATLAB m-Coding** in the **Edit Text** box upon clicking the **Click Me!** control by the user. Here is how you can do it. Hover the mouse over the **Click Me!** control area and then right-click the mouse; you will be prompted by the shortcut menu as shown in Fig. 11.9. Choose the **View Callbacks** item followed by the **Callback** submenu item.

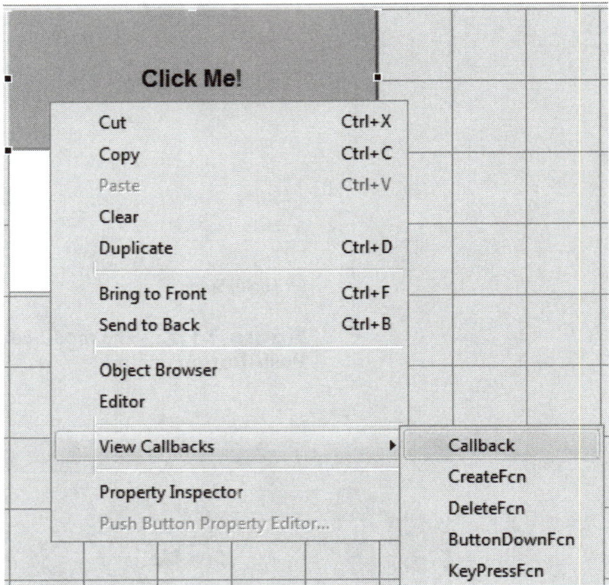

Figure 11.9 The shortcut menu upon right-clicking the mouse while it hovers over the control area of the **Click Me!** button. Select the **View Callbacks** item followed by the **Callback** submenu item.

MATLAB will bring to the front the **newgui.m** file and will pinpoint to the location of the **Callback** function that defines the behavior of GUI upon clicking the push button. Insert the following code (Fig. 11.10) beneath the main function (function pushbutton1_Callback).

Notice that the first two lines of code define two strings, as the entire greeting statement is long and thus is split into two parts for easy code tracking, the third line of code recombines them as one string, and the fourth line of code sets the string of the **Edit Text** box to be equal to the entire welcome statement.

NOTE The **Edit Text** box can be resized in terms of length and width during the design phase so that it can accommodate the welcome statement. Keep in mind that this type of control accepts only one line statement. To output multiple lines, the designer needs to use the ListBox control.

Figure 11.11 shows how the GUI responds to clicking the **Click Me!** button.

```
function PushBut1_Callback(hObject, eventdata,
handles)
% hObject    handle to PushBut1 (see GCBO)
% eventdata  reserved - to be defined in a future
version of MATLAB
% handles    structure with handles and user data (see
GUIDATA)
str1='Sit well with MATLAB GUI   ';
str2='Farewell to MATLAB m-Coding';
str3=[str1,str2];
set(handles.edit1,'String',str3);
```

Figure 11.10 The insertion of four lines of code under the **Callback** function of the **Click Me!** button. The added code will set the **String** property of the **Edit Text** box to be: **Sit well with MATLAB GUI & Farewell to MATLAB m-Coding** after the user clicks the push button.

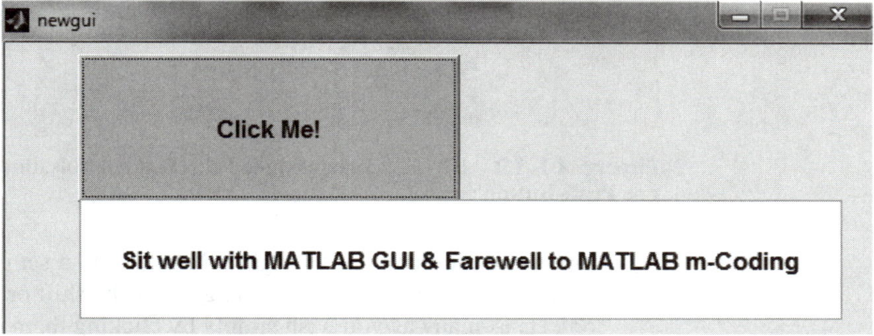

Figure 11.11 GUI shows the greeting statement after clicking the **Click Me!** button.

11.4 Graphical User Interface for an Ideal Gas Volume Calculation

Suppose that we would like to calculate the ideal gas volume of a substance while inputting both the pressure in atmosphere units and the temperature in degrees Centigrade. Let us design a GUI that handles this. Details on how to add controls to a blank GUI and how to customize them are shown in Sec. 11.3. However, more features will be explored here that are not covered in the previous section. Figure 11.12 shows IG_GUI populated with three **Edit-Text** controls, three **Static Text** controls, and one **Push-Button** control. The first two **Edit Text** boxes will be used for inputting temperature and pressure, and the third one for outputting the calculated molar volume (L/mol).

Here it is worthwhile to mention the alignment of controls and the grid/no grid option. For the first point, you can *highlight* the control objects to be aligned and choose from the top menu the alignment tool (labeled with a **black** box in Fig. 11.12) where you may pick the alignment style (i.e., vertical vs. horizontal,

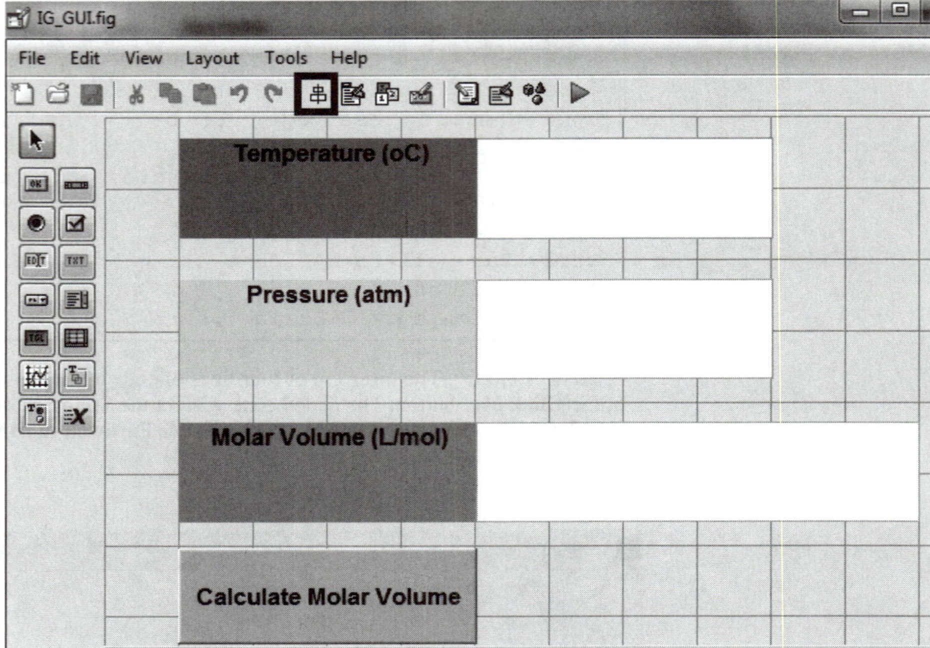

Figure 11.12 IG_GUI contains three **Edit-Text** controls, three **Static Text** controls, and one **Push-Button** control.

top vs. bottom, and left vs. right). To highlight a set of objects, click on the first object by mouse and while pressing either the shift or ctrl key you may add more objects as many as you wish simply by clicking them one at a time.

For the grid vs. no grid option, go to the **Tools** menu then to the **Grids and Ruler** submenu, where you will be able to select or deselect the grid, in addition to the other grid-related attributes. Figure 11.13 shows the code to be inserted by the developer under the **Callback** (pushbutton1_Callback) function that defines the behavior of GUI upon clicking the **Calculate Molar Volume** push button.

```
function pushbutton1_Callback(hObject, eventdata, handles)
[Tempstr]=get(handles.edit1,'String');
Temp=str2num([Tempstr]);
[P_str]=get(handles.edit2,'String');
pressure=str2num(P_str);
[Vol]=0.082057*(Temp+273.15)./pressure;
Volume=num2str(Vol);
set(handles.edit3,'String',Volume);
```

Figure 11.13 MATLAB receives the temperature and pressure values from the first two **Edit-Text** boxes and converts them into numbers via the (**str2num**) function. The volume is then calculated based on the ideal gas law. Finally, the value is converted back into string (via the **num2str** function) so that it can be shown in the third **Edit-Text** box.

Well, because MATLAB in essence deals with matrices (pancake) and vectors (corn flicks) as its favorite meal, then values of temperature and pressure can be entered here as row vectors. Figure 11.14 shows typical results for the calculated ideal gas molar volume expressed in L/mol. A set of three values of T and P was entered twice and the corresponding molar volumes were calculated for each of the two sets. Notice how the molar volume, $\tilde{V}$, increases with T (Fig. 11.14, *top*) and decreases with P (Fig. 11.14, *bottom*).

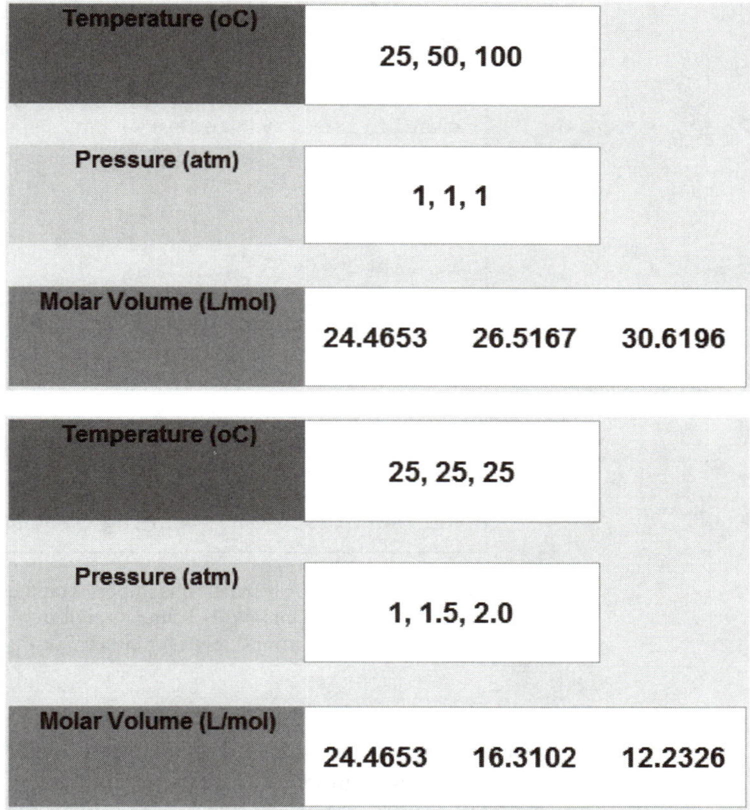

Figure 11.14 The calculated ideal gas molar volume is shown in the third **Edit-Text** box as a function of input temperature and pressure. Three isobaric ($P = 1$ atm) temperature values were entered to see the effect of increasing temperature on volume (*top*); on the other hand, three isothermal ($T = 25°C$) pressure values were entered to see the effect of increasing pressure on volume (*bottom*).

The developer may impose, however, a restriction on the type of input to be plugged in the first two **Edit-Text** boxes for T & P values, which will tell the user that there is something wrong with entering the data for either the temperature or pressure. For example, the following code inspects the input string to verify that the input temperature and pressure are finite (i.e., not NAN, where NAN stands for not a number in the MATLAB language). The code is shown in Fig. 11.15.

```
flag2=1.0;
[Tempstr]=get(handles.edit1,'String');
Temp=str2num(Tempstr);
if isempty(Tempstr)
   flag2=0;
elseif isempty(Temp)
    flag2=0;
end
[P_str]=get(handles.edit2,'String');
pressure=str2num(P_str);
if isempty(P_str)
   flag2=0;
elseif isempty(pressure)
    flag2=0;
elseif pressure==0
    flag2=0;
end
if flag2==0
    mystr2='No proper entry was found';
    mystr3=' for either T or P';
    mystr4=[mystr2,mystr3];
   set(handles.edit3,'String',mystr4);
         return
end
[Vol]=0.082057*(Temp+273.15)./pressure;
Volume=num2str(Vol);
set(handles.edit3,'String',Volume);
```

Figure 11.15 A restriction is imposed on the type of input to either *T* or *P*. The flag (flag2 variable) will change its sign from 1 to zero should the user attempt to enter an invalid input for either *T* or *P*.

Here, a restriction is imposed on the type of input for either *T* or *P*. A blank entry, alphabetic (not numeric), or zero-pressure value will alarm the user that there is something wrong with entered data for either *T* or *P* or both.

Figure 11.16 show samples of output where an attempt to enter an invalid input data type for either *T* or *P* is made. The volume should give inf (∞) value if $P = 0$. However, if $P = 0$ that strictly means there are no molecules of a substance such that they exert pressure on the wall as a result of their collision with and bouncing off the wall.

11.5 Graphical User Interface for Evaluating Volume-Related Properties of a Pure Substance

Suppose that we would like to calculate the following volume-related properties of a pure substance (not necessarily an ideal gas)—the gas molar volume, liquid molar volume, isothermal compressibility factor (kappa, κ), thermal expansion

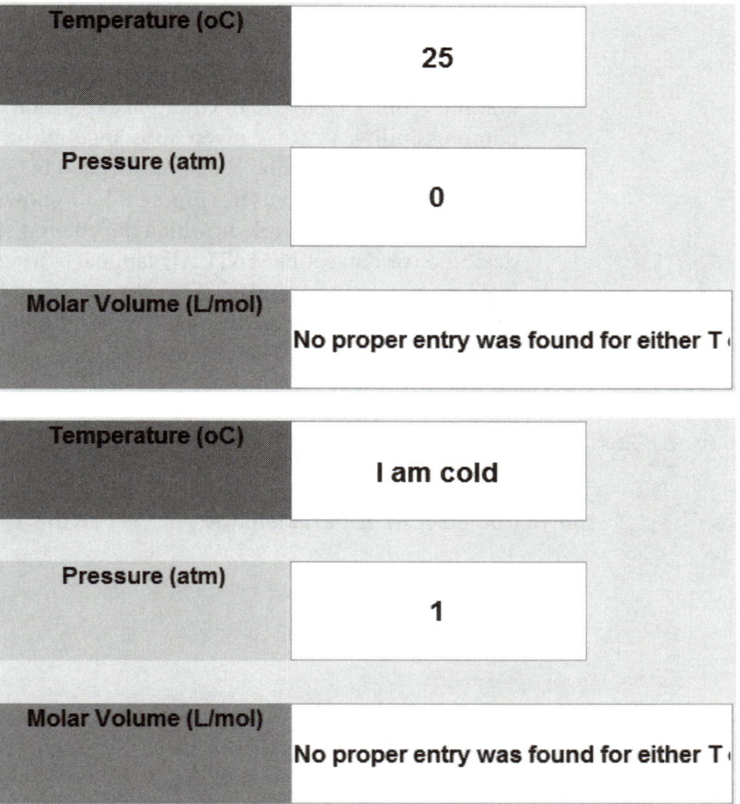

Figure 11.16 Samples of results in which an invalid data entry is deliberately made. A zero absolute atmosphere (*top*) and alphabetic input T (*bottom*) will provoke the program (watching dog) to warn the user that there is something fishy going on.

coefficient (alpha, α), and compressibility factor (Z)—while inputting both the pressure in atmosphere and temperature in Kelvin. The calculation is based on the van der Waals' equation of state. Let us design a GUI that handles such a duty. Details on how to add controls to a blank GUI and how to customize them are shown in previous sections. However, more features are explored here that have not been covered before. Given that a database exists in the form of an Excel sheet, which lists the name of a substance, its chemical formula, its critical pressure, and its critical temperature, we would like the user to search for the substance of interest via a keyword that is based either on name or chemical formula. The search results shall be presented and the user will then decide on the substance of interest via selecting the corresponding serial number assigned to that substance. Therefore, the Excel sheet will be read by and stored in MATLAB as a dataset. A dataset, in general, is an $n \times m$ matrix with n rows or records and m columns. Once the user makes his or her choice, MATLAB will fetch all needed thermodynamic critical properties of the assigned substance. Refer to Sec. 10.2, which shows how we can exploit the MATLAB built-in root finding algorithm to calculate the three roots of a nonlinear cubic equation and how to use the van der Waals' equation of state to estimate the molar volume of the gas and liquid, the compressibility factor, Z, the thermal expansion coefficient, α, and the isothermal compressibility, κ, for a given substance at the given absolute pressure, P, and temperature, T. The same code is embedded in GUI design (i.e., in the background without bothering the user). Figure 11.17 shows the GUI designed for this purpose. It shows an example in which the entered search keyword is "water" and the database (or dataset in MATLAB language, as explained in Sec. 9.4) serial number assigned to water is 188. This number must be entered in the dedicated **Edit-Text** box. Moreover, the absolute temperature and pressure must also be entered in the dedicated **Edit Text** boxes. Upon clicking the **Calculate Properties**

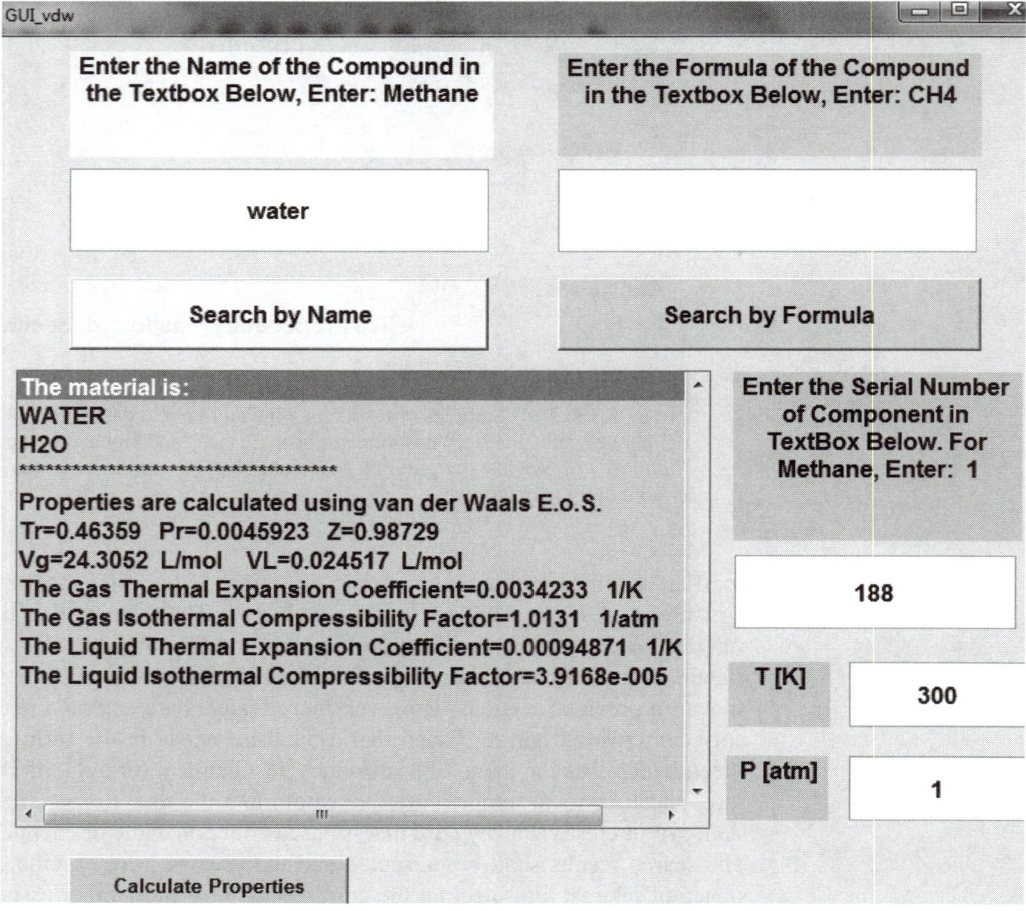

Figure 11.17 GUI for calculating volume-related thermodynamic properties of a pure substance based on the van der Waals' equation of state. The user searches for the substance of interest and after entering the database-assigned serial number, pressure, and temperature, the program will calculate the requested properties and present them in the list box as shown.

```
function GUI_vdw_OpeningFcn(hObject, eventdata, handles, varargin)
global PureComp myvar myvar2

% MATLAB generated comments were removed for easy code tracking

PureComp = dataset('XLSFile','PureComponentData1.xlsx');
myvar=cellstr(PureComp,'Compound');
myvar2=cellstr(PureComp,'Formula');
```

Figure 11.18 Defining the dataset for which MATLAB will import the data from Excel sheet. Two string-type global cell vectors are further defined that contain the names and molecular formulas of 188 chemical species. They are made global so that they can be referenced in other submodules in **GUI_vdw.m**, the counterpart M-file of **GUI_vdw.fig**.

push-button, the calculated properties of water at the given conditions of pressure and temperature will be shown in the reserved list box.

Figure 11.17 obviously demonstrates how easy it is for the user to interact with GUI forms to carry out the same task as laid out in **Van_Der_Waals.m** (Fig. 10.5). Nevertheless, the way to interact with the GUI and present the results is more plausible and at the same time much easier to use than in the **Van_Der_Waals.m** scenario.

Figures 11.18 through 11.21 show only the code that must be inserted by the developer into the auto-generated M-file (**GUI_vdw.m**). Other codes are automatically obtained and there is no need to present them here.

11.6 Saving Results to a File

To simplify the process of saving data or results to a file, I will give here an example in which upon clicking the push button some calculations will be performed, results are shown on the screen via a list box, and finally a dialog box will pop up to ask the reader to enter the name of a file where the data can be stored. Body mass index (BMI) is a measure of body fat based on height and weight that applies to both adult men and women. The user will be required to enter his or her weight and height using the engineering (American) or metric (SI) system of measurements. BMI categories are:

- Underweight = <18.5
- Normal weight = 18.5-24.9
- Overweight = 25-29.9
- Obesity = 30 or greater

Figure 11.22 shows the GUI for calculating BMI upon entering the weight and height either in American or metric units and prompting the user to save the results to a file of his or her choice.

Figure 11.23 shows the code behind clicking the **Calculate BMI (SI)** push button, in which, in addition to showing the results in the dedicated list box, the application will prompt the user to enter the name of the file to which the results will also be saved.

```
function pushbutton1_Callback(hObject, eventdata, handles)
global PureComp myvar
global CompName
tempstr=get(handles.listbox1,'String');
if iscellstr(tempstr)
set(handles.listbox1,'String','');
end
CompName= get(handles.edit1,'String');
matchStart = regexpi(myvar, CompName);
tempindex=1;
flag=0;
% myindx(1,1)=' ';
for idx=1:size(matchStart,1)
    if cell2mat(matchStart(idx,1))>=1
        myindx(tempindex,1)=idx;
        tempindex=tempindex+1;
        flag=1;
    end
end
if flag==0
    mystr2='No Entry was Found!; Good Luck and Try Again.';
   set(handles.listbox1,'String',mystr2);
%     display ('No Entry was Found!; Good Luck and Try Again.')
    return
end
mystr2=[];
for idex2=1:size(myindx,1)
    value=myindx(idex2,1);

mystr=[value;PureComp.Compound(value);PureComp.Formula(value);
'**********************************'];
    mystr2=[mystr2,mystr];
    set(handles.listbox1,'String',mystr2);
end
set(handles.edit1,'String','');
```

Figure 11.19 The response of GUI for clicking the **Search by Name** push button. It will take in the name (i.e., the search keyword) entered by the user and search for all possible records that contain such a string. The results will be shown in the list box. If no entry is found, then a pop-up message will alert the user that his or her entry is not found.

The code for clicking the **Calculate BMI ($)** push button will be the same as that shown in Fig. 11.23, except that it is for calculating BMI as a function of input parameters, expressed in the American (engineering) system of units (Fig. 11.24).

11.7 Deployment of MATLAB-Based Applications

Once the developer is happy with his or her GUI, he or she can deploy it with the MATLAB core platform, which is needed to run the application without having to install MATLAB itself. The MATLAB core platform is called MATLAB

```
function pushbutton2_Callback(hObject, eventdata, handles)
global PureComp myvar2
global FormName
tempstr=get(handles.listbox1,'String');
if iscellstr(tempstr)
set(handles.listbox1,'String','');
end
FormName= get(handles.edit2,'String');
matchStart = regexpi(myvar2, FormName);
tempindex=1;
flag=0;
% myindx(1,1)=' ';
for idx=1:size(matchStart,1)
    if cell2mat(matchStart(idx,1))>=1
        myindx(tempindex,1)=idx;
        tempindex=tempindex+1;
        flag=1;
    end
end
if flag==0
    mystr2='No Entry was Found!; Good Luck and Try Again.';
   set(handles.listbox1,'String',mystr2);
%     display ('No Entry was Found!; Good Luck and Try Again.')
    return
end
mystr2=[];
for idex2=1:size(myindx,1)
    value=myindx(idex2,1);

mystr=[value;PureComp.Compound(value);PureComp.Formula(value);
'************************************'];
    mystr2=[mystr2,mystr];
    set(handles.listbox1,'String',mystr2);
end
set(handles.edit2,'String','');
```

Figure 11.20 The response of GUI for clicking the **Search by Formula** push button. It will take in the formula (i.e., the search keyword) entered by the user and search for all possible records that contain such a string. The results will be shown in the list box. If no entry is found, then a pop-up message will alert the user that his or her entry is not found.

Component Runtime (MCR). Any application designed in MATLAB has to be augmented by MCR so that the user can run it as a stand-alone program or application.

At the command prompt, write:

```
>> deploytool
```

MATLAB will launch its deployment tool starting with the first window, as shown in Fig. 11.25. For example, we would like to create a console application for body

```
function pushbutton3_Callback(hObject, eventdata, handles)
global PureComp
CompNum= str2double(get(handles.edit4,'String'));
Tabs= str2double(get(handles.edit5,'String'));
Pabs= str2double(get(handles.edit6,'String'));
Pcr=PureComp.Pcr(CompNum);
Tcr=PureComp.Tcr(CompNum);
Rgas=0.08206;
aterm=0.42748*Rgas^2*Tcr^2.5/(Pcr*Tabs^0.5);
bterm=0.08664*Rgas*Tcr/Pcr;
Vols=roots([Pabs, - Rgas*Tabs, (aterm - Pabs*bterm*bterm - Rgas*Tabs*bterm),
-aterm*bterm]); % Finds all roots
GasVol=max(Vols(find(imag(Vols) == 0))); % finds largest real root for the
% gas volume
LiqVol=min(Vols(find(imag(Vols) == 0))); % finds smallest real root for the
% liquid volume
Zval=Pabs*GasVol/(Rgas*Tabs);
Tr=Tabs/Tcr;
Pr=Pabs/Pcr;
term1=Rgas*Tabs/(GasVol-bterm)^2;
term2=2*aterm/(GasVol^3);
alphaG=(1.0/GasVol)*(Rgas/(GasVol-bterm))/(term1-term2);
kappaG=(1.0/GasVol)*(1.0/(term1-term2));
term3=Rgas*Tabs/(LiqVol-bterm)^2;
term4=2*aterm/(LiqVol^3);
alphaL=(1.0/LiqVol)*(Rgas/(LiqVol-bterm))/(term3-term4);
kappaL=(1.0/LiqVol)*(1.0/(term3-term4));
mystr=['The material is:';PureComp.Compound(CompNum);PureComp.Formula(Comp
Num);'**********************************'];
mystr1=['Properties are calculated using van der Waals E.o.S.'];
mystr2=['Tr=',num2str(Tr),'   Pr=',num2str(Pr),'   Z=',num2str(Zval)];
mystr3=['Vg=',num2str(GasVol),' L/mol ',' VL=',num2str(LiqVol),' L/mol'];
mystr4=['The Gas Thermal Expansion Coefficient=',num2str(alphaG),'  1/K'];
mystr5=['The Gas Isothermal Compressibility Factor=',num2str(kappaG),'  1/atm'];
mystr6=['The Liquid Thermal Expansion Coefficient=',num2str(alphaL),'  1/K'];
mystr7=['The Liquid Isothermal Compressibility Factor=',num2str(kappaL),
'  1/atm'];
mystr8=[mystr;mystr1;mystr2;mystr3;mystr4;mystr5;mystr6;mystr7];
set(handles.listbox1,'String',mystr8);
```

Figure 11.21 The heart of GUI where the essential steps are carried out upon clicking the **Calculate Properties** push button. The gas molar volume, liquid molar volume, isothermal compressibility factor (kappa, κ) for both phases, thermal expansion coefficient (alpha, α) for both phases, and the compressibility factor (Z) are all estimated for a given selected substance while inputting both the pressure in atmosphere units and temperature in Kelvin. The calculations are based on van der Waals' equation of state.

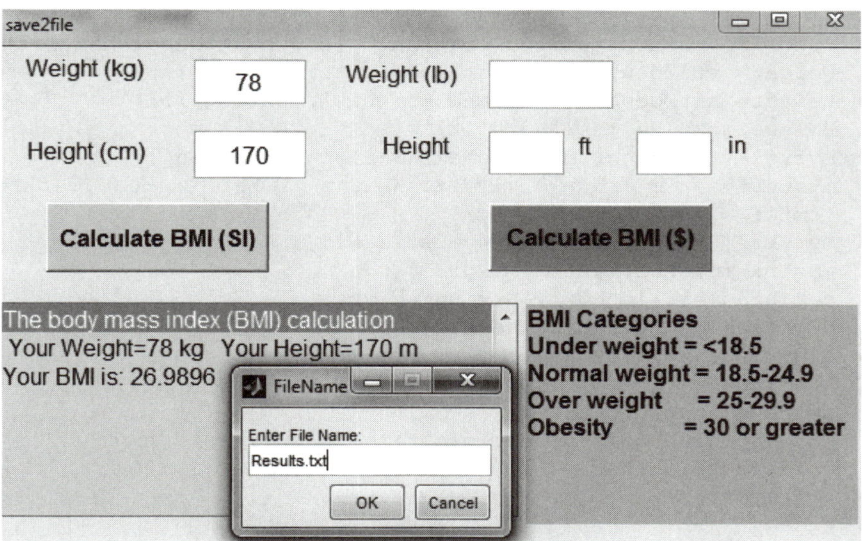

Figure 11.22 Calculation of body mass index (BMI) using either the American or metric system of measurements and prompting the user to save the results to a file of his or her choice.

mass index (BMI) shown in Sec. 11.6. Name the project BMI with .prj extension and choose the proper location.

The deployment tool window of BMI.prj is shown in Fig. 11.26.

The BMI-related files must be added. Click on the **[Add main file]** hyperlink (the middle black box in Fig. 11.26) so that we can add the **save2file.m** file. Then click on **[Add files/directories]** hyperlink (the bottom black box in Fig. 11.26) to add the **save2file.fig** (GUI) file. Figure 11.27 shows the result of file addition.

NOTE It is worth mentioning here that all logistic (resource) files must be added to the deployment project as shared resource files if you, as a developer, refer to them by any of your main M-files, found under the category Main File in the deployment project. For example, in Sec. 11.5, MATLAB imports critical properties of a substance from an Excel sheet; therfore, such an Excel sheet must be added here as a shared resource file.

Click on the **Package** tab (the top black box in Fig. 11.26), and you will see Fig. 11.28. If you want to deploy the application to a user who has already the same MATLAB version that you use to design the GUI and the deployed executable file, then *there is no need to add* MATLAB Component Runtime (MCR) file. However, if the user lacks the same version of MATLAB that you possess, as a developer, then *you will need to create the application, including MCR.*

Upon the addition of MCR, the **Package** tab becomes as shown in Fig. 11.29. Notice that the package size will be about 400 MB, which is a huge size for handling such a light duty as calculating BMI. Consequently, MATLAB will not compete with standard programming tools like Microsoft Visual Studio in terms of

```
function pushbutton1_Callback(hObject, eventdata, handles)
Weight= str2double(get(handles.edit1,'String'));
Height= str2double(get(handles.edit2,'String'));
BMI=Weight/(Height/100.0)^2;
mystr1=['The body mass index (BMI) calculation'];
mystr2=[' Your Weight=',num2str(Weight), ' kg', '  Your Height=',num2str(Height),
'cm'];
mystr3=['Your BMI is: ',num2str(BMI)];
mystr8={mystr1;mystr2;mystr3};
set(handles.listbox1,'String',mystr8);
prompt = {'Enter File Name:'};
dlg_title = 'FileName';
num_lines = 1;
def = {'Results.txt'};
answer = inputdlg(prompt,dlg_title,num_lines,def);
if isempty(answer)
   return;
else
fname=char(answer(1))
nrow=size(mystr8,1);
fid = fopen(fname, 'w');
for row=1:nrow
    fprintf(fid, '%s %d %d %d\n', mystr8{row,:});
end
fclose(fid);
end
tempstr={'Data were saved to Data File:', fname};
msgbox(tempstr, 'FILE Saving!!!');
```

Figure 11.23 The GUI response upon clicking the push button **Calculate BMI (SI)**. The program receives the input parameters, converts them into numeric values, calculates BMI, presents the results on the screen via the listbox, prompts the user to enter the name of the file for data saving, opens the file for writing, and finally informs the user about the success of the file saving process.

```
Weight1= str2double(get(handles.edit3,'String'));
Height1= str2double(get(handles.edit4,'String'));
Height2= str2double(get(handles.edit5,'String'));
Weight=Weight1*0.45359;
Height=(Height1+Height2/12.0)*0.3048;
BMI=Weight/(Height)^2;
```

Figure 11.24 The rest of the code is the same as that shown in Fig. 11.23, except that the code for calculating BMI using the American (engineering) system of units is shown here.

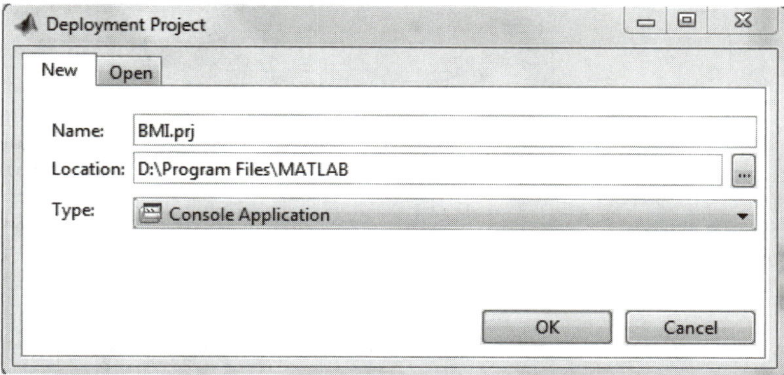

Figure 11.25 The first pop-up window that appears upon launching the MATLAB deployment tool.

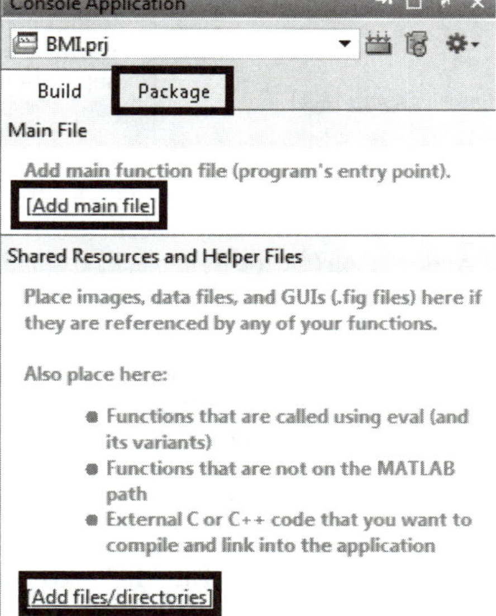

Figure 11.26 The deployment tool window for a BMI.prj (console application).

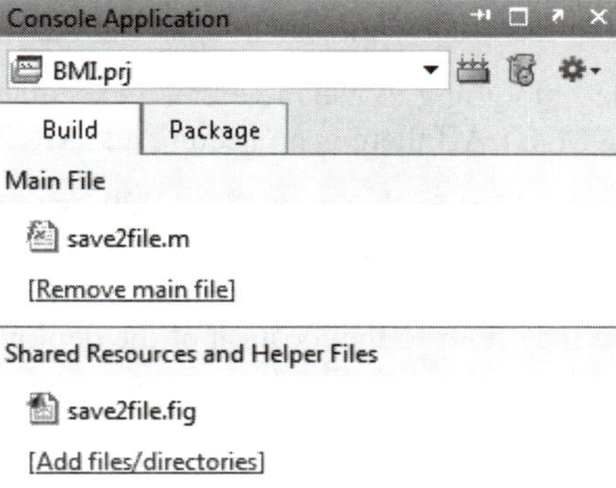

Figure 11.27 The deployment tool window for BMI.prj (console application) with the addition of the main (**save2file.m**) file and the shared resource **save2file.fig** (GUI) file.

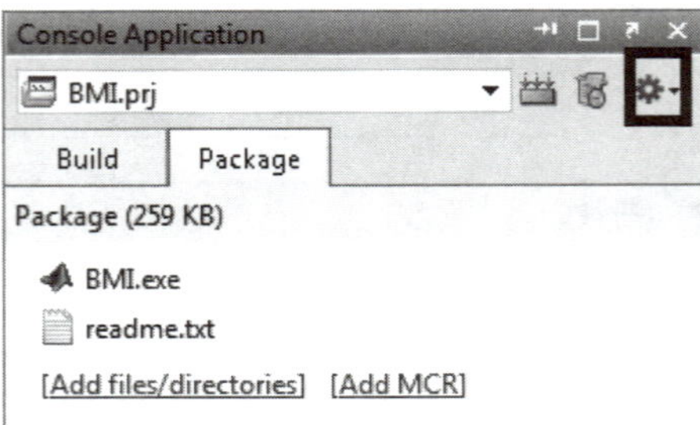

Figure 11.28 The **Package** tab allows the developer to add the MATLAB core platform; that is, the MATLAB Component Runtime (MCR).

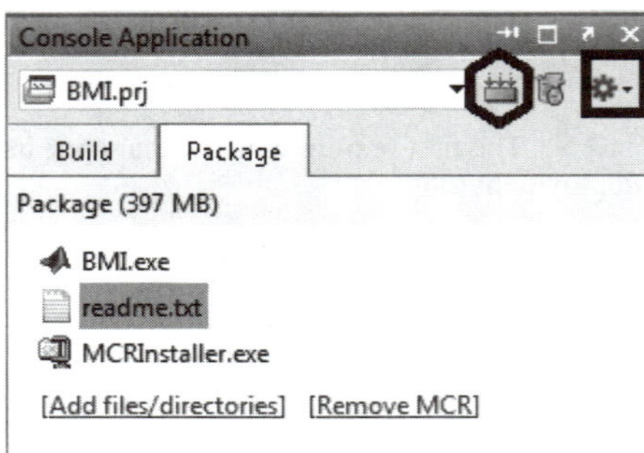

Figure 11.29 The addition of the MATLAB MCR package to the deployment project so that it will be deployed with the application itself to the end user.

minimum file size requirement to handle such a tiny duty. Using MATLAB for this is like using a cannon to kill a tiny insect.

Click on the **Actions** icon (the top right black box in Fig. 11.29) for further inspection of other features that can be either enabled (selected) or disabled (deselected). Choose the **Settings** submenu item, and the **Project Settings** window is shown in Fig. 11.30. It shows the application name and the location of the files to be deployed. You can change such settings if you wish. Click on other tabs and see other features. At this stage, there is nothing to do or add. Select embed CTF archive into the Application. Regarding this option, each application or shared library you produce using the MATLAB compiler has an associated component technology file (CTF) archive. The archive contains all the MATLAB-based content (MATLAB files, MEX-files, etc.) associated with the component. The MATLAB compiler also embeds a CTF archive in each generated binary. The CTF houses all deployable files. All MATLAB files encrypt in the CTF archive using the advanced encryption standard (AES) cryptosystem. In brief, this is to preserve royalties of MATLAB from being re-used by coders or programmers unless they build upon MATLAB-based byte technology.

NOTE Since I am going to use the created executable file on the same laptop that I am using MATLAB there is no need to add MCR.

Click on the **Build** icon shown in Fig. 11.29, which is surrounded by a black hexagon. The building process will start and you will notice the messages generated by the MATLAB compiler. Once the building project process is successfully completed, you may refer to the location of the deployed file and run the executable version of the application file. If there is any error that renders the building process unsuccessful, then you may save the error and warning messages to a log file for future analysis.

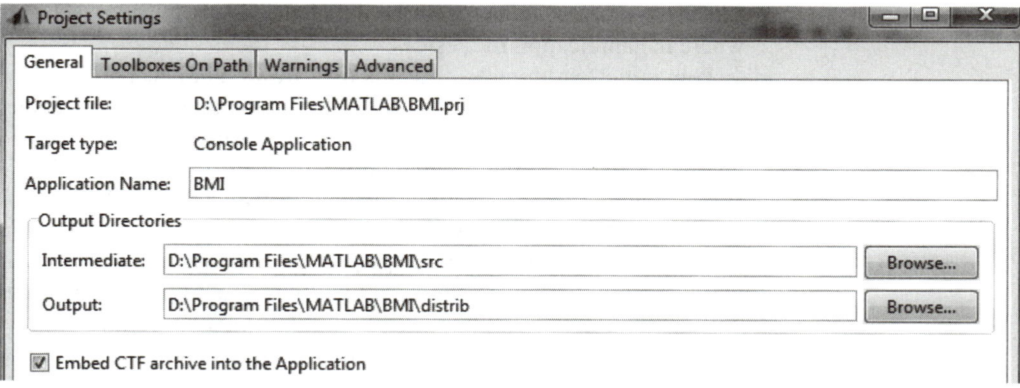

Figure 11.30 **Project Settings** window in which it shows the application name and the location of the files to be deployed. You can change such settings if you wish.

Close the entire MATLAB application and go to the location of the **Output** folder (as shown in Fig. 11.30) and run the executable file (BMI.exe) created by the MATLAB compiler. A DOS "blank" screen will pop up, followed by the same GUI we created during the design phase and that was uploaded as a shared resource during the deployment phase. Figure 11.31 shows the results of the

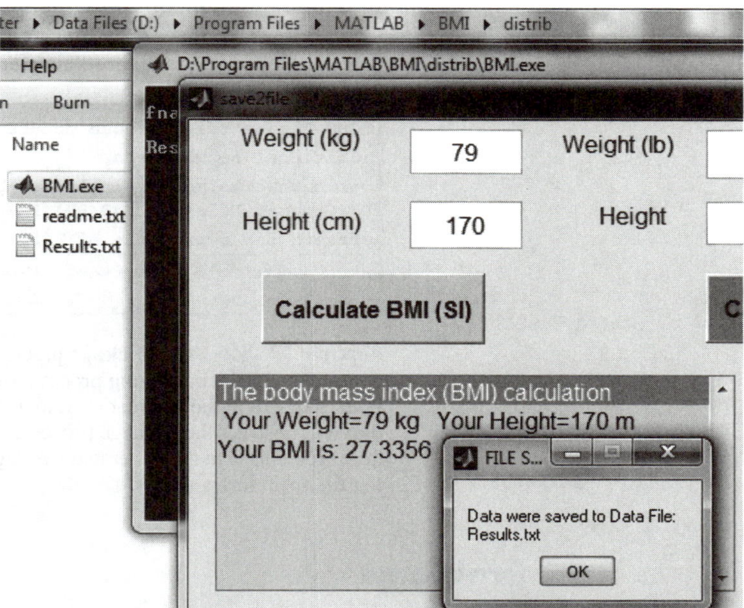

Figure 11.31 Upon running the executable file (BMI.exe), a DOS (blank) screen and the designed GUI will pop up successively. The user may interact with GUI as a stand-alone application. The program responded to user's input data, calculated BMI, presented data on the screen, and finally saved the results to a file.

execution process. Notice that the Results.txt file is created in the root directory where the application file resides.

One important thing to notice is that my weight became 79 instead of 78 kg_f (Fig. 11.22), which is expected as I am approaching the completion of this textbook; thus I am happy, relaxed, and possibly overweight.

Finally, to deploy the application as a package, usually with MCR, click on the **Package** icon (the icon that lies top right between the black square and hexagon in Fig. 11.29). MATLAB will prompt you to save the executable package file in a destination folder and the **Package** process window will show up until the packaging process is complete, as shown in Fig. 11.32. You will notice that MATLAB will create the executable package file (BMI_pkg.exe) that upon running will reveal MCR, if it is added during the deployment stage, and the executable file for GUI.

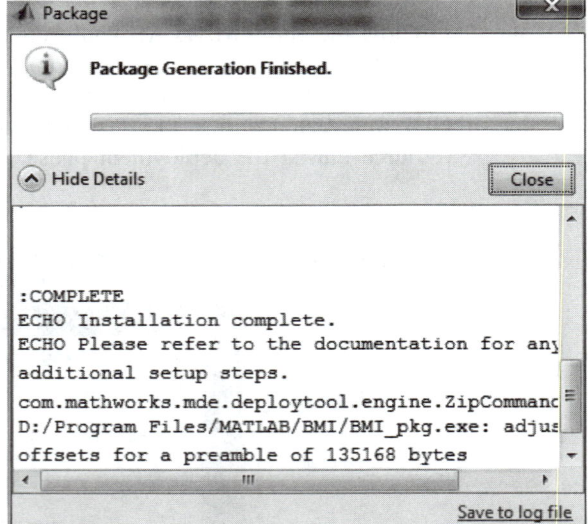

Figure 11.32 The **Package** process window shows the completion of the packaging process itself via creating BMI_pkg.exe in the destination folder. If there is an error that will obstruct the package process, you will notice it here. You can save the generated messages/warnings to a log file for future analysis.

11.8 Problems

11.1 Design a GUI such that it accepts the first name and last name of the user and it returns the following greeting: "Hi first name last name!"

11.2 Design a GUI such that it accepts the *three parameters* of a quadrature: $(ax^2 + bx + c = 0)$ and it returns the *two roots* using $r_{1,2} = \dfrac{-b \pm \sqrt{b^2 - 4ac}}{2a}$.

11.3 Design a GUI such that it accepts the *four parameters* of a polynomial of third degree ($ax^3 + bx^2 + cx + d = 0$) and it returns the *three possible roots*.

11.4 Design a GUI such that when clicking a push button the GUI will present the user with the current date and time. See online help for **datestr(now)** and **date**.

11.5 Design a GUI such that when clicking a push button the GUI will present the user with his or her age in years. See online help for **Birthdate = datenum(Y,MO,D)** and **CurrentDate=floor(now)**. Of course, the user is supposed to enter his or her birthday so that the GUI can calculate his or her age in years.

Open-Ended Questions

11.6 Design a GUI that enables the user to convert from the engineering (American) to metric (SI) system of units for the following physical quantities: length, mass, density, volume, force, energy, speed, and power. Refer to any standard textbook in chemical engineering to pick up the proper unit(s) in the US and SI systems and the required conversion factor from and to the SI system.

11.7 Design a GUI that enables the user to enter some critical properties of a substance while outputting volume-related and other thermodynamic properties of your choice. Use an equation of state other than the van der Waals equation. Examples are Peng-Robinson, Peng-Robinson-Stryjek-Vera, Redlich-Kwong, or Soave-Redlich-Kwong. Notice that you need to reevaluate the volume-based properties and the cubic equation in light of what you select as the equation of state. (See Sec. 11.5 regarding the design of the GUI based on the van der Waals' equation of state.)

Index

Note: Page numbers followed by *f* denote figures.

T